CALCIUM *in* CELL CYCLES *and* CANCER

Second Edition

James F. Whitfield, Ph.D.
National Research Council of Canada
Institute for Biological Sciences
Ottawa, Canada

CRC Press
Boca Raton New York London Tokyo

Library of Congress Cataloging-in-Publication Data

Whitfield, James F.
Calcium in cell cycles and cancer / author, James F. Whitfield. — 2nd ed.
p. cm.
Rev. ed. of: Calcium, cell cycles, and cancer, ©1990.
Includes bibliographical references and index.
ISBN 0-8493-4420-4
1. Calcium—Physiological effect.. 2. Cell cycle. 3. Calcium-binding proteins—Physiological effect. 4. Cellular signal transduction. I. Whitfield, James F. Calcium, cell cycles, and cancer. II. Title.
[DNLM: 1. Calcium—metabolism. 2. Cell Transformation, Neoplastic—metabolism. 3. Cell Cycle—physiology. 4. Cyclic AMP—metabolism. QV 276 C1432 1995]
QP535.C2W53 1995
612′.01524—dc20
DNLM/DLC
for Library of Congress 94-41530
CIP

Direct all inquiries to CRC Press, Inc., 2000 Corporate Blvd., N.W., Boca Raton, Florida 33431.

International Standard Book Number 0-8493-4420-4
Library of Congress Card Number 94-41530
Printed in the United States of America 2 3 4 5 6 7 8 9 0
Printed on acid-free paper

Preface to the First Edition

When I was a graduate student in microbiology at the University of Western Ontario in the early 1950s, I discovered two things: inorganic cations (Ca^{2+}, K^+, Na^+) condense bacterial chromatin, and the structure and function of the bacterial nucleoid are controlled by the cellular K^+ and Na^+ transport mechanisms.[1] Thus began my association with inorganic cations. But my long affair with Ca^{2+} did not begin until 1956 when Ray Rixon, thinking about the possible effects of this chromatin-condensing action of cations on radiation-induced chromosome damage, asked a simple question: could the survival of irradiated rats be increased by raising their plasma Ca^{2+} concentration with the Ca^{2+}-mobilizing parathyroid hormone? Indeed, it did, and significantly so.[2]

During the next 13 years, Ray Rixon, Alan Perris, and I ranged through the empty (except for a small band of egg activists) world of Ca^{2+} and proliferation with no map, very primitive tools, and funny cells such as bone marrow cells and thymic lymphoblasts. But they were enough for us to discover that Ca^{2+} controls the events leading to DNA replication. We were surprised and not a little dismayed by this discovery because, while no one would have been surprised if this unsophisticated cation condensed chromosomes, no one in this pre-Ca^{2+}-binding protein era would believe[3] that it could in any way selectively regulate so sophisticated a function as the initiation of DNA replication, which everyone knew was catalyzed by Mg^{2+}- and zinc-dependent enzymes. Even more controversial was our finding that bradykinin, growth hormone, parathyroid hormone, prolactin, TPA, and vasopressin stimulated DNA replication in thymic lymphoblasts by a Ca^{2+}-dependent mechanism. Happily, it turns out that we were right and had sown the seeds of the great things to come.

Up to the end of the 1960s, I had given no thought to Ca^{2+} and cancer; we were far too busy defending and digesting the fact that Ca^{2+} mediated the DNA-synthetic response to many growth factors and hormones. Then came Dr. Sam Balk at the beginning of January 1970 to ask me whether the Ca^{2+}-dependent, proliferative control mechanism we had discovered might be altered or absent in his model cells: chicken embryo fibroblasts neoplastically transformed by the Rous sarcoma virus. I did not know, but I thought it would be well worth finding out. The experiment I suggested to answer this question had far-reaching consequences for cell biology. I suggested that he change his model from chicken embryo fibroblasts in serum-containing medium to adult fibroblasts in plasma-containing medium because we had found that factors such as bradykinin, which are generated by the blood clotting used to produce serum, enabled our thymic lymphoblasts to initiate DNA replication even in low-Ca^{2+} medium. Sam took my suggestions back to Rockefeller University and the rest is history.[4] Fibroblasts transformed by the Rous sarcoma virus proliferated in low-Ca^{2+} medium which did not support the proliferation of uninfected, untransformed cells.[4] Thus, transformation had indeed profoundly affected the Ca^{2+}-dependent controls.[4] My concern about serum factors was also right. Serum contained a mitogen which was not present in plasma, and that mitogen turned out to be PDGF, the platelet-derived growth factor.[5]

Over the years, the world of Ca^{2+} signals and proliferative control has grown immensely, and the tools have become far more elegant than our humble cellular "shovels" of the 1950s and 1960s. Now we know that eukaryotic cells have devised a sophisticated system of signals, channels, and a large family of Ca^{2+}-binding effector proteins to produce and relay the Ca^{2+} signals that coordinate a vast array of cellular functions and even talk to genes. It has also become certain that Ca^{2+}-dependent mechanisms controlling cell-cycle transit and differentiation can change dramatically during the several stages of carcinogenesis.

Another great regulator, cyclic AMP, was discovered in the 1950s. Unfortunately, an ill-founded notion that cyclic AMP negatively regulates proliferation emerged during the 1960s and early 1970s and still thrives in the minds of many. But cyclic AMP and Ca^{2+} actually work together to trigger various stages of the cell cycle,[6,7] which is why cyclic AMP figures so prominently in a monograph about Ca^{2+}, cell cycles, and cancer.

Hopefully, this book provides a useful account of one veteran's view of the vast, very complex, and rapidly unfolding story of Ca^{2+}/cyclic AMP signals and the control of cell proliferation. Like all such books, it undoubtedly has shortcomings and errors which will annoy some readers. For these I apologize.

I must thank the past and present members and collaborators of my group, Ubaldo Armato, Al Boynton, Balu Chakravarthy, Jon Durkin, Doug Franks, Richard Isaacs, Len Kleine, John MacManus, Alan Perris, Leda Raptis, Marianna Sikorska, Sabine Swierenga, Roy Walker, Ben Tsang, and Tony Youdale who have contributed so much to the establishment of Ca^{2+} and cyclic AMP as major mitogenic mediators. I also thank Dr. Doug Dewar of the Canada Institute for Scientific and Technical Information for searching the literature for me, Ruth Baker, Lynda Boucher, Chris Gobey, Kathy Hamelin, Anne Jamieson, and Evelyn Monson for processing the words, and Dave Gillan and Robert Richards for preparing the illustrations.

Finally, I dedicate this work to my wife Barbara and to my friend and colleague of many years, Ray Rixon, who started it all.

J.F. Whitfield
Ottawa, Canada

REFERENCES

1. **Whitfield, J.F. and Murray, R.G.E.,** The effects of the ionic environment on the chromatin structures of bacteria, *Can. J. Microbiol.*, 2, 245, 1956.
2. **Rixon, R. and Whitfield, J.F.,** The radioprotective action of parathyroid extract, Int. *J. Rad. Biol.*, 3, 363, 1961.

3. **Baserga, R.,** *The Biology of Cell Reproduction*, Harvard University Press, Cambridge, MA, 1985.
4. **Balk, S.D., Whitfield, J.F., Youdale, T., and Braun, A.C.,** Roles of calcium, serum, plasma and folic acid in the control of proliferation of normal and Rous sarcoma virus-infected chicken fibroblasts, *Proc. Natl. Acad. Sci. U.S.A.*, 70, 675, 1973.
5. **Ross, R.,** The platelet-derived growth factor, *Hdb. Exp. Pharmacol.*, 57, 133, 1981.
6. **Boynton, A.L. and Whitfield, J.F.,** The role of cyclic AMP in cell proliferation: A critical assessment of the evidence, *Adv. Cyclic Nucleotide Res.*, 15, 193, 1983.
7. **Dumont, J.E., Jauniaux, J.-C., and Roger, P.P.,** The cyclic AMP-mediated stimulation of cell proliferation, *Trends Biochem. Sci.*, 14, 67, 1989.

Preface

Since the publication of the first edition of this book in 1990, teams of signaling molecules sticking to autophosphorylated velcroceptors and cyclin-dependent "PSTAIRE" protein kinases have emerged as star players in the starting and driving of the cell cycle. But Ca^{2+} and cyclic AMP still work alongside these "new boys on the block" to trigger the various stages of the cell cycle. Unfortunately, progress in finding out what Ca^{2+} and the cyclic nucleotide do has been slow mainly because they can do so many things. Nevertheless, the Ca^{2+} spikes and oscillations that I suspected marked the initiation of the cycle, DNA replication, prophase, and anaphase have now actually been seen and there are more, direct demonstrations of the importance of cyclic AMP in the build-up to DNA replication.

This is not just an updated version of its predecessor. It has been completely rewritten and re-formatted into a more concise, and hopefully more durable, up-to-date narrative overview of the most likely roles of Ca^{2+} in the proliferation, differentiation, and death of five model cell systems. I must thank Lynda Boucher and Josée Levesque for heroically translating my extraterrestrial handwriting into this final form, Geoff Mealing and Lyne Séguin for drawing the illustrations, Tom Devecseri for photographing them, and Diane Kharouba, Maureen Wong, and Mike Nadon at the Canada Institute for Scientific and Technical Information for searching the literature for me over the last years. Finally, as before, I dedicate the book to my wife Barbara and to Ray Rixon who started it all.

J.F. Whitfield
Ottawa, Canada

The Author

James F. Whitfield, B.Sc., Ph.D., obtained his B.Sc. in microbiology from McGill University, Montreal, in 1951 and his M.Sc. and Ph.D., also in microbiology, from the University of Western Ontario in 1955. He started his postgraduate career wrestling with the genetics of bacteriophage λ at Atomic Energy of Canada at Chalk River, Ontario, but, he soon became interested in radiation biology. In 1962, he left Canada to start a cellular radiation biology group at the European Joint Research Center at Ispra, Italy. In 1965, he returned to Canada to start a Group that pioneered study of the key roles of Ca^{2+}, Ca^{2+}-binding proteins, and cyclic nucleotides in the proliferation of normal and cancer cells.

Table of Contents

Chapter 6

Chapter 7

1 Origins

I. THE VITAL MUDS

In the beginning in the hot early Earth were colloidal clays consisting of crystals, which were very "low-tech" inorganic catalyst-genes that could crudely replicate their structures and, most importantly, their structural faults, which supplied the variations needed for the selection and evolution of more effectively self-replicating forms.[1–4] In time, the clay genes began trapping solar energy for the proto-photosynthetic production of ever-increasingly elaborate organic molecules on their surfaces with which they could control their immediate environments.[1–3] A major breakthrough came when the increasingly "organified" clay "organisms" encased themselves in membranes or walls, which enabled them to establish and maintain their own optimal microenvironment. Thus, they could prevail against drought and dispersal, and invade hitherto untenable niches. The next great leap forward came when the proto-organisms began storing information for the organic supports for crystal replication in the structures of polypeptides and especially organic polyphosphates in which lay the seeds of the future RNA and then DNA genes. But eventually the organic genes predominated and pushed the simple clay genes aside and seized control of the now all-organic organisms, which were furiously inventing and perfecting new "high-tech" metabolic machinery and storing the information for this machinery in new RNA and DNA genes.

Thus, had simple, "low-tech" silicon "chips" evolved into "high-tech" prokaryotes equipped with rigid proteoglycan walls and membranes loaded with ATP-generating proton pumps, ion and nutrient transporters, and the attachment sites for the replication origin and terminus of their novel, one-replicon chromosome.

II. THE PROKARYOTES

The prokaryotes that emerged from this epochal "genetic takeover" about 3.0 billion years ago[1–4] invented one of the great signal mediators, cyclic AMP, to initiate and orchestrate such vital things as gene expression, DNA replication, cell wall elongation, and cell division.

The prokaryotes generate a cyclic AMP surge in response to a falling ATP level to signal food shortage to the appropriate genes.[5] They invented a cyclic AMP-binding protein known as CAP ("catabolite gene activator protein") to relay selectively the cyclic AMP signal to target genes with a specific cyclic AMP-CAP-binding nucleotide sequence in their 5′-regulatory elements. The products of these cyclic AMP/cyclic AMP-CAP-responsive genes enable the cells to switch over to alternative nutrients.[5]

Cyclic AMP by itself starts DNA replication in the bacterium *Escherichia coli* by displacing ADP from inactive ADP-*dna*A protein complexes associated with the membrane/cell wall attachment site for the chromosomal origin sequence, *ori*C, at one pole of the cell.[6–10] This enables the formation of active ATP-*dna*A protein complexes which bind to four 9-mer repeats ("*dna*A boxes") to form an *initial complex* consisting of negatively supercoiled *ori*C-DNA wrapped around a central core of 20 to 40 ATP-*dna*A monomers.[6,6a] The ATP-*dna*A monomers "melt" three AT-rich 13-mers to produce a "bubble", or *open complex*, into which complexes consisting of *dna*B unwinding enzyme and *dna*C protein can move to prime replication.[6,6a] The replication origin then replicates and the sister origin is moved, as bidirectional replication progresses, by DNA helicase to an attachment site at the opposite pole of the cell.[7,8] The still-unreplicated replication terminus detaches from the cell wall and is pulled to the equator of the cell by a simultaneous DNA gyrase-induced supercoiling of the two replicated chromosomes.[7,8] The replicated chromosomes are decatenated and separated at their termini by topoisomerase II (as we shall see in the next chapter replicated mammalian chromosomes must also be decatenated by topoisomerase II), the terminal attachment site divides, and the two new termini reattach to the cell wall.[7,8] While all of this is going on, the cyclic AMP/cyclic AMP-CAP signal also drives cell wall elongation and division by activating PBP2, a transpeptidase that initiates new elongation sites, and by stimulating the expression of the *fis*A gene, whose product is involved in septum formation between the new terminus attachment sites of the cell wall.[11]

The emerging prokaryotes did not need receptors or use Ca^{2+} signals to activate genes, trigger DNA, or operate their flagellar motors.[12] But there were hints of the great signaling systems to come when some *Streptococcus* species learned to use a Ca^{2+} transient to inactivate a surface-associated growth inhibitor, nisin, to start DNA replication and proliferation,[13–15] and cyanobacteria such as *Phormidium uncinatum* learned to use a brief Ca^{2+} surge through voltage-sensitive channels to reverse their gliding motion when entering a dark zone (the photobic response).[16] Most importantly, the prokaryotes laid the foundation of the spectacular biotechnological breakthrough to come by inventing various ways (Ca^{2+}/H^+ or Ca^{2+}/Na^+ antiporters, Ca^{2+}-ATPase pumps) of ridding themselves of the *dangerous* Ca^{2+} ion to prevent it from precipitating phosphates and activating endonucleases, phospholipases, and proteinases which could damage and kill them.[17–19] Thus was created the very steep, inwardly directed Ca^{2+} gradient which had enormous signaling potential if it could ever be *safely* harnessed.

III. THE GREAT CALCIUM-DRIVEN REVOLUTION

Something momentous happened around 1.5 billion years ago. Specially primed Gram-positive actinomycetes lost their ability to make murein and with it their rigid cell walls.[7] This "decorsetting" presented the cells with a host of formidable dangers. They would become susceptible to swelling and lysis from osmotic pressure fluctuations; their once-quiet surfaces would start evaginating and invaginating; and the chromosome attached to this seething surface would be enormously stressed, damaged, and rearranged, which would trigger an evolution-accelerating genetic upheaval.[7,8] To survive this loss and eventually to prevail they had to have already invented prokaryotic, actin-like proteins; elastic gels which, by extending throughout the cell and attaching to the cell membrane, could prevent the cell being torn apart by excessive swelling. Then they developed special proteins and a signaling mechanism at first simply to enable the gel to respond actively to swelling, but later to trigger a host of new activities.[7,8,20] They strengthened their membranes by learning to make and insert sterols and glycoproteins into them. They also invented microtubules to be dynamic reinforcing rods which stabilize the cell membrane, oppose the contracting action of the actin gel, and replace key binding sites which were lost with the cell wall. Thus, the *neomura* ("new walls") had replaced the ancient rigid exoskeleton with a cytomusculature and dynamic endocytoskeleton which endowed them with an enormous range of new activities.

The loss of the cell wall disrupted the tight coupling of chromosome replication to cell division and exposed the still surface-attached chromosome to the now roiling surface. The proteins of the chromosome attachment site evolved into a MTOC (microtubule-organizing center), a proto-centrosome, in order to reinforce and stabilize this vital membrane site.[7] However, the new cytomusculature and mobile surface were the engines of endocytosis and phagocytosis, and one of the things phagocytosed was a vesicle or cisterna bearing the chromosome attachment site and its associated microtubles.[7] The internalized vesicle, with its MTOC, MTOC-attached chromosome, and ribosome-binding sites then expanded to enclose the chromosome and form the first proto-eukaryotic nucleus.

As time went on, more genes were loaded into the chromosome and more replication origins and replicons had to be created in order to be able to replicate the chromosome in a reasonable time. This called for radical nuclear restructuring. Obviously, the old attachment site of the original origin at the MTOC was no longer adequate for a multireplicon chromosome. Therefore, a novel structure, the nuclear matrix, was invented to provide the additional attachment sites for the new origins.[21] The old origin attachment site became the centrosome MTOC, the ancient *ori* sequences became the chromosomal telomere, and the replication terminus became the kinetochore that binds to the mitotic spindle microtubules and participates in moving replicated chromosomes to the spindle poles of the dividing cell.[7] The new "super-high tech" eukaryotic nucleus, with its cell wall-replacing lamina and matrix and rapidly replicable multireplicon chromosomes, was born. And despite the extensive nuclear remodeling, DNA replication was still coupled, though very differently so, to cell division.

The *neomurans* also changed and expanded the role of cyclic AMP in controlling the more elaborate cell functions. No longer did a cyclic AMP surge mean a food shortage and signal the need to switch over to alternative nutrients. Instead, it came to signal nutrient *adequacy* and a readiness to start the program of events leading to the initiation of DNA replication and mitosis or meiosis. But cyclic AMP no longer participated directly in the initiation of replication as it did via the *dna*A-like proteins of the ancestral prokaryotes. With these changes came a radical restructuring of the cyclic AMP signaling mechanism. Instead of binding to, and activating, regulator proteins which targeted various cyclic AMP-responsive genes, the cyclic AMP-binding, signal-sensing proteins became the regulatory subunits of the cyclic AMP-dependent protein kinases (PKAs), which respond to a cyclic AMP signal surge by dissociating into cyclic AMP-R subunit complexes and active C subunits which phosphorylate gene transcription enhancers or suppressors, enzymes, and structural components. In other words, the *neomuran* cyclic AMP signal was now relayed to many different targets as bursts of protein kinase activity. Because of this change, a cyclic AMP signal could trigger the phosphorylation and either the stimulation or inhibition of many different enzymes and modulators. Therefore, cell cycle cyclic AMP signal surges had to be strictly limited in size and duration in order to stimulate key events but not inhibit downstream events. Thus, the replication-triggering origin-unwinding activity had to become cyclic AMP-independent.

At first the *neomurans* did not have mitochrondria or chloroplasts.[22] But they soon got them in a dramatically direct way. Unlike the stiff-walled prokaryotic ancestors who could only ingest small molecules, the *neomurans* were *phagotrophic*, that is, they could ingest entire bacteria. Among their prey were purple sulfur bacteria, which were phagocytosed and converted into mitochondria, and cyanobacteria, which were phagocytosed and converted into chloroplasts.[22]

The key event that triggered the emergence of the eukaryotes was the loss of the ability to make murein, a major cell wall component.[6] But this was not enough by itself to start such a revolution. Undoubtedly many prokaryotes lost their walls and promptly blew up and died: there were no large eukaryotes in which to safely shelter while they reduced their size and fortified their membranes with a skeletal structure as would happen much later with the mycoplasma.[6] There were two keys to the successful exploitation of this event: an already functioning protocytoskeleton/cytomusculature and a signaling mechanism to operate it. The successful *neomurans* had harnessed the ancestral inwardly directed Ca^{2+} gradient and developed gated Ca^{2+} channels to produce the Ca^{2+} signal surges to operate the new machinery. Thus, Ca^{2+} signals drove the phagocytoses that internalized the chromosome replication apparatus and ingested the bacterial ancestors of the modern chloroplasts and mitochrondria. Ca^{2+} took over from proton loops to activate the new kinds of ATP-powered motors that had been invented to replace the old prokaryote motors and drive microtubular cilia and flagella, move chromosomes, and elongate the spindle during mitosis.[12,24–26] Of course Ca^{2+}, and the endonucleases, proteinases, and phospholipases it stimulates, were as dangerous as ever,[27–29] but the *neomurans* ingeniously solved this problem by inventing the first of a large family of Ca^{2+}-binding effector proteins with such high affinities for the ion that they could selectively activate specific targets in response to Ca^{2+} signal surges that were well below the lethal level.[29] They even recruited the deadly Ca^{2+}-dependent phospholipases into a signaling mechanism which breaks minor membrane phospholipids such as phosphatidylinositol-4,5-*bis*-phosphate, PtdIns(4,5)P_2, down into products such as inositol-1,4,5,-*tris*-phosphate, Ins(1,4,5)P_3, which mobilizes Ca^{2+} from storage vesicles and directly or indirectly opens Ca^{2+} channels in the plasma membrane, and diacylglycerols (DGs), which collaborate with Ca^{2+} to stimulate some membrane-associated PKCs activity.

Eventually, the glycoproteins that first served simply to strengthen and protect the naked surfaces of the wall-less *neomurans* evolved into elegant receptors through which cells could talk to each other and form functioning multicellular societies. These receptors consist of external ligand-binding domains, transmembrane domains to relay the ligand-binding signal from the external domain to the internal cytoplasmic domain, which by activating signal-coupling G-proteins, or by becoming active, autophosphorylated-velcroized protein tyrosine kinases (YPKs) that collect and activate teams of signaling molecules, send the signal on to various parts of the cell. Thus evolved the wonderfully networking systems that are operated by Ca^{2+} and cyclic AMP to start and drive the cell cycle.

REFERENCES

1. **Cairns-Smith, A.G.,** *Genetic Takeover and the Mineral Origins of Life*, Cambridge University Press, Cambridge, 1982.
2. **Cairns-Smith, A.G.,** *Seven Clues to the Origin of Life*, Cambridge University Press, Cambridge, 1985.
3. **Cairns-Smith, A.G. and Hartman, H., Eds.,** *Clay Minerals and the Origin of Life*, Cambridge University Press, Cambridge, 1986.

4. **Shapiro, R.,** *Origins: A Skeptic's Guide to the Creation of Life on Earth,* Summit Books (Simon and Schuster, Inc.), New York, 1986.
5. **Ullman, A. and Danchin, A.,** The role of cyclic AMP in bacteria, *Adv. Cyclic Nucleotide Res.*, 54, 915, 1988.
6. **Kornberg, A. and Baker, T.A.,** *DNA Replication*, 2nd ed., W.H. Freeman, New York, 1992.

6a. **Skarstad, K. and Boye, E.,** The initiator protein DnaA: evolution, properties and function, *Biochim. Biophys. Acta,* 1217, 111, 1994.

7. **Cavalier-Smith, T.,** The origin of eukaryote and archaeobacterial cells, *Ann. N.Y. Acad. Sci.*, 503, 17, 1987.
8. **Cavalier-Smith, T.,** Bacterial DNA segregation: its motors and positional control, *J. Theor. Biol.*, 127, 361, 1987.
9. **Hughes, P., Landoulsi, A., and Kohiyama, M.,** A novel role for cAMP in the control of the activity of the *E. coli* chromosome replication initiator protein, dnaA, *Cell*, 55, 343, 1988.
10. **von Meyenberg, K. and Hansen, G.G.,** The regulation of chromosome replication, in *Escherichia coli* and *Salmonella typhimurium, Cellular and Molecular Biology*, Vol. 2, Neidhardt, F.C., Ed., American Society for Microbiology, Washington, D.C., 1987, 1555.
11. **Vincente, M.,** The control of cell division in bacteria, in *The Microbial Cell Cycle*, Nurse, P. and Streibloba, E., Eds., CRC Press, Boca Raton, FL, 1984, 30.
12. **Harold, F.M.,** *The Vital Force: A Study of Bioenergetics*, W.H. Freeman, New York, 1986.
13. **Hurst, A.,** Apparent destruction of nisin by the producer organism before initiation of growth in *Streptococcus lactis, Nature*, 219, 403, 1968.
14. **Hurst, A. and Kruse, H.,** The correlation between changes in absorbancy, calcium uptake, and cell-bound nisin activity in *Streptococcus lactis, Can. J. Microbiol.*, 16, 1205, 1970.
15. **Hurst, A. and Lazarus, W.,** Calcium uptake during growth of *Streptococcus lactis, Nature*, 219, 404, 1968.
16. **Hader, D.-P.,** Signal perception and amplification in photomovement of prokaryotes, *Biochim. Biophys. Acta*, 864, 107, 1986.
17. **Devés, R. and Brodie, A.F.,** Active transport of Ca^{2+} in bacteria: bioenergetics and function, *Mol. Cell. Biochem.*, 36, 65, 1981.
18. **Fletcher, J.,** Importance of calcium in microbiological systems, in *The Role of Calcium in Biological Systems*, Vol. 1, Anghilieri, L.J. and Tuffet-Anghilieri, A.M., Eds., CRC Press, Boca Raton, FL, 1982, 229.
19. **Silver, S.,** Calcium transport in microorganisms, in *Microorganisms and Minerals*, Weinberg, E.D., Ed., Marcel Dekker, New York, 1977, 49.
20. **Schliwa, M.,** *The Cytoskeleton: An Introductory Survey,* Springer-Verlag, Vienna, 1986.
21. **Cavalier-Smith, T.,** Evolution of the nuclear matrix and envelope, in *The Nuclear Envelope and the Nuclear Matrix*, Maul, G.G., Ed., Alan R. Liss, New York, 1982.
22. **Cavalier-Smith, T.,** The simultaneous symbiotic origin of mitochondria, chloroplasts, and microbodies, *Ann. N.Y. Acad. Sci.*, 503, 55, 1987.
23. **Campbell, A.K.,** *Intracellular Calcium: Its Universal Role as Regulator,* John Wiley & Sons, Chichester, England, 1983.

24. **Pratt, M.M.,** ATPase in mitotic spindles, *Int. Rev. Cytol.,* 87, 83, 1984.
25. **Satir, P.,** Switching mechanisms in the control of cell motility, *Mod. Cell Biol.*, 4, 1, 1985.
26. **Campbell, A.K.,** Intracellular calcium: friend or foe?, *Clin. Sci.*, 72, 1, 1987.
27. **Orrenius, S. and Bellomo, G.,** Toxicological implications of perturbation of Ca^{2+} homeostasis in hepatocytes, in *Calcium and Cell Function*, Vol. 6, Cheung, W.Y., Ed., Academic Press, Orlando, 1986, 185.
28. **Trump, B.F. and Berezesky, I.K.,** The role of calcium in cell injury and repair. A hypothesis, *Surv. Synth. Pathol. Res.*, 4, 248, 1985.
29. **Whitfield, J.F., MacManus, J.P., Boyton, A.L., Durkin, J.P., and Jones, A.,** Futures of calcium, calcium-binding proteins, cyclic AMP and protein kinases in the quest for an understanding of cell proliferation and cancer, in *Functional Regulation at the Cellular and Molecular Levels*, Corradino, R.A., Ed., Elsevier/North-Holland, New York, 1982, 61.

2 Calcium, Cyclic AMP, and Velcroceptor-Started Cycles

I. INTRODUCTION

The journey of a cell and its descendants to cancer begins with the mutation or deletion of the first of a series of genes that somehow subverts the receptor-operated mechanisms that depend on Ca^{2+} and/or cyclic AMP to control proliferation and differentiation. The mutant cell and its increasingly mutant descendants may escape from the tyranny of receptors and "do their own proliferative thing" by spontaneously expressing normally receptor-controlled key cell cycle genes. Or they may respond to phantom voices and signals from their own mutant or virally derived, spontaneously signaling receptors or receptor couplers that drive them relentlessly to proliferate regardless of whether or not they should. Or they may have lost the ability to receive or respond to signals from neighbors or substrates to stop cycling and start differentiating.

Before 1950, the cell cycle, that is, the life history of a proliferatively active cell from birth to mitosis, was thought to consist of a featureless interphase or growth phase which ended in a wonderfully vivid division phase during which nuclear division by mitosis or meiosis was followed by cytokinesis or cytoplasmic division and the emergence of two newborn daughter cells.[1] But no one knew when the chromosomes were replicated, although some logically (for those times) believed that they were replicated during prophase. Eventually, it was found that the chromosomes are replicated during interphase, not mitosis, and that most cell cycles have a first prereplicative gap, or G_1 phase, when the cell's DNA is not yet replicating and cannot yet be labeled with a radioactive precursor such as [^{3}H]-thymidine; a replication, or S, phase when DNA can be labeled; a second gap, or G_2, postreplication phase when the DNA can no longer be labeled; and finally mitotic, or M, phase (Figure 1).[1] However, some cells such as cleaving frog *(Xenopus laevis)* eggs and Chinese hamster V79–8 cells have no G_1 or G_2 phases in their cycles because they maintain pools of the basic parts for chromosome replication and mitosis, and achieve the properly

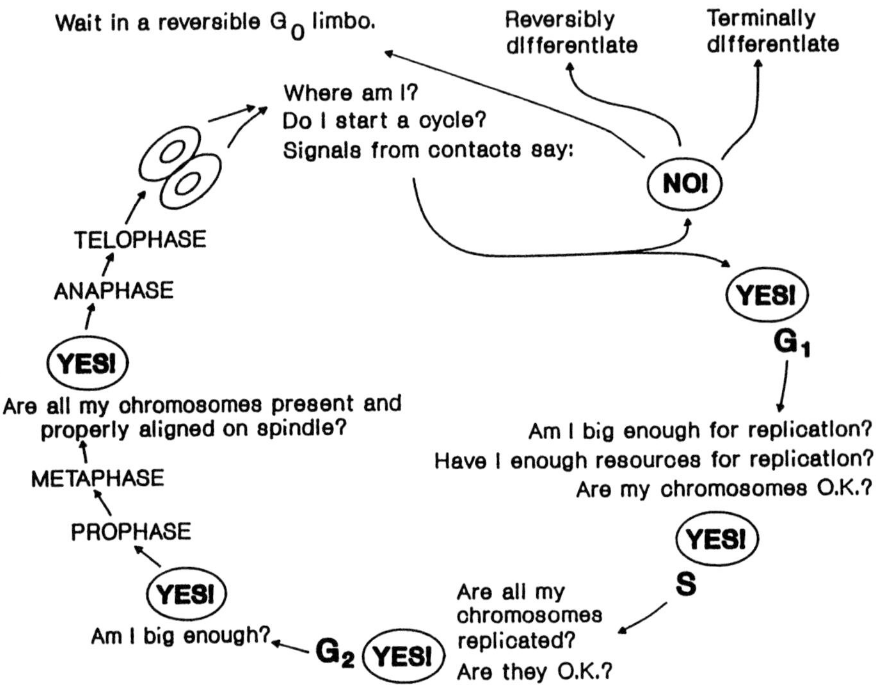

FIGURE 1 The cell cycle, its several checkpoints, and the questions the cell must ask at each checkpoint.

timed initiations of replication and mitosis by sequentially making and destroying stage-specific cyclin-dependent PKs.[1]

The prereplicative G_1 and postreplicative G_2 phases are simply two time delays: the first delay extends between the cell's birth at cytokinesis, or its stimulation from proliferative quiescence, and the initiation of chromosome replication; and the second delay extends between the completion of chromosome replication and the onset of mitosis. The variability of the lengths of the G_1 phases of the cells in an actively proliferating cell population is much greater than the variabilities of the other three phases.[1] This G_1 variability is not due to genetic differences between cells, because even the lengths of the G_1 phases of genetically identical sister cells are different, probably because sister cells are not the same size at birth and therefore are unequally endowed with resources. The length of G_1 is the time that the cell needs to accumulate critical amounts of the things needed for chromosome replication and assemble the G_1-specific engines that drive the build-up and start replication. Therefore, the bigger and more lavishly endowed a cell is at its birth, the shorter will be the G_1 build-up. It is essential to note that from the earliest times logic and observation have indicated that the operation of the engine that activates the chromosome-replicating machinery is growth-sensitive: as soon as the cell passes a certain minumum size threshold it pushes the replication button.

Before venturing into the S phase or mitosis, a normal eukaryotic cell must run down a checklist of prerequisites to avoid such potentially disastrous things as running out of parts for chromosomes in the middle of the S phase, replicating damaged chromosomes, producing daughters with incomplete, or otherwise aberrant, chromosomes, or producing "wee" or midget daughters that are too small to start a cycle. Therefore, before starting chromosome replication, the cell must be sure that it is big enough, that it has enough components for a round of replication, and that its chromosomes are fit to replicate (Figure 1). Then, before starting mitosis, the cell must again be sure that it is big enough and that its chromosomes have all been replicated (Figure 1). And, finally, the cell must make sure that all of its chromosomes are lined up on the equator of the metaphase spindle before giving the signal that starts the anaphase movement of chromosomes to the spindle poles and gets rid of the last of the cell cycle engines (Figure 1). Thus, eukaryotic cells have set up major checkpoints at the G_1/S and G_2/M boundaries and armed them with devices that prevent the G_1 or G_2 cell cycle engines from starting replication or mitosis until all is in order and may even command the cell to self-destruct if a problem such as damaged chromosomes cannot be repaired.

In this chapter we see how Ca^{2+}, cyclic AMP, and the newly found cyclin-dependent PKs[2,3] drive the different parts of the cycle. Here we focus on animal cell cycles, particularly those that are started by signals from teams of signalers sticking to autophosphorylated protein tyrosine kinase (YPK) velcroceptors.

II. THE CYCLE

A. The Proliferative Wakening: From Quiescence to G_1 Build-Up

Our story begins with a proliferatively "switched-off" or quiescent cell. There are two kinds of reversible quiescence. First is the so-called "G_0" quiescence which is seen in cells such as yeast and cultured animal cells when they are starved, deprived of growth factors, detached from a substrate, or exposed to an inhibitor of mRNA translation before passing the point-of-no-return (the yeast cells' "START" point; the animal cells' "*restriction*" point) when the cell cycle engine has been built and started, and enough replication-related parts have been accumulated to commit the cell to a round of chromosome replication.[1,3] The cell responds to starvation or the silencing of growth factor receptors by activating an ancient pleiotypic shutdown program that includes the destruction of the cell cycle engine and switches the cell into the G_0 state.[1,2] Second is the *functional* quiescence in which a well-fed cell, instead of growing and making the G_1-specific and mitotic cell cycle engines, makes things for its particular structure and function and networking with other cells.

The starved or growth factor-deprived "G_0" cultured cells shut down the expression of the cell cycle genes by reducing their transcription and/or increasing the destruction of their transcripts while turning on genes encoding quiescence factors which inhibit the initiation of DNA replication, but do not affect on-going DNA replication. One of these quiescence factors may be a protein that collects on the outside of serum-deprived quiescent human neonatal foreskin fibroblasts and prevents proliferatively active fibroblasts from initiating DNA replication.[1] Another quiescence factor is a 57-kDa protein, statin, which collects around the nuclei of quiescent cells.[1] Significantly, some neoplastic cells (SV-40-transformed IMR-90 human fibroblasts, adenovirus 5-transformed human kidney cells, and HeLa human cervical adenocarcinoma cells), which do not need external growth factors to proliferate, do not trigger the cycle-shutdown program in response to serum deprivation and make something that overrides the quiescence factors and enables shut-down normal cells (e.g., IMR-90 human fibroblasts) to initiate DNA replication.[1] A functionally equivalent quiescence factor in budding yeast, *Saccharomyces cerevisiae*, is the 238-amino acid ARD-1 (arrest defective) protein, without which the cell cannot shut down into a G_0 state or sporulate in response to a lack of nutrients and thus may kill itself by trying to go through a round of chromosome replication and budding (division) while starving.[1]

The silencing of growth factor receptors stops cell growth and the G_1 buildup by striking at all aspects of RNA and protein metabolism as well as the mitochondrial energy supply. An important key to understanding how a cell switches in and out of the cycling mode is the fact that the mRNAs from genes encoding growth-regulating proteins are "weak", low-priority messages which owe their low place in the translational totem pole to extensive secondary structure in capped, 5′-upstream untranslated regions which must be "melted" by growth factor-activated translation initiators before they can bind to 43-S pre-initiation complexes and be translated (Figures 2 and 3).[4–10] The shutdown program blocks the processing of the mRNA transcripts of cell cycle genes, demobilizes the "weak" mRNA messages for growth-promoting peptides, reduces the total mRNA content, decreases ribosomal RNA synthesis, increases ribosomal RNA degradation, increases the degradation (by lysosomes), reduces the flow into the nucleus of short-lived non-histone proteins such as transcription factors, and lowers mitochondrial activity.[1] Only the "strong" messages (with their low 5′-secondary structure) for basic housekeeping and quiescence factors are translated while the number of polyribosomes dramatically drops as the cell shunts a large part of its growth factor receptor-regulated, "weak" mRNA into a nonpolysomal pool.[1,5,9] For example, when African green monkey kidney cells (Vero cells) shut down, they drop their mRNA content to about 25% of the level in proliferating cells and the fraction of polyribosome-associated mRNA from 75 to 35%, which together reduce the protein-synthetic

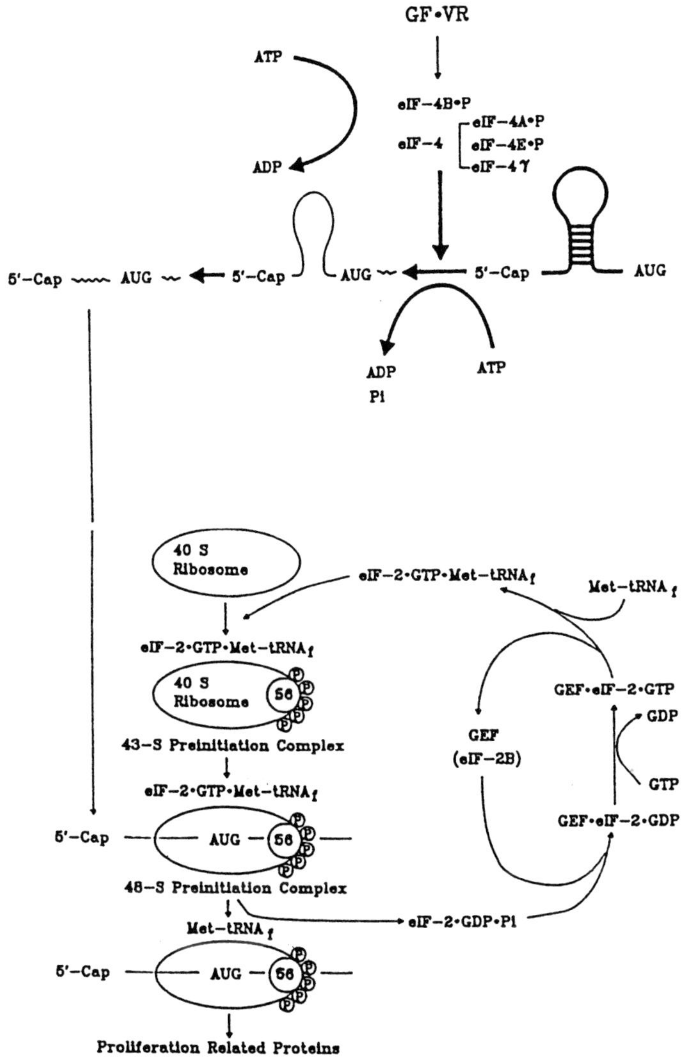

FIGURE 2 The stimulation of the translation of the 5′ highly structured RNA messages for growth-related proteins in response to a signal from a growth factor's velcroceptor (GF.VR)

activity of the quiescent cells, to only 12% of the activity in proliferating cells.[1] While functional "weak" mRNAs for proliferation-related peptides drop dramatically, functional "strong" mRNAs for housekeeping and other peptides continue to be translated.[1,5,7] The drop in the translatability of the growth-promoting mRNAs is due to at least two things. First, when the growth factor-activated PKs fall silent, the translation rate-limiting eukaryotic initiation factor-4 (eIF-4) components are dephosphorylated and inactivated.[4–10] Thus, there are fewer active, phospho-eIF-4 complexes to bind to the 5′-caps

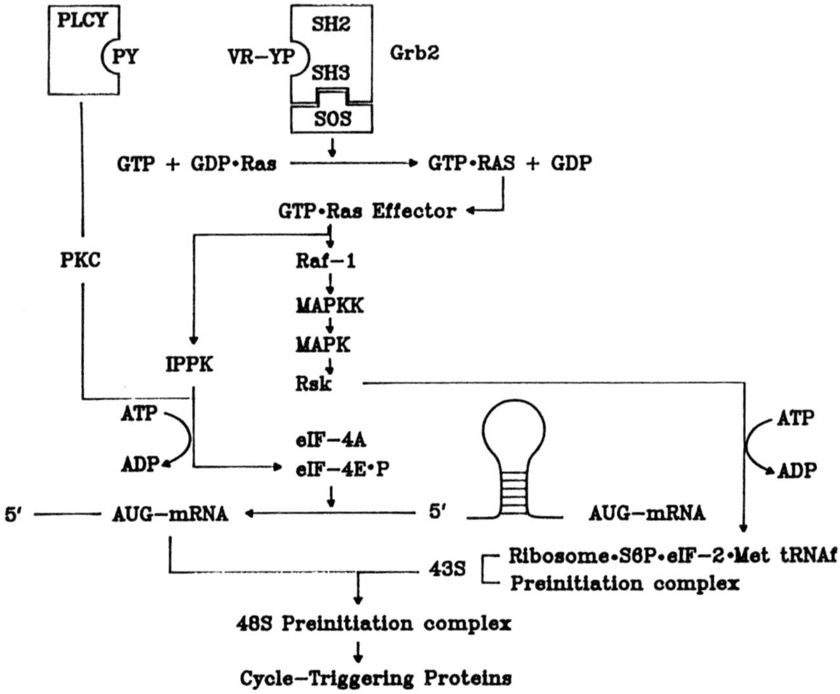

FIGURE 3 The Ras-mediated cascade of PK activities triggered by an activated, autophosphorylated velcroceptor (**VR**) that stimulates the translation of translationally "weak" (low-priority) growth-related mRNAs. Things begin when the Grb2 adaptor protein binds with its SH2 pocket to the velcroceptor and the mSos GEF protein[16a] attaches to Grb2's SH3 domain. This brings the mSos protein into contact with membrane-associated GDP·Ras complexes, which it then converts into active GTP·Ras complexes by stimulating the exchange of the complexes' GDP for GTP. A key event in the GTP·Ras-triggered events is the phosphorylation and activation of eIF-4 complexes, whose eIF-4A helicase component then melts the characteristically very extensive secondary structure of the "weak" growth-related mRNA 5′ UTRs. This enables the eIF-4 complex to promote the association of the capped 5′ region of the messages with primed, S6-phosphorylated 43-S preinitiation complexes. It should be noted that the Ras-induced pathway can be bypassed partially when eIF-4 is phosphorylated by PKCs stimulated by DGs from PtdIns(4,5)P_2 hydrolyzed by velcroceptor-activated PLC-γ_1 (**PLCY** in the diagram) or the potent PKCs activator, TPA.

(m^7GpppX) of growth-regulated mRNAs and enable the 50-kDa eIF-4A helicase components of the eIF-4 complexes to melt the messages' excessively structured (bulges, helices, loops, pseudoknots) 5′-UTRs and thus enable them to bind to the eIF-2·GTP·Met-tRNAf·40-S ribosome preinitiation complexes (Figures 2 and 3).[4–10] Second, the silencing of the growth factor receptors and Rsk STPKs result in the dephosphorylation and inactivation of 40-S ribosome S6 protein components that are needed to form the eIF-2·GTP·Met-tRNAf·40-S ribosome preinitiation complexes (Figures 2 and 3).[7]

An additional component of the shutdown mechanism under some conditions is an increase in the level of phosphorylated (on the S^{51} residue) eIF-2α. Phospho-eIF-2α binds tightly to eIF-2B, a guanine nucleotide exchange factor (GEF), that maintains translational activity by catalyzing the regeneration of eIF-2·GTP complexes from eIF-2α·GDP complexes released from 48-S preinitiation complexes (Figure 1).[7] eIF-2B is present in only limiting amounts and locking it into complexes with phospho-eIF-2α stops translation.[7] The level of phospho-eIF-2α is controlled by PKs and protein phosphatases. In starving yeast *(Sa. cerevisiae)* cells, uncharged lysyl-tRNA accumulates and binds to GCN2 protein which activates the protein's PK domain. GCN2 then phosphorylates eIF-2α.[4] By contrast, the increased level of phospho-eIF-2α in amino acid-starved hepatocytes is due to the inactivation of a protein phosphatase instead of activation of an eIF-2α PK, although amino acyl tRNA synthetases in animal cells do seem to have eIF-2α PK activity.[4] The importance of phospho-eIF-2α in the proliferative quiescence of cultured animal cells deprived of growth factors in an otherwise complete medium or cells such as hepatocytes in an intact adult liver is unknown. But it seems likely that in these cases it is the level of phospho-eIF-4E that is the key limiting factor.

Quiescent cells of the commonly used established cell lines all seem superficially to be in an identical G_0 state, because they all take about the same long time (10 to 15 h) to start replicating DNA after being stimulated by activated growth factor receptors.[1] But this is in fact quite wrong. Some of them retain considerable traces of the G_1 build-up, such as ciliated centrioles, peak levels of calmodulin and cyclic AMP, and monophosphorylated histone, all of which must actually be eliminated by a "reset" mechanism before the G_1 build-up can start again.[1] Depending on how much of the G_1 build-up they manage to maintain, the continuing pattern of gene expressions, and exactly what is missing or suppressed, different quiescent cells will respond differently to various factors. Thus, for example, the decidedly freakish quiescent Swiss Albino 3T3 mouse cell (unfortunately one of the most popular model cell lines for the study of cell cycle control), responds to things such as high pH, potent PKCs-stimulating phorbol esters, cyclic AMP and cyclic AMP elevators (e.g., cholera toxin, PGE_1) that stimulate other quiescent cells such as BALB/c3T3 mouse cells and T51B rat liver cells *only* when they are arrested near the end of the G_1 build-up, but *never* when they are in a pre-G_1 G_0 quiescence. Another example is the ability of cytoplasmic or nuclear microinjection of adenovirus E1A protein, a *trans*-acting protein transcription enhancer, to trigger a slow G_1 build-up in Swiss albino 3T3 mouse cells but not in human T98-G glioma cells.[1] Yet another example is the ability of nuclear microinjection of a 10 m*M* *Tris* (pH 7.5), 20 m*M* KCl buffer to activate quiescent human WI-38 cells, but not Swiss albino 3T3 mouse cells.[1]

A signal to G_0 cells in quiescent growth factor-deprived cultures from receptors for a growth factor such as epidermal growth factor (EGF) or one of the platelet-derived growth factors (PDGFs) starts the cells growing to a state

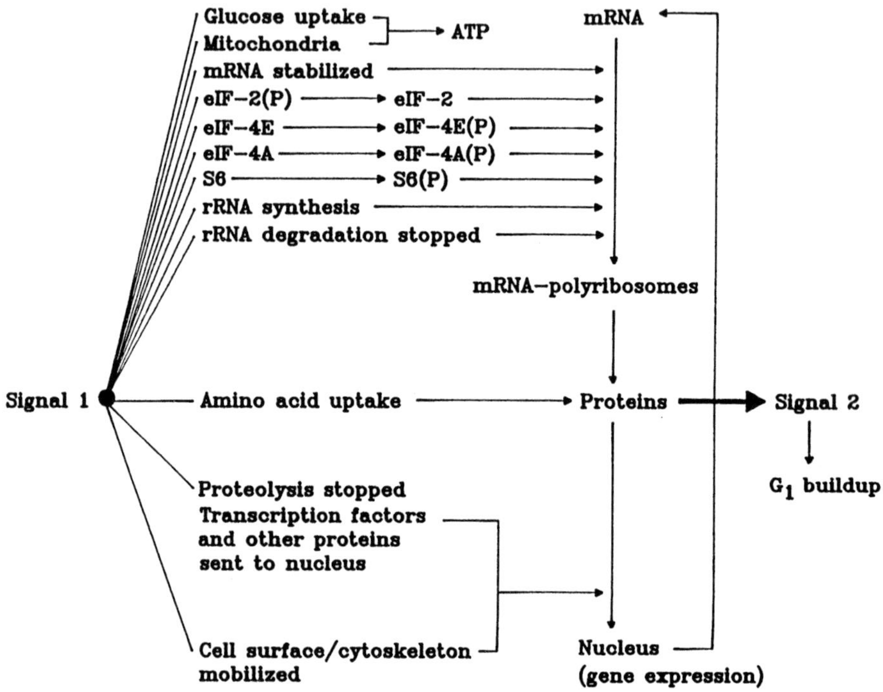

FIGURE 4 The first signal to quiescent, growth factor-deprived cells from velcroceptors (such as those for EGF or the PDGFs) triggers a dramatic cascade of events that mobilizes and activates transcription factors and sends them to the nucleus, repositions the genes for optimal response to the incoming transcription factors, increases the translatability of growth-related mRNA, increases ribosomal RNA production, increases amino acid uptake and protein synthesis, and increases the number and activity of the mitochondria to provide the ATP needed to fuel the growth of the cell into a state of competence to respond to a second signal to start the actual G_1 build-up to DNA replication.

of competence in which they can generate and respond to a second signal to start the actual direct build-up to replication. This first "wake-up" signal triggers a dramatic cascade of events (Figure 4). Mitochondrial activity surges upward to provide the ATP needed to fuel growth.[1] The cytoskeleton is activated as indicated by ruffling of the cell surface, and the nucleus-specific actomyosin motor is activated to reposition the chromatin to enable the cell cycle genes to respond to incoming transcription factors and PK and PPase modulators of resident, *in situ* nuclear transcription factors.[11a] However, at the heart of this cascade is the mobilization and translation of the pools of normally poorly translatable growth-regulated mRNAs for proteins such as c-Fos, c-Fra's, c-Max, c-Myc, and ODC, which are major players in the switching of the cell into the cycling mode.[1–11] To make these, and all of the other proteins needed for growth, the cell stops degrading ribosomal RNA, and starts making more of it.[1] Lysosomal protein degradation stops and a large number of housekeeping proteins and new proteins such as the c-Myc and c-Fos transcription factors and nonhistone chromatin-decondensing proteins are now rapidly shipped to the nucleus to restructure the chromatin and stimulate gene expression.[1] Putrescine-ODC activator complex subunits of RNA polymerase I are sent specifically to the nucleolus to stimulate ribosomal RNA production.[1]

What is this potent "wake-up call"? It is the individualistic, cell type-dependent part of the cycle in which the cell, according to its own particular make-up, collects enough of the necessary things and grows to the critical size to be able to satisfactorily answer the first questions, pass the first checkpoint (Figure 1), and start replicating its chromosomes. Different kinds of cells use a variety of differently coupled receptors and a smaller number of different signal pathways leading away from these receptors into the depths of the cell to get to the same tightly preprogrammed chromosome-replication and mitotic phases of their cycles. I will not attempt to cover the vast, review-studded literature on this subject as I did in the previous edition of this book. Indeed, I will often refer to this first edition for the extensive pre-1990 literature. Instead, in this chapter, I will focus especially on what seems to be the basic logic underlying growth-triggering receptor tyrosine kinase (YPK) receptors, such as the monomeric EGF and PDGF receptors and the heterodimeric insulin and insulin-like growth factor (IGF)-1 and -2 receptors, which I will call "velcroceptors", the hookers of the receptor world (Figure 5).[12,13]

Ligand binding to the extracellular parts of the monomeric velcroceptors causes receptor molecules to dimerize and activate their cytoplasmic protein tyrosine kinase (YPK) domains (Figure 5).[12,13] The activated YPK domain of each receptor component then phosphorylates tyrosine residues on the other's cytoplasmic portion. This velcroizes the receptor homodimer which means that the dimer can hook onto a variety of co-signaling molecules by thrusting its phosphorylated tyrosines deeply into the co-signalers' so-called SH2 (*S*arc *H*omologous 2) pockets where the phosphate groups are grabbed and held by the terminal guanido groups of an arginine residue at the bottom of the pocket (Figure 5).[14–16] The logic of velcroization is beautiful. It enables a receptor to hook-collect a "velcrocluster" or team of signaler molecules at the cell surface where they can coordinately be phosphorylated by the YPK receptor and interact with themselves and other membrane-associated molecules to switch on several differently targeted signal mechanisms. In other words, velcroization enables one receptor to coordinately activate a cluster of mediators, the velcrocluster, and operate several different signal pathways. The kinds of SH2-bearing co-signalers in the velcrocluster on a particular velcroceptor homodimer (or heterodimer in the case of the IGF-1 or insulin velcroceptor) depends on the amino acid sequences flanking the velcroceptor's pY residues and surrounding the co-signalers' SH2 pockets.[14–16] Each velcroceptor dimer hooks a particular SH2-bearing co-signaler with each of its pY residues. Thus, the pY^{739} residue of the β-PDGF receptor, with its associated pYNAPY sequence, hooks GAP;

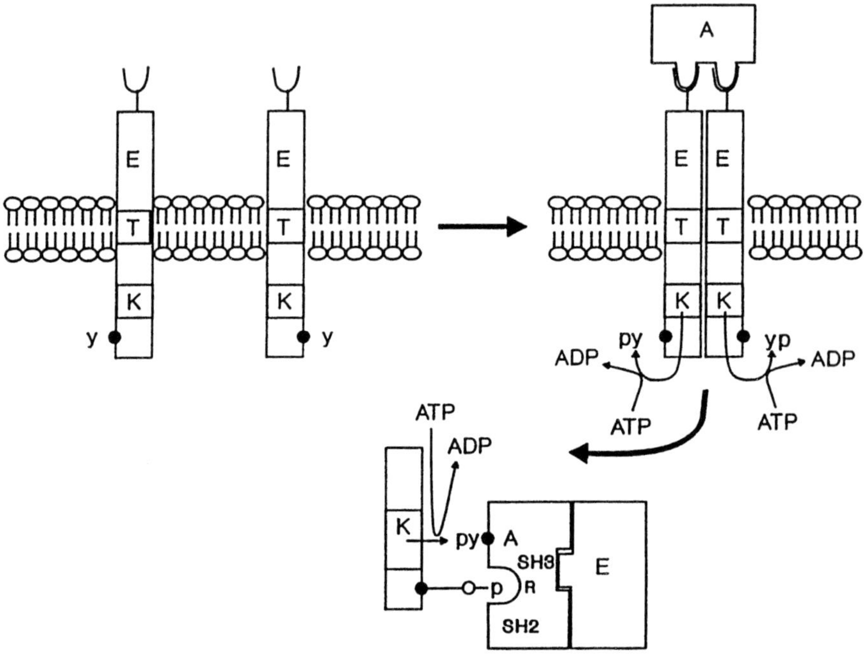

FIGURE 5 Velcroceptors and how they work. The *upper two diagrams* illustrate the functional anatomy of an integral velcroceptor. It consists of an external, agonist (**A**)-binding region (**E**), a transmembrane region (**T**), and a cytoplasmic region. The cytoplasmic region contains a YPK **K** domain and **Y**-bearing motifs specific for certain signal enzymes and adaptors. The binding of agonist (**A**) activates the cytoplasmic **K** domain and dimerizes the receptor. The **K** domain of one of the paired receptor molecules phosphorylates the various **Y** bearing motifs on the other receptor's cytoplasmic region. The paired receptors are thus "velcroized". Their **pY**-bearing motifs then selectively hook certain enzymes or adaptors by their specifically "motifed" **SH2** pockets. The *lower diagram* illustrates the hooking and activation of an adaptor (**A**) such as the Grb2 protein. The receptor **pY** sticks its phosphate group into the adaptor's **SH2** pocket. The result is a hooked, reconfigured adaptor whose **SH3** domain binds a specific enzyme (the mSos GEF in the case of hooked Grb2). A signaling enzyme such as PLC-γ_1 does not need an adaptor to associate with a velcroceptor because it has its own specifically "motifed" **SH2** pocket. It hooks onto an appropriately flanked receptor **pY** residue with its own **SH2** pocket and is then directly phosphorylated by the receptor YPK.

the receptor's pY^{1021}, with its associated pYIIPY sequence, hooks PLC-γ_1; pY^{708} and pY^{719}, with their associated pYMDMS and pYVPML sequences, hook the 85-kDa regulatory subunit of PtdIns-3′-K;[13a] and the pY^{579} residue, with its associated pYCVD sequence, hooks Src-family kinases.[14] Thus, each type of velcroceptor dimer hooks a different team of SH2-bearing signalers.

There are two kinds of SH2-bearing co-signaler adaptors and enzyme effectors (Figure 5).[15,16] Examples of adaptors are the 85-kDa regulatory subunit of PI3′K, which, when hooked by an activated velcroceptor, such as one of the PDGF receptors, activates the enzyme's mitogenically essential 110-kDa catalytic subunit, which produces PtdIns(3,4)P_2 and PtdIns(3,4,5)P_3 that may be involved in actin reorganization, PKCs activation, vesicular trafficking, and receptor downregulation.[13a] Other examples are Sem-5 and Grb2, which are involved in activating Ras proteins (Figure 3).[15–16a] An important example of an enzymic SH2-bearing effector is PLC-γ_1 which triggers PtdIns(4,5)P_2 hydrolysis.[15,16]

Co-signalers also have the evolutionarily more ancient SH3 (*S*arc *H*omologous *3*) domain.[15,16] This domain enables a hooked co-signaler or adaptor to bind to, and activate, other signaling components. Thus, for example, the Grb2 adaptor is in the cytoplasm of unstimulated cells where it is complexed through its SH3 domain with mSos, a guanine nucleotide exchange promoter, but when one of the cell's velcroceptors is activated the complex shifts to the cell membrane when the Grb2 component hooks onto the autophosphorylated velcroceptor with its SH2 pocket (Figure 3).[14] This brings mSos to membrane-associated inactive GDP·Ras complexes which it activates by stimulating the replacement of GDP by GTP (Figure 3).[14,16a]

To awaken a quiescent, cultured G_0 cell, we must get it to express the first of its cell cycle genes. To do this, we must make it do two things — make and then activate the transcription factors that will stimulate the expression of these genes. According to the currently popular scheme (Figure 3), things begin when Grb2 adaptors hook onto autophosphorylated velcroceptors. The receptor-bound, altered adaptors bind and activate mSos molecules which activate plasma membrane-bound Ras proteins by stimulating them to replace their attached GDP with GTP. The now-active membrane-bound GTP·Ras complexes interact with the N-terminal regulatory domain of the 74-kDa Raf-1 STPK.[17,17a] This interaction activates Raf-1 PK by causing it to form a detergent-insoluble complex with a membrane protein.[17a] The now membrane-bound, activated Raf-1 PK then separates from the Ras protein.[17a] The activated Raf-1 PK phosphorylates and activates a dual-function STYPK, the 45-kDa MAPKK (*M*itogen-*A*ctivated *P*rotein *K*inase *K*inase), which, in turn, activates (by phosphorylating T and Y residues) the 44-kDa MAPK (or ERK1) STPK and the 42-kDa MAPK (or ERK2) STPK that activate the 90-kDa Rsk (or S6) STPK which phosphorylates five S residues of the S6 subunits of 40-S ribosomes.[7,17–19b] These five S6 serine residues are on the "platform" of the 40-S ribosome which is involved in codon:anticodon binding. This S6 phosphorylation triggers a conformational change that increases the ability of the 40-S ribosome to bind messages.[7] GTP·Ras also activates another STPK, possibly IPPK (*I*nsulin-*S*timulated *P*rotamine *P*rotein *K*inase), which phosphorylates the S^{53} residue of eIF-4E to produce the phospho-eIF-4E needed to form complexes with the eIF-4A helicase that mobilize growth-regulated messages by melting the extensive secondary structure of their capped 5′-UTRs and binding them to the primed, S6-phosphorylated, 43-S preinitiation complexes.[4–11] The Ras/Raf-1 activated MAPK network also reaches into the nucleus to stimulate a transcription factor, c-Myc, and thus immediate-early genes whose products start the cell cycling.[16,19–21] The MAPKs directly or indirectly phosphorylate N-terminal S^{63} and S^{73} residues in the gene-activating region of c-Jun, a member of the BZip (conserved basic [B] and leucine zipper [Zip] domains for DNA-binding and dimerization, respectively) transcription factor family.[20] This promotes the gene-activation ability of c-Jun after it homodimerizes with another c-Jun or a member of the Fos family (c-Fos, c-Fos B, c-Fra 1, c-Fra 2).[20] The c-*fos* gene is transiently stimulated by signal-activated factors such as p67SRF or p62TCF which operates by binding to c-*fos*'s SRE (*S*erum-*R*esponse *E*lement).[20] The reconfigured phospho-c-Jun protein dimerizes with another c-Jun or one of the Fos family, usually c-Fos itself, to form so-called AP-1 transcription factor complexes which stimulate the transcription of early cell cycle genes.[20] The AP-1 complexes are reinforced by the arrival of another BZip transcription factor, the ubiquitous c-Myc,[20] from the cytoplasm where its handicapped message has been de-structured and mobilized.[4–11] c-Myc, unlike the multipurpose c-Jun/c-Fos (AP-1) complexes, is dedicated to stimulating the expression of proliferogenic genes when it homodimerizes or heterodimerizes with a relative, c-Max. We shall see further on that c-Myc rather ominously also stimulates the expression of apoptogenic genes at the same time it stimulates the proliferogenic genes.[22,22a] The apoptogenic gene products will cause the cell to self-destruct unless other signals are given.

Mobilizing fewer than 10 key growth-related messages is enough to start some quiescent cells cycling. This has been dramatically demonstrated with NRK rat kidney fibroblasts infected with mutant RSV or KSV viruses which, respectively, produce thermolabile, membrane-associated v-Sarc, a velcroizing nonreceptor YPK, and a mutant, activated v-K-Ras G protein.[1] The persistent signaling from these viral proteins drives their host cells to cycle indefinitely without help from serum growth factors.[1] Inactivating these mitogenic proteins by raising the temperature to a nonpermissive level (40 or 41°C) causes the cells in serum-free medium to stop cycling and switch into a G_0 quiescence. Adding serum or activating PDGF velcroceptors stimulates the serum-starved cells to make RNA and protein and start cycling. By contrast, reactivating the v-Sarc or v-K-Ras proteins by dropping the temperature induces the cells to make only a select few key proteins to trigger cycling without increasing total RNA or protein syntheses.[1] It would be reasonable to expect that the membrane-associated v-Sarc YPK,[12] like the velcroceptors, activates normal membrane-associated Ras proteins which selectively, and perhaps exclusively, stimulate the enzyme that phosphorylates and activates eIF-4E, which collaborates with the other components of the eIF-4 complex to increase the translatability of the 5′-capped, growth-regulated messages without affecting the translatability of other messages or the translation machinery. Reactivated v-K-Ras protein would simply bypass the v-Sarc step in this scheme by replacing v-Sarc-activated cellular Ras proteins.

One of the enzymes hooked, phosphorylated, and activated by the velcroceptors is the pivotal signal enzyme PLC-γ_1[23] which may, or may not in some cases, be essential for mitogenesis.[13a] This enzyme's target is a minor membrane phospholipid (9% of the total phospholipids), PtdIns(4,5)P_2. Normally, PtdIns(4,5)P_2 is linked to profilin, a protein that protects the lipid from PLC-γ_1.[24] But, when PLC-γ_1 is hooked, phosphorylated, and activated by the velcroceptor, it can knock profilin aside and attack PtdIns(4,5)P_2.[24,25] This assault breaks PtdIns(4,5)P_2 down into Ins(1,4,5)P_2 (or IP_3) and a first wave of diacylglycerols (DGs).[26–28]

The first wave of DGs can stimulate 8(α, β1, βII, γ, δ, ϵ, η(L), θ, but not ζ, λ) of the 10 currently known PKCs.[26,29] At low concentrations they can stimulate any inactive PKCs that the cell might be keeping in its membrane, and at higher concentrations with the help of an Ins(1,4,5)P_3-induced Ca^{2+} surge they can also induce the translocation of PKCs from the cytosol to activation sites in the plasma membrane.[30] The first burst of membrane-associated PKC activity phosphorylates a host of membrane proteins or proteins near the membrane. It phosphorylates and inhibits some nearby velcroceptors such as the EGF/TGF-α receptor and enzymes such as PLC-β or PLC-γ_1.[1,28] It stimulates the nonreceptor c-Sarc YPK, transcription factors such as c-Fos and c-Jun, and the release of the NF-κB transcription factor from its inhibitor.[1,31] It also activates eIF-4E.[7,10] However, the first wave of DGs crashes within seconds because the PKCs it has stimulated have phosphorylated PLC-γ_1 on S^{1248} (or PLC-β on S^{887}) which shuts the enzyme off and thus stops further generation of the intrinsically ephemeral DGs.[28] But the PKCs stimulate two other phospholipases, PLD and PLA_2.[27,28,32] PLD now breaks phosphatidylcholine (PtdCh) into PA (phosphatidic acid) and choline. The PA is dephosphorylated into DGs by PAP (phosphatidic acid phosphohydrolase).[27,28,32] Thus is started a second large, long-lasting wave of DG production from a far more plentiful membrane phospholipid.[28,30] Meanwhile, the PKC-stimulated PLA_2 generates arachidonic acid, which can leave the cell and serve as a paracrine factor and be the progenitor of a shower of potent signalers, the eicosanoids, some of which (e.g., PGEs) can stimulate adenylyl cyclase and thus trigger the cyclic AMP/PKAs signal mechanism with yet another set of cytoplasmic and nuclear targets.[1,28,33] Arachidonic acid can also bind to and inactivate the Ras-Gap (GTPase-activating protein) and activate the Ras-GIP (GTPase-inhibiting protein) proteins, which results in a surge of active GTP·Ras complexes that can trigger the Raf-1/MAPK cascade and the IPPK-induced phosphorylation of eIF-4E, as well as stimulate PtdCh-PLC, which would contribute to sustaining the second DG wave and thus prolonging PKCs activity to drive the proliferogenic process.[11,28]

The transient surge of Ins(1,4,5)P_3 that accompanies the first wave of PtdIns(4,5)P_2-derived DGs does a very important thing — it binds to 260-kDa receptors on Ca^{2+} storage vesicles, known as calciosomes, which may bud from, or remain as specialized regions of, the endoplasmic reticulum (ER, Figure 6).[33a-33d] The calciosomes are also linked to the plasma membrane by actin microfilaments and membrane-inserted ankyrin adaptors, which serve as signal wires to tell membrane Ca^{2+} channels when the stores are empty and to

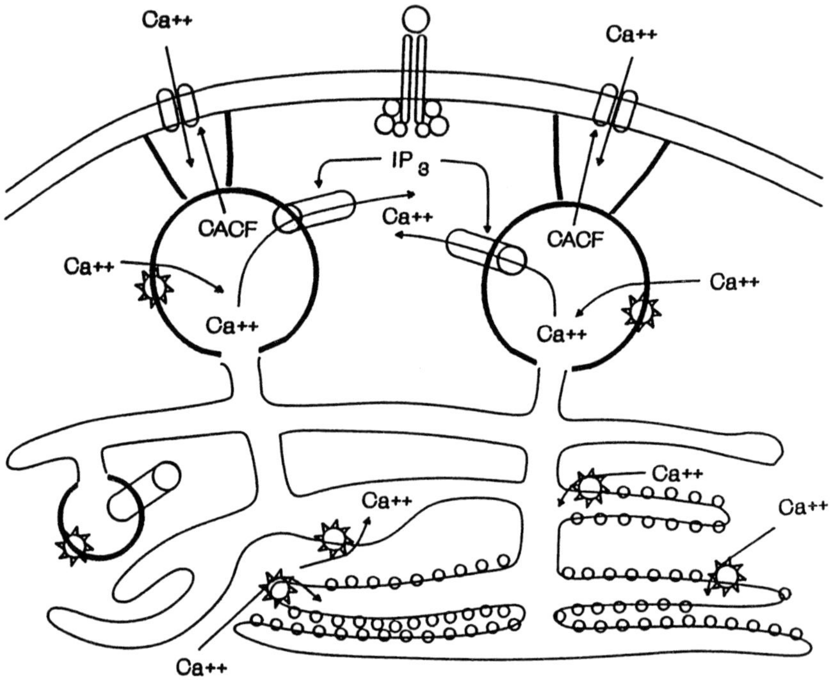

FIGURE 6 The sources of the internal Ca^{2+} surges that follow the activation of a velcroceptor such as the HGFR. The HGF-dimerized HGFR is auto-velcroized and collects and activates a team of signaling enzymes, one of which is PLC-γ_1. PLC-γ_1 breaks membrane PtdIns(4,5)P_2 down into DGs and IP_3[Ins(1,4,5)P_3]. IP_3 binds to, and activates, its receptor/channel on membrane-associated Ca^{2+} storage vesicles (calciosomes), which are specialized, receptor signal-responsive parts of the ER, and are linked to the cell membrane by actin microfilaments attached to the cell membrane by membrane-inserted ankyrin adaptors. The stores release their Ca^{2+} through the IP_3-opened channels. Their emptying closes the IP_3 channels, stimulates the release of the CACF factor, and activates thapsigargin-inhibitable ATPase pumps, which promptly refill the stores. Thus, the initial Ca^{2+} surge rapidly subsides, but CACF is a Ca^{2+} channel activator which has opened channels in the cell membrane. This CACF-induced Ca^{2+} influx stops the decline and holds the Ca^{2+} concentration at a lower, but still suprabasal, level. It should be noted that the (○) symbols represent ER-attached ribosomes.

open and admit enough external Ca^{2+} to refill the stores.[34–37] The empty calciosomes may also release a factor that opens Ca^{2+} channels.[38,39]

Is a Ca^{2+} surge needed to start resting G_0 cells cycling in response to velcroceptor signals? There are several reasons to believe that the answer to this question is ... yes! A Ca^{2+} surge is a common part of the response to various mitogens.[1] Moreover, "newborn" (i.e., just emerged from telophase) C127 mouse cells briefly generate a train of Ca^{2+} transients, that is, they start singing a Ca^{2+} song, when starting the G_1 build-up (Figure 7).[40] Confluent or serum-deprived BALB/c3T3 mouse cells, human foreskin fibroblasts, and the

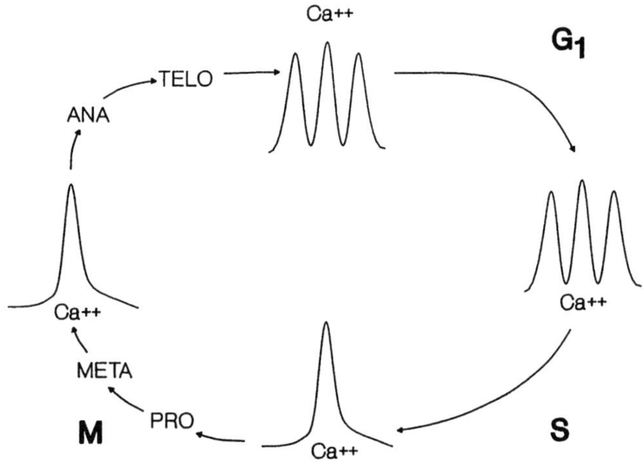

FIGURE 7 The Ca^{2+} signals that trigger various parts of the cycle.[40] When a newborn C127 mouse cell emerges from telophase and decides to start a cycle, it generates a train of Ca^{2+} oscillations (its Ca^{2+} song). When it nears the end of the G_1 build-up it starts singing again. It generates another train of Ca^{2+} oscillations starting at the nuclear periphery and spreading throughout the cell. When the chromosomes are replicated, the cell generates a Ca^{2+} spike to help initiate prophase and another Ca^{2+} spike to trigger anaphase movements.

notorious Swiss albino 3T3 mouse cells promptly empty their loaded calciosomes in response to addition of serum, growth factors such as EGF and PDGF, and mitogenic hormones.[1] The mitogenic signal from PDGF velcroceptors triggers a rapid, large (eightfold) rise in both nuclear and cytoplasmic Ca^{2+} in quiescent BALB/c3T3 cells. Preventing the Ca^{2+} surge by loading the cells with the Ca^{2+}-buffer Quin 2 prevents them from starting to cycle.[41] Emptying the internal Ca^{2+} stores of DTT1MF-2 smooth muscle cells using an inhibitor of the calciosome-filling or refilling ATPase pumps (Figure 6), such as thapsigargin, keep the cells in a G_0 quiescence, whereas switching the pumps back on and refilling the stores enables the cells to resume cycling.[42] Because of their ample internal Ca^{2+} stores, BALB/c3T3 cells and Swiss albino 3T3 cells in low-Ca^{2+} medium can respond to various velcroceptor signals by generating the necessary Ca^{2+} surge and reach a late G_1 stage despite the external Ca^{2+} shortage, but then at this late stage they must start singing another Ca^{2+} song (Figure 7) and cannot start replicating DNA without external Ca^{2+} or in the presence of a Ca^{2+} channel blocker such as verapamil.[1] By contrast, serum-deprived or confluent T51B rat liver cells and WI-38 human fibroblasts do not have enough stored calcium to trigger the early pre-G_1 events without help from external Ca^{2+}.[1] If they are stimulated with fresh Ca^{2+}-deficient (e.g., 0.02 m*M* Ca^{2+}) medium-serum they will not trigger the early events, but they are ready to do so if the external Ca^{2+} concentration is raised to a normal level (e.g., 1.5 m*M*) at any time during the next 2 d. Evidently, serum factors are powerless under these conditions, and nothing happens without Ca^{2+}: key starting events must await the arrival of Ca^{2+} and the internal Ca^{2+} surge. The importance of Ca^{2+} at the start of the cycle as well as at the end of the G_1 build-up is indicated by the observation that depriving growing cultures of C3H10T1/2 mouse fibroblasts and WI-38 human fibroblasts of Ca^{2+} blocks cells at either the start or at the end of the G_1 build-up, but not at stages in-between.[1]

How do Ca^{2+} and the other products of a mitogenic velcroceptor start cells cycling? The short answer is, I don't know because we are dealing with mind-boggling complexity and ignorance. One of the key players in the first wave of events is the minor membrane phospholipid, PtdIns(4,5)P_2, the ultimate source of the functionally obscure, but seemingly important, PtdIns(3,4)P_3, as well as the Ca^{2+}-mobilizing Ins(1,4,5)P_3[13a,26–28] and the first surge of DGs. In the quiescent cell, the 80-kDa MARCKS (*M*yristoylated *A*lanine-*R*ich *C*-*K*inase *S*ubstrate) protein is bound to its membrane receptor and cross-links or bundles actin filaments into a rigid cortical meshwork which prevents secretory granules from contacting the plasma membrane.[33,44,45] PtdIns(4,5)P_2 is shielded from attack by PLC-γ_1 by anchoring the Ca^{2+}-binding gelsolin and profilin to the cell membrane. When the velcroceptors of a growth factor, such as the AA or BB isoform of PDGF, are activated and then capture and phosphorylate PLC-γ_1, the phosphoenzyme can attack PtdIns(4,5)P_2 and break it down into DGs and Ins(1,4,5)P_3. This detaches gelsolin and profilin from the cell membrane.[25] The PKCs stimulated by the brief first wave of DGs phosphorylate the MARCKS protein, which is released from the cell membrane and can no longer cross-link actin filaments.[43,45] At the same time, the Ca^{2+} released from Ins(1,4,5)P_3-stimulated calciosomes binds to the liberated gelsolin which has attached itself to the debundling actin filaments, and the detached profilin binds to the actin monomers.[43–45] Also at the same time, Ins(1,4,5)P_3 might reach the ER-derived nuclear envelope, activate receptors on its outer membrane, and thus release Ca^{2+} from a pool in the space between the two membranes (Figure 8).[46–48] The surging Ca^{2+} and the multipurpose Ca^{2+}-binding/activated signaler, CaM, activates a formidable array of enzymes that includes some adenylyl cyclase isoforms, calcineurin (a PPase), calpain (a protease), CaMPKII, CaMPKIII (eEF-2 kinase), and MLCK (myosin light-chain kinase). The nuclear Ca^{2+} content rises and the resulting nuclear Ca^{2+} surge and Ca^{2+}•CaM complexes stimulate the chromatin-shifting, matrix-associated actomyosin, as well as CaMPKII which supplements the gene-stimulating actions of AP-1 complexes, c-Myc homodimers, and c-Myc/cMax heterodimers by phosphorylating other transcription factors bound to a CaMRE (*CaM*-*R*esponsive *E*lement) in the regulatory regions of certain genes.[1,11a,46] CaMPKII may also stimulate other genes by phosphorylating the CREB (*C*yclic *A*MP-*R*esponsive *E*lement *B*inding) protein transcription factors that are the targets of the PKAs and bind to the genes' CREs (*C*yclic *A*MP-*R*esponsive *E*lements).[46,48] Thus, the Ca^{2+} spike or oscillations will activate immediate-early cell cycle genes, such as c-*fos,* which have a CRE.[49] The Ca^{2+} surge or oscillations may also stimulate

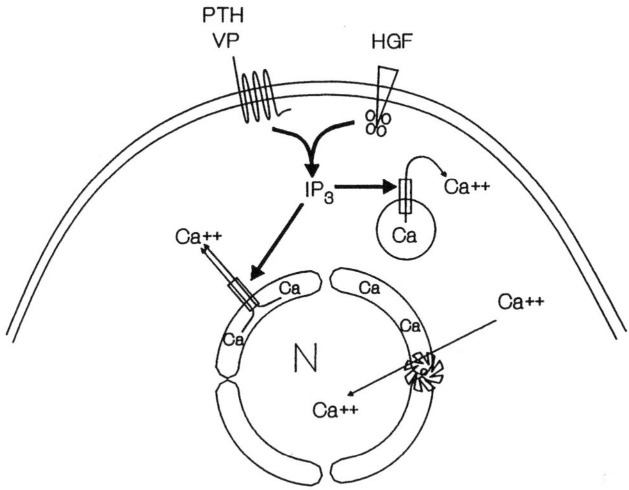

FIGURE 8 PLCs activated by receptors such as HGF velcroceptors or the 7-TM (or serpentine) PTH or VP1 receptors generate [$IP_3Ins(1,4,5)P_3$] by hydrolyzing membrane PtdIns(4,5)P_2. The IP_3 binds to, and opens, specific receptor channels on cytoplasmic ER-derived calciosomes and on the outer surface of the ER-derived nuclear envelope. Released Ca^{2+} is pumped into the nucleus by an ATPase pump where it triggers the several Ca^{2+}•CaM-dependent and other Ca^{2+}-dependent reactions discussed in the text.

the release of Rel-family transcription factors which have been bound by ankyrin repeats to the cytoskeleton and membrane proteins.[31] The combination of Ca^{2+}-induced disruption of the cortical actin meshwork, activation of MLCK and cortical myosin II, the consequent contraction of cortical actomyosin filaments, and an associated activation of microtubule-associated motors, starts the velcroceptor-activated cell twitching, ruffling its surface, and pumping transcription factors and other things into its actively reconfiguring nucleus.[1,2–5,11a,50]

Another thing started by Ca^{2+} in collaboration with the Grb2 adaptor and its attached Ras-activating mSos protein (guanine nucleotide exchange protein) sticking to the velcroceptor (Figure 3) is the mobilization of the low-priority growth-regulated messages. The Ins(1,4,5)P_3-triggered Ca^{2+} surge increases the binding of eIF-2 and eIF-3 translation-initiation factors to the 40-S ribosomes.[1,51] The Ca^{2+} surge also stimulates CaMPKIII (eEF-2 kinase), which is highly specific for the T^{56} residue of the eEF-2 peptide chain elongation factor.[7] This T^{56} phosphorylation stops the eEF-2-GTP-driven translocation of peptidyl-tRNAs from the A-site to the P-site on the 80-S ribosome.[7] This would promote the cell's switch-over from quiescent housekeeping to active cycling by making peptide chain elongation, rather than initiation, the limiting factor.[7] This, along with the Ras/IPPK-activated phospho-eIF-4E (Figure 3) would facilitate access of the 5′-UTR-destructured, previously disadvantaged, growth-regulated messengers vis-à-vis their former competitors to the ribosomes by causing a brief build-up of initiation factors.

The Ca^{2+}- and Ras-dependent eIF-4E-induced mobilization of the low-priority growth-regulated messages looks as if it might be the most important cycle-triggering event in quiescent immortalized cells, while many other events, such as PtdIns(4,5)P_2 breakdown and the first surge of PKCs activity may not be essential.[13a] This is strongly indicated by the ability of microinjected eIF-4E to trigger a dose-dependent stimulation of DNA synthesis by quiescent (because of serum deprivation) NIH 3T3 mouse cells and Rat-2 fibroblasts.[4–6,8] Moreover, overexpressed eIF-4E also eliminates the need for adhesion to a solid substrate to proliferate (i.e., the overexpressing cells can form colonies even while suspended in semisolid soft agar medium) and makes the cells tumorigenic in nude mice.[4–6,8] By contrast, quiescent primary, hence still "mortal", fibroblasts cannot be stimulated by overexpressing eIF-4E: they must express c-Myc or another immortalizing transcription factor, such as E1A, to respond.[4–6,8] This, plus the fact that expression of an activated Ras protein increases phosphate turnover on eIF-4E S^{53} residues,[52] suggests an answer to why the appearance of an activated Ras mutant is often associated with carcinogenesis in a variety of tissues and why Ras must collaborate with Myc to transform cells. It also applies to the Ras-mediated proliferogenic action of the Rous sarcoma virus's v-Src YPK or the polyoma virus's membrane-associated middle T protein, which operates by binding and activating the host cell's normal c-Src YPK.[1,12] Thus, the expression of a mutant activated Ras in a "mortal cell" with no, or only a subthreshold, pool of growth-regulated messages from c-Myc's client genes would not stimulate or transform the cell. But when the cell has been immortalized and contains a suprathreshold load of low-priority growth-regulated messages, either an activated Ras protein or overexpressed eIF-4E would stimulate and dysregulate cell proliferation.

But what about the PKCs? I have argued in the first edition of this book and elsewhere that the DGs-stimulated PKCs were not invented to stimulate proliferation.[1,53–55] As we shall see in Chapters 4 and 5, PKCs actually *stop* the proliferation and stimulate the differentiation of many normal cells. Quiescent, immortal BALB/c3T3 cells do not need PKCs to respond to the mitogenic signal from their PDGF velcroceptors. This is indicated by the fact that down-regulating (eliminating) PKCs by prolongedly treating the cells with TPA did not prevent PDGF from stimulating DNA synthesis in experiments by Diliberto et al.[56] And mutation of the PDGF receptor's PLC-γ_1-binding site eliminates PLC-γ_1 binding to the autophosphorylated, velcroized receptor, but it does not affect the binding of other key components, such as the 85-kDa regulatory component of PtdIns-3′K, to the receptor or eliminate the mitogenic response.[13a] Moreover, neither the Ca^{2+} surge nor the PKCs-stimulating DGs resulting from the activation of fibroblast growth factor (FGF) velcroceptors are needed for the mitogenic response of L6 myoblasts to FGF.[57,58] Thus, L6 myoblasts with mutant FGF receptors that cannot induce PtdIns(4,5)P_2 breakdown (because they have a F^{766} residue instead of the autophosphorylatable Y^{766} needed to hook and activate PLC-γ_1) initiate DNA synthesis and proliferate normally in response to FGF. Thus, Ca^{2+} and PKCs may be meant to stimulate other things such as differentiation and migration while it is some of the other signalers that can still hook onto the mutant FGF velcroceptor that are parts of the proliferogenic mechanism. However, there are many examples of potent PKC stimulators, such as TPA, being able by themselves to stimulate some quiescent immortalized and neoplastic cells to start cycling. The reason for this difference might be found in the size of the cell's pool of growth-regulated messages. PKCs are not normally used to activate eIF-4E,[7] but they can be made to phosphorylate eIF-4E and mobilize the growth-regulated messages.[5,7,8,10] Thus, massively activating PKCs with the potently nonphysiological TPA can bypass the normal Ras/Raf-1-triggered STYPK and STPK cascade and mobilize the cycle-starting messages. Therefore, PKC activators such as TPA, like microinjected eIF-4E, can stimulate message-preloaded quiescent immortal cells such as Swiss albino 3T3 mouse cells, but they cannot induce mortal cells to cycle because these cells need other mediators to stimulate the production of mitogenic messages. However, potent PKC activators can also stimulate a phosphatase that enhances the ability of AP-1 complexes to stimulate cell-cycle genes by dephosphorylating c-Jun's C-terminal phospho-serines (e.g., S^{243}) which increases the protein's ability to bind to DNA.[20]

In summary, the main goal of the several mediators fanning out from the activated growth factor velcroceptors is to load the G_0 cell's nucleus with transcription factors that can stimulate selectively the expression of a set of genes whose products will induce the cell to start making the parts of its cell cycle engine. This is done in four ways: mobilize any preexisting, low-priority messages for transcription factors; reduce the turnover of these messages and their products; stimulate the genes coding for the transcription factors; and increase the effectiveness of the transcription factors by phosphorylating or dephosphorylating key residues. However, the ubiquitous proliferation-specific c-Myc BZip transcription factor does something potentially deadly as it helps trigger a cycle — it activates genes that unless they are turned off or the actions of their products are blocked by subsequent signals from cycle-progression-stimulating factors such as the Bcl-2 protein, will initiate a suicide mechanism known as apoptosis (from the Greek meaning to fall off like leaves in the Autumn, or like hair from a balding pate) which will be mentioned often later in this and other chapters.[22,22a] Thus, the c-Myc-induced apoptogenic program is meant to kill the cell in order to avoid generating clones of dangerous mutant cells should the G_1/S checkpoint monitor find something irreparably wrong with the cell's chromosomes (Figure 1).

The other kind of reversible quiescence, the functional quiescence of cells such as hepatocytes in the intact adult rat liver, is not due to a lack of nutrients, but it is due to subthreshold levels of a group of proliferogenic hormones, growth factors, and eicosanoids in the portal blood. Unlike the sluggish G_0 cultured cells, the nonproliferating hepatocytes in the liver are very busy cells which make, secrete, detoxify, and otherwise dispose of, a lot of things.[1] Each

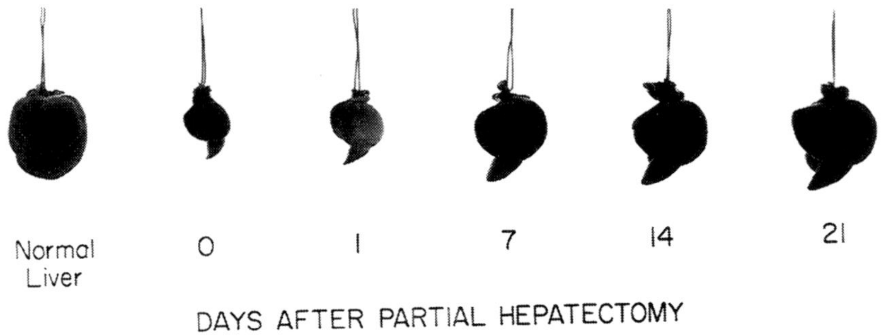

FIGURE 9 The growth of the liver remnant and the restoration of the normal tissue mass by 21 d after 70% HPX in 200-g male Sprague-Dawley rats.

hepatocyte makes very large amounts of protein, much of which it exports, some of which it uses for self-maintenance, and the rest of which is wasted.[1] Despite this intense activity, hepatocytes do not grow and proliferate, because they do not express their cell cycle genes, whose products such as the cyclin-dependent PKs of the cell cycle engine would switch them into the proliferative mode. However, we do not know whether they maintain a pool of the low-priority, handicapped growth-regulated messages to enable them to start cycling in response to an emergency. Such an emergency occurs when there is a sudden, life-threatening loss of cells.

Removing 70% of a young, male Sprague-Dawley rat's liver somehow urges the remaining liver cells to start growing and preparing to initiate the G_1 build-up to chromosome replication and mitosis (Figure 9). Within an hour of 70% partial hepatectomy, HPX, the hepatocytes increase gene transcription and primary transcript processing and start pumping messages out of the nucleus, which produces an early surge of messages into polyribosomes and a steady rise in the total concentration of polyribosome-associated mRNA that goes on for 16 h until it stops just before the onset of DNA replication.[1] The switching signal does not trigger the de-differentiation of the remaining hepatocytes: this would kill the animal. On the contrary, unlike cultured immortalized cells, they must actually step up their normal activities in order to compensate for their lost comrades and maintain essential body-wide ("federal") functions while at the same time preparing to replace the lost cells as fast as possible. Therefore, most of the wave of new messages being produced and pumped into the cytoplasm are for function-related, rather than proliferation-related components. But among these messages are the crucial, but poorly competitive, low-priority, 5′-UTR-handicapped messages from the cycle-starting immediate-early and early cell cycle genes. First to come are the general-purpose mode-shifting c-*fos* messages between 10 and 15 min, which peak between 15 and 20 min and then vanish by 2 to 3 h.[1] Then come the proliferation-specific, c-*myc* messages which peak at 1 h after HPX just at the height of the flow of messages into the polyribosome fraction and then drop to a lower, but still suprabasal level. There must be a burst of CaMKIII activity, the GTP·Ras-induced kinase cascade, and eIF-4E activation to bring these messages to the head of the translation line.

What triggers the first response? HPX sets off a general alarm throughout the body, perhaps by the remnant releasing an injury protein that stimulates the production of endocrine/paracrine mitogen(s).[58a] This triggers the release of signalers from the anterior and posterior pituitary, lung, parathyroid glands, thyroid C cells, and the remnant's own Kupffer cells and sinusoid endothelial cells.[1,53–55] Hepatocytes have receptors for calcitonin, glucagon, glucagon-related intestinal peptides (GRP-glucagon-37 or oxyntomodulin), hepatocyte growth factor (HGF), parathyroid hormone (PTH), and vasopressin (VP1), all of which surge into the portal blood during the first 15 min to 3 h after HPX. The parathyroid gland's chief cells are stimulated to release PTH prolongedly and maximally by the large drop (1 mg/100 ml of plasma) in the blood Ca^{2+} concentration caused by the operation and is probably triggered by trauma-induced release of calcitonin from the thyroid gland and interleukin-1 (IL-1) from the Kupffer cells in the liver remnant.[1,53–55] PTH is a general wound-response hormone that affects the differentiation, functions, and proliferation of cells, including bone marrow cells, dermal and kidney fibroblasts, keratinocytes, osteoblasts, spleen lymphocytes, thymic lymphoblasts, and, most importantly at this point, hepatocytes.[1] Indeed, the post-HPX, hypocalcemia-induced PTH release might be partly responsible for the stimulation of thymic lymphoblast proliferation that follows HPX.[59] However, neither the PTH receptors nor the hepatocyte's VP1 receptors are velcroceptors: they are 7-TM or serpentine receptors.[60–64] These receptors pass, or "snake", in and out of the plasma membrane seven times (hence 7-TM or transmembrane domains) and couple to their target signaling enzymes, a PLC (probably a PLC-β) in the case of the high-affinity PTH receptor and the VP1 receptor and both a PLC and adenylyl cyclase in the case of the low-affinity PTH receptor, by G proteins (G_s for adenylyl cyclase and $G_{q/11}$ for the PLC) (Chapter 4, Figure 6).[60–64] Hepatocytes do not need the signals from PTH receptors to start cycling. But the large surge of vasopression in the portal blood which peaks at 2 h after HPX and the signals from VP1 receptors must be directly or indirectly needed, because the regeneration response is greatly impaired in Brattleboro rats which do not make the hormone.[1] Another early signaler is prostaglandin E_2, an adenylyl cyclase stimulator whose accumulation peaks about 1 h after HPX. Indeed, arachidonic acid, the eicosanoid precursor, which is released, along with its various metabolites, from the liver remnant's Kupffer cells, can be a potent hepatocyte activator.[1] It is one of the two things, HGF is the other, that can stimulate the proliferation of the EGF-insensitive, deeply quiescent, mature hepatocytes that make up about 70% of the hepatocytes in primary neonatal rat liver cultures.[1] A prompt surge of prolactin (PRL), probably another response to the surgery-induced drop in the blood Ca^{2+} concentration, could be another mitogenic signaler.[65,66] According to some, PRL, like PTH and vasopressin, directly or indirectly stimulates PtdIns(4,5)P_2 breakdown into Ins(1,4,5)P_3, which releases Ca^{2+} from internal stores, and DGs, which stimulate PKCs, and it may also be translocated into the hepatocyte nucleus where it stimulates nuclear PKCs that directly activate AP-1 transcription factor complexes as well as the expression of the c-*myc* gene.[65–67] According to others, PRL does not stimulate PtdIns(4,5)P_2 breakdown, but rather it may stimulate the breakdown of PtdCh, which does not yield the Ca^{2+}-mobilizing Ins(1,4,5)P_3, but does yield the PKCs-stimulating DGs.[68,69] The full range of mechanisms that are activated by the PRL receptor is unknown. This receptor, which may be both nuclear and surface, belongs to a family that includes the growth hormone receptor and the β(p75) component of the IL-2 receptor (to be discussed in Chapter 6), which are not homodimerizing, PLC-γ_1-stimulating velcroceptors, but rather are modular velcroceptors which activate G proteins and recruit nonreceptor Src-family YPKs to compensate for their lack of YPK cytoplasmic domains.[69]

These many HPX-induced members of the "Portal Signaling Symphony" can together generate a Ca^{2+} surge as well as bursts of PKA and membrane-associated PKCs activities. Ca^{2+} and PKCs alone or together cannot stimulate hepatocyte proliferation, nor can cyclic AMP/PKAs.[1] However, it is known that the cyclic nucleotide or an adenylyl cyclase activator is an essential part of cocktails that stimulate hepatocyte proliferation in the intact rat liver.[1] But something is missing — the potency of the coordinated activation of a velcroceptor-hooked team of Ras-activators, PI3′K, and other things that can activate eIF-4A and eIF-4E and the MAPK cascade with their potent effect on the production, translatability, and translation of cycle-starting messages. Is a velcroceptor involved in starting regeneration? The answer is yes! Indeed it has been suggested that 82-kDa hepatocyte growth factor (HGF), consisting of a 69-kDa α-subunit and a 34-kDa β-subunit, is in fact the main hepatocyte proliferogen which activates a heterodimeric velcroceptor (the product of the c-*met* proto-oncogene) consisting of an extracellular 50-kDa-subunit which is linked by disulfide bonds to a membrane-spanning 145-kDa β-subunit whose cytoplasmic domain has a YPK motif.[13,70–74] During the first 3 h after HPX, the putative injury protein from the remnant recruits HGF from the lungs as well as the remnant's own Kupffer cells, sinusoid endothelial cells, and particularly its Ito cells.[58a,72,73] Thus, HGF is both an endocrine and paracrine proliferogen (in fact, it is both a mitogen and motogen or motility stimulator). The HPX-mobilized HGF binds to the 500 to 600 high-affinity (K_d = 20 to 30 p*M*) velcroceptors on each hepatocyte's surface.[13,73] Like all velcroceptors, the HGF receptors dimerize and each dimer hooks onto its specific velcrocluster such as GAP, Grb2, PI3′K, and PLC-γ_1.[58a]

Clearly, HPX unleashes a swarm of signalers that converge on the hepatocyte surface facing the sinusoid and the portal blood and redundantly stimulate all of the major signal mechanisms. HGF velcroceptors and vasopressin's 7-TM VP1 receptors might start the hepatocytes singing a Ca^{2+} song, because both trigger $[Ca^{2+}]_i$ oscillations in cultured hepatocytes.[74,75] HGF triggers "sinusoidal wave Ca^{2+} oscillations" which are rapid and superimposed on a suprabasal Ca^{2+} level, while vasopressin triggers "baseline spikes" which occur at a lower frequency than sinusoidal oscillations and rise from, and fall to, the basal level.[74,75] Many mechanisms have been proposed to explain these oscillations, but the simplest ones that are consistent with the evidence have been presented by Putney and Bird.[75] The sinusoidal oscillations passively accompany DG or PLC/PKCs oscillations. Thus, activation of PLC-γ_1 by HGF velcroceptor stimulates the generation of Ins(1,4,5)P_3 and DGs from PtdIns(4,5)P_2. The Ins(1,4,5)P_3 binds to its receptor/channels on calciosomes (Figure 6). This releases Ca^{2+} from the calciosomes. Meanwhile, the DGs stimulate membrane-associated PKCs, which feed back onto, and inhibit, PLC-γ_1 and thus stop the generation of Ins(1,4,5)P_3 and DGs. The Ca^{2+} stores are refilled by the calciosomes' ATPase pumps when the stores are empty and the Ins(1,4,5)P_3 level falls. But when the DG level falls, the PKCs activity falls, the PLC-γ_1 is dephosphorylated and reactivated, and PtdIns(4,5)P_2 breakdown starts again and causes a second release of Ca^{2+} from the refilled calciosomes. The release of Ca^{2+} from the stores opens Ca^{2+} channels in the plasma membrane by sending signals along actin "wires" connecting the calciosomes to the Ca^{2+} channels, or by releasing a channel-activating coupling factor (Figure 6). This coupling of calciosomes to channels is meant to replace Ca^{2+} lost from the cell through membrane Ca^{2+}-ATPase pumps or taken up by the mitochondria. Vasopressin-induced baseline spiking also begins with Ins(1,4,5)P_3-induced emptying of the Ca^{2+} stores. As the cytosol Ca^{2+} concentration rises above a certain level, the Ins(1,4,5)P_3 receptor/channel is shut off by Ca^{2+} and/or Ca^{2+}·Cmd, as well as by the emptying of the calciosome lumen. Now the calciosome ATPase pump is activated and refills the calciosomes. As the cytosol Ca^{2+} drops, the cytosol Ca^{2+} concentration reaches the level for optimal activity of Ins(1,4,5)P_3-receptor complexes and the refilled calciosomes empty again. Of course, some of the released Ca^{2+} will be lost with each calciosome emptying because some of it is either pumped out of the cell by the Ca^{2+}-ATPase pumps in the plasma membrane or taken up by the mitochrondria. But membrane Ca^{2+} channels are signaled to open and each Ca^{2+} spike is followed by a surge of external Ca^{2+} through these channels into the cell. It seems that the $[Ca^{2+}]_i$ oscillations originate from a specific locus in the hepatocyte and spread out from there in waves. The starting point in a given hepatocyte is the same regardless of the stimulator and receptor, which indicates a specific site for the signal transduction mechanism. Whether $[Ca^{2+}]_i$ oscillates and the hepatocytes start singing as they switch into the cycling mode in the real world of a liver remnant is unknown and perhaps unknowable. But why would they bother? Putney and Bird[75] doubt the importance of sinusoidal oscillations such as those triggered by HGF, because they are usually seen only at subphysiological temperatures. But the baseline spiking triggered by vasopressin does occur at physiological temperatures. It can accurately tell a cell about the concentration of the agonist and enable the cell to discriminate between noise and signal even at very low agonist concentrations. Increasing the agonist concentration shortens the delay before the first spike (i.e., the time required for the concentration of Ins(1,4,5)P_3 to reach a critical level) and increases the frequency of spiking, but does not affect the amplitude of the spikes. The spike frequency would be reflected in the rate of build-up of at least one key signal enzyme, CaMKII.[76] Thus, each spike produces Ca^{2+}•CaM complexes which can bind to the enzyme's approximately 10 α- and β- or β^1-subunits and separates each subunit's autoinhibitory domain from its catalytic site. Each Ca^{2+}•CaM-activated subunit does two things — autophosphorylates and phosphorylates its many substrates. When a Ca^{2+} spike crashes down, Ca^{2+} dissociates from CaM, which in turn should promptly dissociate from, and deactivate, the subunit. But this does not happen. The autophosphorylation of each activated subunit's T^{286} residue has greatly increased (about 1000-fold) the affinity for CaM, which now needs several hundreds of seconds, rather than the expected fraction of a second, to dissociate from the subunit. This CaM-trapping increases the impact of brief Ca^{2+} spikes on the enzyme and the cell. It also enables the multimeric enzyme to detect and reflect the agonist level. Thus, at a low frequency of spiking, each spike will provide only enough Ca^{2+}•CaM to partially activate the multimeric enzyme. As each spike crashes, these activated subunits have enough time before the next spike to lose their CaM, to be dephosphorylated, and to be deactivated. As the spike frequency rises, the subunits will have increasingly less time to lose their CaM between spikes and at a certain frequency, the enzyme's multimers will actually *recruit* CaM released from other proteins and the enzyme will become maximally activated when all of its 10 subunits have trapped a CaM.

We must not fail to note that Ins(1,4,5)P_3-responsive cytoplasmic Ca^{2+}-storage vesicles are not the only participants in cellular Ca^{2+} surging and oscillation. If Ins(1,4,5)P_3 produced at the hepatocyte's surface reaches the nucleus, it will bind to receptors on the nuclear membrane and release Ca^{2+} from a pool between the two nuclear membranes (Figure 8). Also, hepatocytes have another kind of Ca^{2+}-storage vesicle (the type II vesicle in Figure 10) which responds directly to cytosolic Ca^{2+} instead of Ins(1,4,5)P_3 through Ca^{2+}-activated ryanodine receptors and can take up Ca^{2+} released from the Ins(1,4,5)P_3-sensitive vesicles and thus can modulate Ca^{2+} influx and contribute to oscillations.[37,75,77]

One of the two goals of the first burst of signaling is to get the cell growing to a proper size and infrastructure for initiating chromosome replication and ultimately dividing. The other goal is to start the production of the first cyclin-dependent PK to stimulate the expression of the transcription factor(s) that coordinately starts the expression of the genes for the various chromosome parts, replication enzymes, and the second wave of cyclin-dependent PKs which will activate the replication complexes when they have assembled on the replication origins and/or replicators.[2,3]

The proliferogenic switch-over begins with a dramatic surge of message production which peaks between 1 and 3 h after HPX and then drops slowly

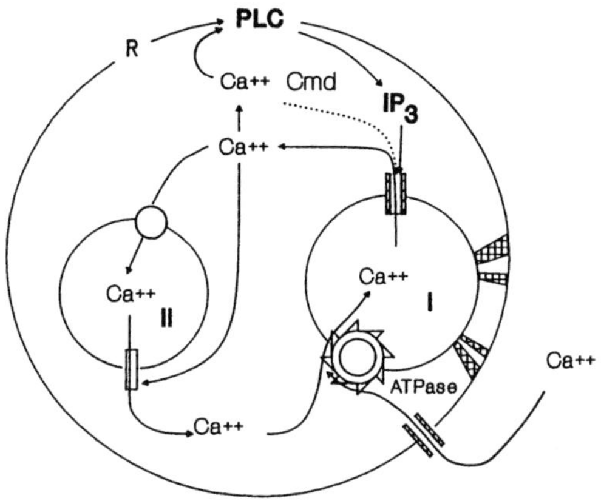

FIGURE 10 A possible interaction between the two kinds of Ca^{2+} storage vesicle. A velcroceptor, **R**, activates PLC-γ_1 which hydrolyzes membrane PtdIns(4,5)P_2 to produce IP_3Ins(1,4,5)P_3. IP_3 binds to its receptor/channels on the first kind (**I**) of vesicle. The vesicles empty through the IP_3-opened Ca^{2+} channels. The released Ca^{2+} would promote PLC-γ_1 activity. It also produces Ca^{2+}•Cmd (calmedin) complexes which, together with the emptying of the vesicles' Ca^{2+} load, closes the IP_3 receptor/channels. The emptying also causes the release of CACF, which opens membrane Ca^{2+} channels and activates the vesicles' ATPase pumps. When the cytosol Ca^{2+} concentration rises to a critical level it triggers the release of Ca^{2+} from the second type (**II**) of vesicle (which has been accumulating the released Ca^{2+}) by activating their ryanodine receptors/channels. This is known as CICR, Ca^{2+}-induced Ca^{2+} release. The IP_3-responsive vesicles refill with Ca^{2+} from ryanodine vesicles (**II**) and Ca^{2+} flowing in through the CACF-activated membrane channels. As the cytosol Ca^{2+} concentration drops, the Ca^{2+}•Cmd complexes dissociate and this, together with the refilling, reactivates the IP_3 receptor/channels on the type I vesicles. These channels can now be reopened by IP_3 to start another Ca^{2+} cycle if the receptor and PLC-γ_1 are still activated.

to the starting level just before the onset of DNA replication between 16 and 18 h.[1] The total concentration of polyribosome-associated mRNA rises steadily to a peak by 16 h and then starts dropping after the onset of replication.[1] There is a burst of ODC activity which starts between 1 and 2 h, peaks sharply at 4 h then falls gradually during the next 2 h.[1] ODC and its product, putrescine, combine to make a 64-kDa subunit of RNA polymerase I and thus contribute to the increased ribosomal RNA synthesis which peaks at 16 h and then drops upon the onset of DNA replication.[1]

During the first 8 h after HPX there is a brief burst of activity of a protamine-phosphorylating, Ca^{2+}•CaM-dependent PK. There is a first cyclic AMP surge which peaks at 2 h and returns to the starting level by 8 h (Figure 11).[1,78] Along with this cyclic AMP surge come parallel rises and falls in the levels of mRNAs. Between 1 and 2 h another PK activity appears and peaks between 8 and 10 h. This is a peculiar, potent, and probably very important, 105-kDa enzyme with the remarkable multiple talents of phosphorylating S and T residues (i.e., form P-O bonds), guanido groups of arginine residues, and

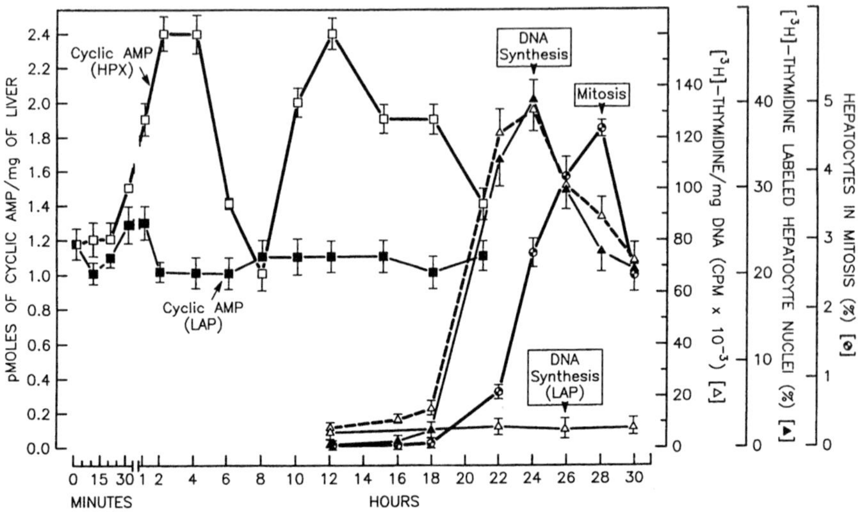

FIGURE 11 Seventy percent partial hepatectomy in 200-g male Sprague-Dawley rats triggered two cyclic AMP surges in the remaining liver cells before they started replicating DNA. There were no cyclic AMP surges or initiation of DNA replication in the still-intact livers of laparotomized (LAP, sham-operated) rats. See Whitfield[1] for original references.

ε-amino groups of K residues (i.e., form phosphoramide bonds).[1] In other words it is a KRSTPK. Its favorite substrate is a chromatin-associated 45-kDa protein which migrates close to H1 histones on gels.[1] Obviously this big multitalented PK must be a powerful tool for the restructuring and repositioning of chromatin needed to make cycle-related genes accessible to signals-induced or -activated transcription factors and it also might directly interact with transcription factors and suppressors such as Rb. It is interesting to note that a 105-kDa chromatin-associated protein, which might be our KRSTPK, is in fact involved in the early restructuring of chromatin in plant lectin-activated human peripheral blood lymphocytes.[1]

What would happen if the blood Ca^{2+} concentration was reduced by removing the parathyroid glands? Would the hepatocytes in the liver remnant of HPX'd parathyroidectomized (PTX) or thyroparathyroidectomized (TPTX) rats still be able to initiate the proliferogenic switch-over? Unfortunately it is practically impossible to reduce the Ca^{2+} concentration in the blood by more than 50% and still have a living rat. PTX or TPTX triggers a rapid drop in the blood Ca^{2+} concentration within the first 24 h and a slower, but ultimately severe, drop in the $1\alpha,25(OH)_2D_3$ concentration by 72 h.[79] If the animals are fed a phosphorus-free diet after TPTX, the concentrations of both Ca^{2+} and $1\alpha,25(OH)_2D_3$ will not drop and the hepatocytes will remain fully responsive to HPX.[79] By 24 h after TPTX the $1\alpha,25(OH)_2D_3$ level in animals given a normal diet is still high, but their blood Ca^{2+} concentration has halved and their hepatocytes cannot initiate DNA synthesis after HPX. In these hypocalcemic animals, only late-G_1 events cannot occur, and just one i.p. injection of 2 mg of $CaCl_2$/100 g of body weight as late as 15 h after HPX enables the remnant's hepatocytes to complete the G_1 build-up during the next 3 to 4 h and start replicating DNA *on time.*[1] Moreover, infusing Ca^{2+} through a jugular vein for only the first 8 h after HPX does not enable the remnant's hepatocytes to initiate DNA synthesis.[1] Thus, something demanding a high level of external Ca^{2+} appears near the end of the G_1 build-up and is needed to complete the build-up to DNA replication. The $1\alpha,25(OH)_2D_3$ deficiency that develops by 72 h after TPTX reduces the necessary enhanced production of the PKA's C subunits. But even at 72 h most of the major first events, except for the surges of PKAs activities, still happen undiminished in the hepatocytes remaining after HPX.[1] Therefore, the early events are unaffected by mild external Ca^{2+} deprivation induced by PTX or TPTX, and the failure to increase the level of PKAs activities.[1] However, the later part of the G_1 build-up is exquisitely sensitive to the external Ca^{2+} concentration and to the failure to accumulate the PKAs C subunits.[1] This does *not* mean that Ca^{2+} is not involved in the proliferative awakening of hepatocytes *in vivo.* It just means that these first events, unlike the later parts of the G_1 build-up, are unaffected by halving the external Ca^{2+} concentration and a lack of signals from low-affinity Ca^{2+} sensors on the cell surface.

B. The G_1 Build-Up

The pre-G_1 hypertrophic, or competence-establishing, phase ends around 8 to 10 h, when the cells are big enough and ready to generate, and respond to, the next signals that will start the actual G_1 build-up, which is focused on replication, and then mitosis. The first surges of cyclic AMP and PKA activity have subsided, but now adenylyl cyclase activity starts rising again and PDE activity drops (Figure 11).[1] Cyclic AMP and PKA activity surge upwards to a second transient peak around 12 h.[1] The cells also begin loading their membranes with PKAs, particularly PKAI, which will be involved in an autocrine mechanism near the end of the G_1 build-up.[1] The G_1 cyclic AMP surge has been retained throughout eukaryotic evolution. It occurs in a very wide variety of cells ranging from diatoms, *Sa. cerevisiae* and *Tetrahymena,* to worms, quail oviduct cells, as well as all normal and many neoplastic human, mouse, and rat cells.[1,53–55,78] Because it has been so carefully preserved throughout evolution from unicellular organisms to animal cells, the G_1 cyclic AMP surge would be expected to be triggered by something ubiquitous and endogenous. But at first glance the evidence seems to indicate a variety of causes ranging from the growth-activated RAS proteins of *Sa. cerevisiae* to exogenous hormones in many mammalian cells *in vivo.*[1,78] It has been suggested that the G_1 cyclic AMP surge in the hepatocytes of the liver remnant in the HPX rat might be triggered by glucagon or glucagon-like proteins (GLPs) from the intestine which peak in the blood around 12 h after HPX.[1,78] Indeed, preventing the GLP-loaded portal blood from reaching the liver by splanchnic evisceration or rerouting the blood through the vena cava prevents the initiation of DNA replication.[1,78] This block can be lifted simply by injecting glucagon into the animal around 8 to 10 h after HPX, but not earlier.[1,78] However, we must not forget that hepatocytes and other cultured cells still generate the G_1 cyclic AMP surge without surging external hormones, although they may express and display hormone receptors at specific points in the cycle. The fact that cells can be stimulated to finish the G_1 build-up by cyclic AMP elevators such as glucagon in the case of hepatocytes does not necessarily mean that this hormone is the physiological cause of the G_1 cyclic AMP surge. Indeed, it may only mimic, and thus enhance or replace, the activity of the actual, ubiquitous, endogenous cyclic AMP elevator. However, it now seems that the surges of adenylyl cyclase activity and cyclic AMP are caused by rises in the level of the membrane-associated adenylyl cyclase-activating $G_{s\alpha}$ G protein while the crashes of adenylyl cyclase activity are caused by rises in the level of the inhibitory G_{ia} protein.[80] These fluctuations appear to be triggered by an endogenous chain of post-translational events set in motion by the first signal.

The G_1 cyclic AMP surge is a key player in the G_1 build-up.[1,78,80a] Delaying or reducing it by injecting an α-adrenergic blocker (phenoxybenzamine or phentolamine) at 8 h after HPX correspondingly delays and reduces the DNA-synthetic response in the remnant.[1,78] Cyclic AMP or the cyclic AMP-elevating glucagon (along with insulin) stimulates the G_1 build-up in EGF-activated hepatocytes in primary neonatal or young rat liver cultures.[1,78] T51B rat liver cells stall in an advanced stage of prereplicative development and either cannot reach the point of generating the G_1 cyclic-AMP surge, or they cannot retain cyclic AMP when stimulated by serum in very low (0.02 m*M*)-Ca^{2+} medium.[1,81] The latter appears to be the case for primary neonatal rat hepatocytes. Thus, EGF-activated neonatal hepatocytes increase cyclic AMP synthesis when in low (0.02 m*M*)-Ca^{2+} medium. Indeed, they make as much cyclic AMP as hepatocytes stimulated by EGF in high (1.0 m*M*)-Ca^{2+} medium, but they cannot generate the two prereplicative cyclic AMP surges or initiate DNA replication, because they dump the additional cyclic AMP into the medium.[81] Adding Ca^{2+} (to a final concentration of 1.2 m*M*) triggers a cyclic AMP transient which is followed by the initiation of DNA replication by the stalled T51B liver cells. Adding cyclic AMP or one of many things (calcitonin, cyclic GMP, glucagon, the PDE-blocking IBMX, hPTH-(1–34), vasopressin) that trigger a cyclic AMP surge, or adding PKAII (*not* PKAI) holoenzyme, or the PKAs' C subunit kicks the stalled T51B liver cells into the S phase.[1,78] In the next chapter we shall see that parotid gland acinar cells in the rat cannot generate the G_1 cyclic AMP surge or initiate DNA replication when *d,l* propranolol is injected at 8 h after i.p. injection of a proliferatively activating dose of isoproterenol, but injecting a mixture of dibutyryl-cyclic AMP and the PDE-blocking theophylline along with the propranolol enables the cells to complete the G_1 build-up and start replicating DNA at the normal time. The importance of cyclic AMP and the PKAs for the G_1 build-up in animal cells has been convincingly proven by Sheffield[82] using mouse mammary cells and oligonucleotides antisense to the translation-start segment of the mRNA coding for the PKAs' C subunit. In these cells, PKA activity rose fourfold to a peak shortly before the onset of DNA replication after the cells were stimulated to start cycling by adding fresh fetal bovine serum to the culture medium. Stopping this PKA surge by adding the antisense oligonucleotides along with the serum prevented the cells from initiating DNA replication. Gadbois et al.[83] have also supported a role for PKAs in the G_1 build-up by showing that a specific PKAs inhibitor, K5720, blocked the cells in logarithmically growing cultures of human diploid dermal fibroblasts midway through their 13-h G_1 phase. We shall see further on that preventing PKAs C subunits from increasing in the regenerating liver remnant by TPTX 3 d before HPX also prevents the initiation of DNA replication.[1] The message seems clear: a surge of PKA activity is needed for the initiation of DNA replication.

What does cyclic AMP do for the G_1 build-up? The answer has probably been found in the budding yeast *Sa. cerevisiae* (Figure 12). To see what it does we must first put the yeast cells to sleep. When these cells are incubated in a nitrogen-rich medium containing only 1.1 m*M* glucose they stop cycling and shut themselves down into a quiescent, G_0, state.

But before waking them up we must get some historical perspective. Both archaebacterial and eubacterial prokaryotes must be able to respond promptly to depletion of the currently favored nutrient by switching on genes encoding enzymes that enable them to use other available nutrients. To do this, they invented a structurally simple enzyme, adenylyl cyclase, to make cyclic AMP from ATP to respond to a fuel crisis.[1,78] They also invented CAP, catabolite activator protein, to bind cyclic AMP, and they coupled appropriate genes to promoters which could bind, and be activated by, cyclic AMP·CAP complexes.[1,78] The promoter-bound cyclic AMP·CAP complexes convert the cyclic AMP signal into a burst of appropriate gene expressions and from these a surge of new enzymes to process alternative nutrients. The first "higher-tech" unicellular eukaryotes changed the way they used cyclic AMP. They no longer used it to signal nutrient stress. Instead, they began using it to signal nutritional adequacy and attainment of the right size to start the G_1 build-up. They learned how to use their monomeric GTP-binding RAS proteins, especially RAS 2, to translate this growth information into a cyclic AMP signal, and they made the first of the sophisticated PKAs with CAP-like, cyclic AMP-sensing R (regulatory) subunits reversibly coupled to a STPK catalytic subunit.[1,78] The R subunit inhibits the PKA's catalytic subunit, but when the R subunit binds cyclic AMP, it dissociates from, and thus derepresses, subunit activity.[84]

And now let's wake up our yeast cell. Adding fructose, glucose, or mannose (to a concentration of 25 m*M*) signals the cell to start growing (Figure 12).[1] Adenylyl cyclase, cyclic AMP, the PKA, and their various handlers have been identified as key players in starting the G_1 build-up to replication by

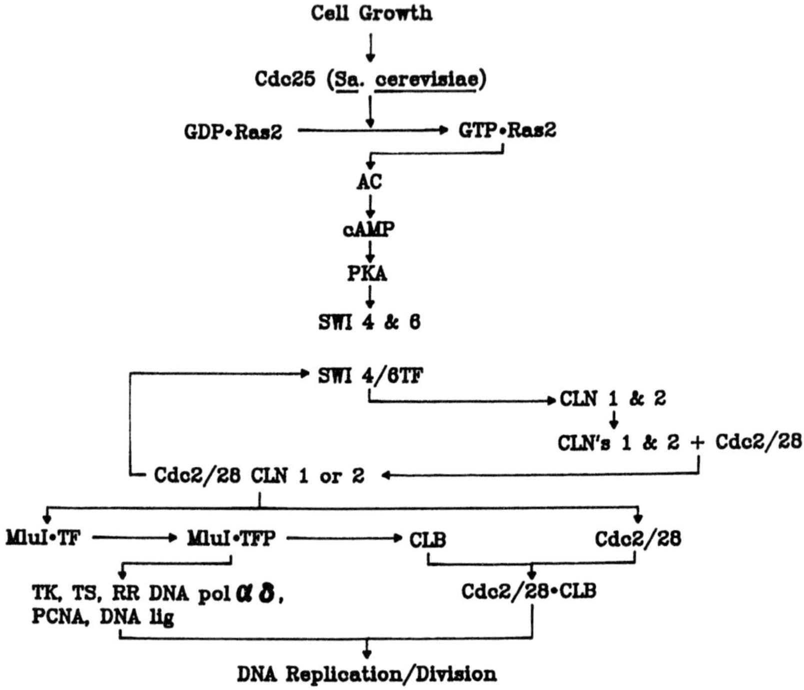

FIGURE 12 Cyclic AMP triggers the **START** of the build-up to DNA replication and division in the budding yeast *Sa. cerevisiae.* Things start when the growing cell activates Cdc25, a GEF protein related to the *Drosophila* Sos and mammalian mSos proteins,[16a] which catalyzes the exchange of GDP for GTP to convert inactive GDP·RAS2 into GTP·RAS2 complexes which stimulate the cyclic AMP-synthesizing adenylyl cyclase (AC). (This Cdc25 GEF protein must not be confused with the Cdc25 dual-function phosphatase of *Sc. pombe* and animal cells that activates Cdc2·cyclin PKs.) The cAMP surge (the equivalent of the G_1 cyclin AMP surge or transient in animal cells) stimulates the yeast cell's single type of PKA. The burst of PKA activity activates (or derepresses) the expression of the *swi* 4 and 6 genes. The resulting SWI4/6 transcription factor complexes (SBF in Figure 13) stimulate the expression of the genes for the CLNs that combine with Cdc2/28 to form the Cdc2/28·CLN PKs that phosphorylate, and thus activate, the MluI·TF (DSCI complex in Figure 13) factors that then coordinately stimulate the expression of the various DNA-replication enzymes as well as the gene for CLB cyclin which replaces the CLNs to produce the Cdc2/28·CLB PKs needed for cell division.

using a panel of conditional (temperature-sensitive or *ts*) mutations.[1] As the yeast cell grows it does a very important growth-determined thing — it expresses its *cdc* 25 gene.[1] The Cdc 25 proteins, which are related to animal cells' Sos and mSos GTP/GDP exchange proteins,[16a] stimulate the exchange of GDP for GTP on RAS 2 proteins to make active GTP·RAS 2 complexes which activate the yeast cell's adenylyl cyclase (Figure 12).[1] The resulting cyclic AMP surge and burst of PKA activity trigger the events leading directly to the initiation of DNA replication and budding. Indeed, this is the so-called START step of the yeast cell's cycle (Figure 12).[1,3] If Cdc 25 *(Sa. cerevisiae),* RAS proteins, adenylyl cyclase, or the PKA should fail to function, the yeast cell will not start the build-up to DNA replication and budding. Ca^{2+} is also involved in the activation of the yeast cell's adenylyl cyclase. Reducing the cell's Ca^{2+} content by incubating it with 10 μM A23187 (a Ca^{2+}-ionophone) and 10 μM EGTA (a specific Ca^{2+}-chelator) in the growth medium reduces the cellular cyclic AMP content and temporarily arrests progression toward the G_1 START step.[1,40] On the other hand, if the cell should continuously generate the PKA signal because of a permanently activated mutant GTP•RAS protein (which cannot shut itself off by autodegrading its bound GTP to GDP) or a high level of free PKA C subunits (because the cell cannot make functional R subunits), it simply cannot switch its cell cycle engine off, even in the face of severe nutritional deficiency such as exists in a nitrogen-deficient medium in which the cell would normally stop cycling and sporulate to avoid killing itself.[1]

The cyclic AMP/PKA signal's job is to stimulate the assembly of STPK complexes of the cell cycle engine each consisting of a Cdc2/28 catalytic component and a CLN1, CLN2, or CLN3 cyclin regulatory subunit (Figures 12 and 13).[2,3,84a] Cdc2/28 is constitutively expressed and is the budding yeast's only representative of a family of small STPKs that have a cyclin-binding PSTAIRE (or closely related) motif near their N-termini and must combine with a cyclin to achieve an active configuration (look ahead to Figure 22).[2,3,85] It seems that the cyclic AMP/PKA signal derepresses the production of the closely related (57% homologous) unstable, CLN1 and -2 cyclins, maybe by phosphorylating and activating the Sit4 phosphatase which, in turn, inactivates an inhibitor of the expression of the *swi* (switch) 4 and 6 genes.[3,86] The products of these *swi* genes, the SWI 4 and SWI 6 proteins, are the parts of a dimeric transcription factor SBF which stimulates the *cln* 1 and 2 genes, when the SWI 4 component of the SBF heterodimer binds to the genes' promoter element containing multiple repeats of the sequence CAGGAAAA, formerly known as CCB elements, but now known as SCB elements.[86] These elements confer a late-G_1 expression time on any gene coupled to them. The accumulating CLNs combine with Cdc2/28 to produce active STPKs. But something else must happen before these active PK heterodimers can START the direct build-up to chromosome replication. Yet another component, Cks1 (Suc 1 in *Schizosaccharomyces pombe*; CksHs1 and CksHs2 in human cells) must now assemble into hexamers.[87] These serve as hubs for the formation of the Cdc2-oligomers that START the G_1 build-up by phosphorylating and activating a protein transcription factor complex, DSCI/MCB, containing the 120-kDa *cdc*10 gene product and SW16, which recognizes, and binds to, the MluI cell cycle box (MCB; -ACGCGT-) in the upstream regulatory regions of a bloc of genes whose products replicate DNA (DNA ligase, DNA primase, PCNA [DNA polymerase-δ's "sliding clamp" accessory factor], polymerases-α and -δ, ribonucleotide reductase, topoisomerase, thymidine kinase, thymidylate

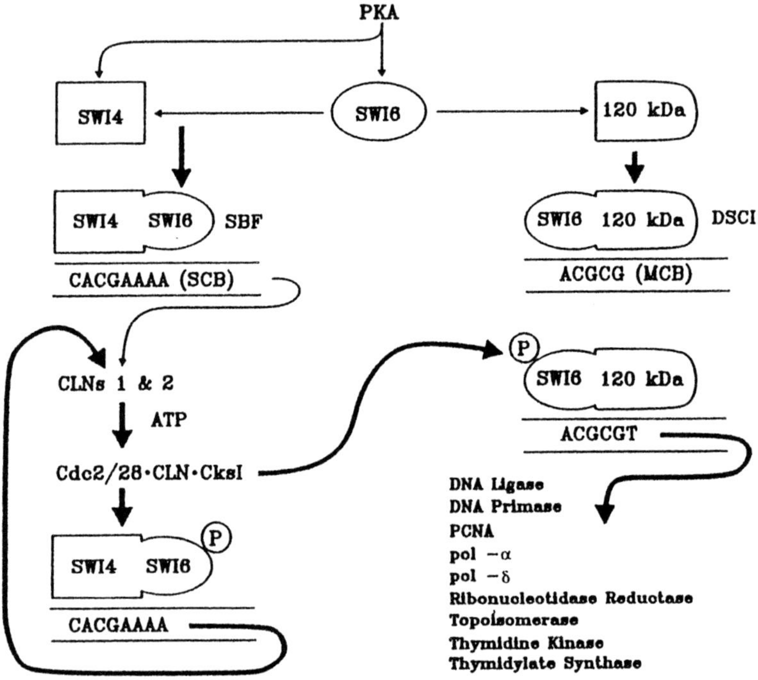

FIGURE 13 The PKA-triggered **START** events that lead to the coordinate expression of DNA replication genes in *Sa. cerevisiae* — a model for the role of cyclic AMP and the PKAs in the G_1 build-up in animal cells? The yeast PKA somehow stimulates the expression of SWI4 and -6. SWI4 combines with SWI6 to produce SWI4/6 complexes (SBF) that bind to the SCB elements of the *cln* 1 and 2 genes. This results in the appearance of CLNs which combine with Cdc2/28 to make Cdc2/28·CLN PKs. The Cdc2/28·CLN PKs combine with Cks1 hexamers to form Cdc2/28·CLN·Cks1 oligomers. These oligomers phosphorylate the SWI6 of the SWI4/6 complexes, which increases the ability of the complexes to stimulate *cln* gene expression and thereby the production of Cdc2/28·CLN PKs. Meanwhile, SWI6 has also joined with a 120-kDa protein to produce MluI or DSCI complexes that bind to MluI or MCB (Mlu cell cycle box) elements of the genes for the several listed DNA-replication enzymes. Phosphorylation of the SWI6 component by Cdc2/28·CLN·Cks1 oligomers activates the MluI or DSCI complexes which stimulate the expression of the DNA-replication genes.

synthase) (Figure 13).[2,3,86–88] The Cdc2/28-CLN complexes, oligomerized on Cks1, enhance the activity of the SWI 4/6 heterodimeric transcription factors by phosphorylating SWI 6, thereby starting a positive-feedback loop that increases the production of more CLN1 and -2 and thus more Cdc2/28-CLN complexes to coordinately start the production of DNA-replicating enzymes by phosphorylating the SWI 6 component of the DSCI/MCB complex on the MCB regulatory elements of the genes coding for these enzymes (Figure 13). It is essential to note that the key coordinating factor is phosphorylated SWI6.

As the yeast cell coordinately makes its various DNA-replicating enzymes, it somehow switches over to expressing another cyclin, replication-related cyclin Bs (CLB5,6) and destroys the CLNs1 and -2 (Figure 11).[2,3,84a,86] The Cdc2/28 catalytic subunits liberated from their union with the CLNs now join up with the CLBs to form oligomerized Cdc2/28-CLB5,6 complexes (Figure 11).[2,3,86] These new complexes start DNA replication by activating the specifically origin-seeking helicase and the assembly of replication complexes on the unwinding origins.

One of the giant steps in cell biology has been the recent discovery of the cell cycle engine which operates in all eukaryotes and is a parade of cyclin surges (CLN3 to CLN1,2 to CLB5,6 to CLB1,2) and crashes and bursts of stage-specific cyclin-dependent "PSTAIRE" STPK activities.[2,3,84a] When *Sa. cerevisiae* cells are cycling in a nutritionally rich growth medium, they maintain high levels of PKA and Cdc2/28 catalytic subunit.[2,3] Then, it is surges of stage-specific, activating cyclins that cue the different parts of the cycle, but it is the growth-triggered burst of PKA activity that starts the cell cycle engine.[1–3]

It must be noted that the cell cycle of *Sc. pombe*, the fission yeast, differs from the unrelated *Sa. cerevisiae* and animal cells in a key respect — it does not use its PKA to trigger the G_1 START. But exposing *Sc. pombe* cells to millimolar concentrations of external cyclic AMP or loading the cells with active adenylyl cyclase by planting *Sa. cerevisiae*'s adenylyl cyclase gene into them, can start the G_1 build-up. Evidently, *Sc. pombe* (perhaps like plant cells) has acquired the knack of using a different kind of protein kinase to start the cycle-driving parade of oligomerized Cdc2/28-cyclin complexes. It must be noted that the Cdc28·CLN1 PK complexes that operate at the G_1/S transition are regulated differently from the Cdc28·CLB1 or -2 PKs that operate at the G_2/M transition. Thus, dephosphorylation inactivates Cdc28·CLNs while Cdc28·CLBs are activated by dephosphorylation at G_2/M. This illustrates the big difference a cyclin makes to the control and targeting of a PSTAIRE-motif STPK.

Animal cells use several different kinds of cyclin-dependent PSTAIRE STPK catalytic subunits and cyclins having different substrate preferences to drive the different stages of the cycle. As a cell such as a hepatocyte grows, something happens at a critical point to activate the trimeric G_s protein, which (like *Sa. cerevisiae's* simpler monomeric RAS 2 protein) activates the more elaborate receptor/channel-like animal cell adenylyl cyclase (Figure 14).[89–91] The first cyclic AMP surge activated PKAIs in the cytoplasm and on the cell surface and sent their C subunits surging into the nucleus to activate a set of genes related to growth and the reconfiguration of the cell into a state of competence to respond to second-stage signals. The next cyclic AMP surge now activates PKAII holoenzymes which had been accumulating and some of which had been tethered by their RII subunits to specific anchor proteins near the nucleus. The cyclic AMP surge releases C subunits from the strategically tethered holoenzymes which then home onto nearby intranuclear targets.[84] Some untethered PKAIIs have entered the nucleus as ternary [cyclic AMP·RII·C] complexes because of their RII subunits' -KKR- NLS (nuclear localization

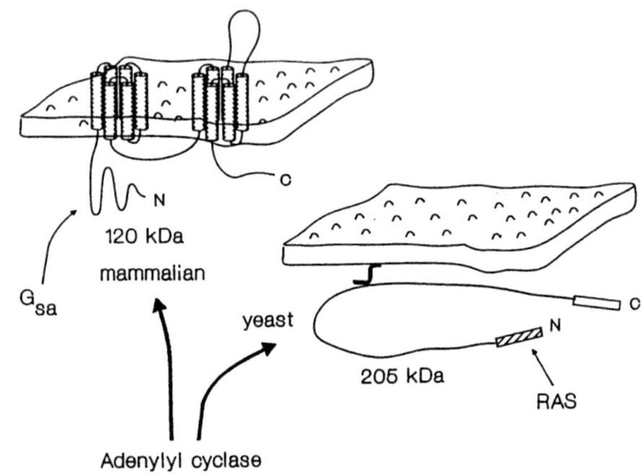

FIGURE 14 The adenylyl cyclases of *Sa. cerevisiae* and mammalian cells. The yeast enzyme is attached by an adaptor to the plasma membrane and it has an N-terminal motif (hatched block) which binds the GTP·RAS2 activator protein. By contrast, the smaller mammalian enzyme is inserted into the membrane and has two clusters of six transmembrane domains. The mammalian enzyme does not bind GTP·Ras proteins, but instead it binds, and is activated by, the GTP·$G_{s\alpha}$ protein.

signal) which qualified them for carriage into the nucleus by the ATP-powered mechanism in the nuclear pore complex, to be discussed further on.[92–94] The special gene-activating power of PKAII is indicated by the fact that a cyclic AMP-responsive somatostatin-CAT fusion gene can respond to a cyclic AMP signal in normal PC12 cells, but it responds only marginally to a cyclic AMP signal in mutant PC12 cells that do not make PKAII but do make PKAI normally.[1,93] The G_1 cyclic AMP surge thus drives RIIβ subunits and associated C subunits toward their target transcription factors bound to specific upstream regulatory regions of cyclic AMP-responsive genes in transcriptionally active nuclear matrix (skeleton)-attached bases of chromatin loops which have been repositioned, exposed, and thus made accessible to transcription factors during the preliminary, competence-inducing hypertrophic stage.[1] These transcription factors recognize and bind to the so-called cyclic AMP-responsive element (CRE) octamer -TGACGTCA- and modify gene function, provided the CRE is flanked by certain sequences.[93–97] One indication of the importance of cyclic AMP-responsive gene activities at this stage of the G_1 build-up is a three- to fivefold increase in the expression of the gene coding for the CRE-βP1 (CRE-binding protein transcription factor), which binds to CREs either as a homodimer or a heterodimer with the c-Jun protein transcription factor, and would certainly be a target of the inrushing PKAII complexes and C subunits.[95,96,98] The 47-kDa cyclic AMP-responsive transcription *repressor* protein, which disappeared after HPX (possibly to derepress some immediate-early genes such as c-*fos*) also now reappears.[97,99] The message is very clear: the growing cell equips itself with the strategically sited PKAIIs and transcription factors needed to respond to the G_1 cyclic AMP surge with a flurry of gene activity. But what are these genes? In *Sa. cerevisiae* the key job of the single PKA is to initiate the production of CLN1 and CLN2 cyclins and thus start the cell cycle engine. The resulting Cdc2/28·CLN1 or CLN2 PK complexes stimulate the expression of DNA-replicating enzymes. We should expect that the similar G_1 PKA signal in animal cells does the same thing — triggers the production of the cyclin-dependent PKs of the cell cycle engine. Thus, the PKA signal might unleash the accumulation to threshold values of a G_1 cyclin, cyclin C or D, G_1 cyclin-dependent, PSTAIRE-motif PK catalytic subunits, e.g., Cdk2 and Cdk4, and transcription factors like *Sa. cerevisiae*'s DSCI/MCB complex (Figures 12 and 13) which are activated by the accumulating Cdk·cyclin PKs and coordinately stimulate the expression of replication-related genes. One of these transcription factors, A-Myb or B-Myb, is normally expressed later in the prereplicative period and stimulates the expression of DNA-polα and DNA-Polδ's PCNA ("sliding clamp") accessory factor.[100]

While we have been focusing on the nucleus-seeking PKA RII and C subunits and the genes they directly or indirectly activate we must not forget that the G_1 cyclic AMP signal is also accompanied by a second large surge of cytoplasmic PKAI activity which must stimulate other things as well as sending a pulse of its C subunits into the nucleus.[1] The signal might stimulate the transport of mRNA through the nuclear pore complexes into the cytoplasm by phosphorylating and activating mRNA transporter proteins.[1] Indeed, in regenerating liver cells the total amount of polyribosome-associated (poly A^+) mRNA peaks at a level that is three times the preoperative starting value along with the G_1 cyclic AMP surge.[1] The processing of this mRNA is promoted by two things, CaM and K-Ras. The first cyclic AMP surge causes the cellular CaM content to double after 8 h. The accumulated CaM promotes the translation of the G_1-specific messages by binding to ribosomal 40-S subunits and stimulating the translation initiating step that follows the binding of the eIF-4 factor to the [eIF-2·GTP·*met* $tRNA_f$·40S subunit] initiation complex.[1] The G_1 cyclic AMP surge is accompanied by, and might even cause, a dramatic burst of c-K-*ras* gene expression in BALB/c3T3 mouse cells and rat hepatocytes.[1,78,101] The surge of membrane-associated Ras proteins would mediate, hence amplify, signals from the activated IGF-1 and insulin YPK velcroceptors that are required to drive the middle and later G_1 development of BALB/c3T3 cells and hepatocytes, respectively. The activated Ras proteins enhance the translation of the G_1-specific messages by stimulating IPPK, which activates eIF-4E translation-initiation factor (Figure 3). But there may be something special about surface-display of IGF-1 velcroceptors. In Swiss albino 3T3 mouse cells they selectively stimulate nuclear PtdIns(4,5)P_2 breakdown with the release of DGs from the inner nuclear membrane and the translocation of PKCs to and/or into the nucleus through the nuclear pores, where they could do many things from promoting the translocation of messages from the nucleus to phosphorylating, and thus modulating, the activity of transcription factors.[102]

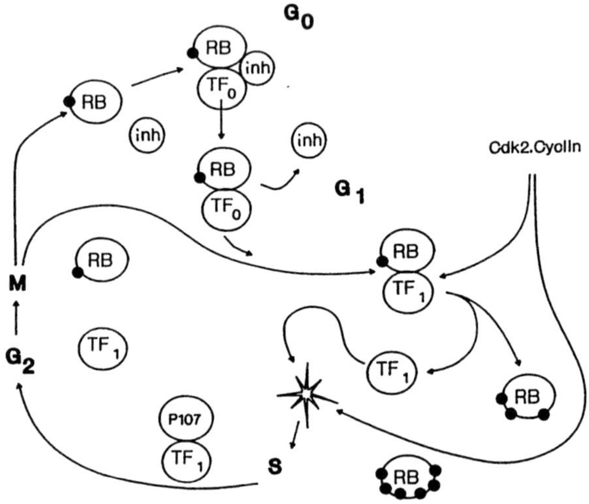

FIGURE 15 A possible interplay between the Rb (**RB**) retinoblastoma susceptibility protein, the Rb-related 107-kDa protein (**P107**), and Cdk2·cyclin PKs in prereplicative development. In quiescent G_0 cells underphosphorylated Rb forms a complex with an inhibitor protein (**inh**) and an early-response transcription factor(s) $\mathbf{TF_0}$. The proliferatively activating signal releases the inhibitor from the complex and RB·TF_0 can contribute to the stimulation of competence-establishing early gene expression. During the G_1 build-up, underphosphorylated Rb binds and suppresses the action of later-stage transcription factors ($\mathbf{TF_1}$) such as E2F on the promoters of genes coding for the replication enzymes. At a critical point in the G_1 build-up Rb is hyperphosphorylated by a Cdk2·cyclin or Cdk4·cyclin D PK and dissociates from the $\mathbf{TF_1}$ transcription factor(s). The derepressed transcription factors stimulate the coordinate expression of replication enzymes. When the replication complexes have collected on the leading replication origins, Cdk2·cyclin A PKs trigger replication and further phosphorylate, and thus "side-track" Rb (**RB**). Eventually, the $\mathbf{TF_1}$(s) binds to the Rb-related 107-kDa protein (**P107**) and is inactivated. This scheme was proposed by Zacksenhaus et al.[103]

One of the targets of accumulating Cdk2·cyclin A or E (or Cdk4·cyclin D) PKs is the 110-kDa Rb (retinoblastoma susceptibility) protein. In its active, underphosphorylated form, Rb binds to a transcription factor such as E2F on the promoters of replication-related genes such as dihydrofolate reductase, thymidylate synthase, and DNA polymerase-α, as well as the *cdc*2 gene whose product will be one of the key players in mitosis.[103–105] When Cdk2·cyclin phosphorylates it, Rb separates from the E2F transcription factor which can then stimulate the expression of the replication-related genes (Figure 15). Perhaps the protein kinase complexes also enhance the activity of the transcription factor as the yeast's Cdc2/28·CLN1 or -2 activates the MluI·TF (Figures 5, 12, and 13).

One of the new proteins that peak after the G_1 cyclic AMP surge (i.e., around 14 to 18 h after HPX) is p53 (or TP53),[106–110] which is *not* shared by yeasts. One might reasonably expect that something which peaks so close to the onset of DNA replication would be a replication promoter. But, No! p53 is actually a supressor which must be inactivated before the cell can start replicating its chromosomes. Why would a cell be so crazy as to load itself with something to suppress chromosome replication while preparing to start replication? Finding the answer to this question is one of the more exciting achievements of recent cell biology.

The p53 that accumulates near the end of the G_1 build-up is underphosphorylated and active.[106,108] p53 has a NLS(-SSSPQPKKKP-) at its C-terminus.[107,111–114] The protein is thus activately transported into the nucleus to carry out its late-G_1 mission.[13] When it has finished, it will be hyperphosphorylated and inactivated by Cdk2•cyclin A[107,114] and casein kinase II.[115,116] Having done its job, p53 leaves the chromosome-replicating S-phase nucleus.[113] But what does it do?

Before a cell takes the big step of starting chromosome replication it must make sure that its chromosomes are at least intact (Figure 1). One of its damage detectors is poly (ADP-ribose) polymerase (PARP) which senses DNA strand breaks.[116a] PARP recognizes the V-shaped configuration of a nicked DNA strand, binds to the nick, and blocks DNA replication until the nick can be repaired. PARP enables repair enzymes to get at the broken strand by relaxing the chromatin superstructure by poly-ADP ribosylating H1 and H2B histones and causing itself to dissociate from the break by auto-poly-ADP ribosylation. The superstructure of the repaired strand is restored by the de-poly-ADP ribosylation of H1 and H2B histones by poly (ADP-ribose) glycohydrolase. It now appears that p53 is another of the cell's damage detectors and replication suppressors that works alongside PARP.

P53 can block chromosome replication directly by binding to the replication initiators encrusting DNA replication origins.[108] But this is probably irrelevant because the protein operates in late G_1 *before* the cell commits itself to chromosome replication.[106] Indeed, p53 will have been phosphorylated, inactivated, and be in the process of being destroyed by the time Cdk2•cyclin A PKs trigger replication.[109] It is a transcription factor which binds to a specific nucleotide sequence and thereby stimulates the expression of genes such as the gene for a 21-kDa protein known as *Cip*1 or WAF1 and the human GADD-45 (*G*rowth *A*rrest *D*NA *D*amage) gene.[117–119] The products of these genes prevent the initiation of DNA replication. The 21-kDa protein stops the G_1 build-up by simply binding and inactivating the cycle-driving Cdk2•cyclin PK complexes.[119] p53 also binds to, and inhibits the expression of, the cyclin A gene which is needed to make the Cdk2•cyclin A PK complexes which, as we shall see further on, activate the DNA replication complexes on replicon origins.[120] p53 is a *censor* or *checkpoint operator* whose job is to check the chromosomes for breaks before the cell plunges irrevocably into the S phase. As we shall soon see, p53 is the late-G_1 equivalent of G_2's "Rad 9" protein that scans the replicated chromosomes for incomplete replication (i.e., free ends) before the cell initiates mitosis. p53's checkpoint activity is indicated by the fact that normal cells make more p53 and stabilize this normally very labile protein when their chromosomes are broken by sublethal ionizing radiation just as they make more p53 in late G_1.[121] p53 and a PK, the so-called DNA-activated STPK (because it is activated by binding DNA), bind to the single-strand ends of broken chromosomes,[122,123] and it has been suggested[122] that the activated PK then phosphorylates key residues of, and thus stabilizes, p53. The stabilized p53 then directly promotes the repair of the broken chromosome and also delays the initiation of DNA replication until the damage is repaired by preventing the appearance of active Cdk2·cyclin A complexes. DNA replication can only begin when the damage is repaired, the DNA-activated PK is silenced, and p53 is destabilized. By preventing the initiation of DNA replication, the p53-induced suppression of the cyclin A gene and expression of the 21-kDa and GADD-45 proteins protect the cell from trying to replicate aberrant chromosomes and malignancy-inducing mutations until the damage is repaired. If the damage cannot be repaired and p53 and its products are not eliminated within a certain time, they will somehow trigger the apoptogenic mechanism to eradicate the cell and with it the dire threat to the body of dangerously delinquent, malignancy-bound clones of mutant progeny.[124]

Now we must go back to the beginning of our story and recall that the c-Myc transcription factor, one of the main cycle starters, did two opposing things — trigger events leading to chromosome replication and apoptogenic events that will kill the cell unless switched off when it reaches the G_1/S checkpoint.[22,22a] It seems reasonable to suspect that p53 is the ultimate linchpin of the c-Myc-initiated apoptogenic program. As we shall see the loss of the p53 safety-check mechanism through deletion or loss-of-function mutation is responsible in large part for the chromosomal and genetic instability that marks the later stages of neoplastic transformation and enables the survival and accumulation of malignant variants. The key message is that the proliferogenic process in a normal cell includes the activation of a safety mechanism designed to kill the cell unless it is switched off by the Cdk2·cyclin PK of the cell cycle engine. It is to the p53-operated mechanism that the cell poses the question "Are my chromosomes fit for replication?" (Figure 1). If the answer is "No", the p53 mechanism tells the cell to fix the problem fast or die.

Near the end of the G_1 build-up, when the cell cycle engine is operating and the various parts of the replication machinery are accumulating, the cell becomes extremely sensitive to external Ca^{2+}. If the external Ca^{2+} concentration falls below a certain level, the cell will be unable to finish the G_1 build-up and start replicating DNA.[1]

Many things happen both in the cell and on its surface at this climactic point in the cell cycle. Around 12 h after HPX (when the cyclic AMP surge has peaked) the hepatocytes in the regenerating liver remnant become more responsive to EGF.[1] Indeed, EGF has been rising steadily in the portal blood and peaks at around four times the normal value between 12 and 18 h before crashing down to the preoperative value after the onset of DNA replication.[1]

The hepatocytes have now (16 to 18 h) begun making autocrine/paracrine transforming growth factor (TGF)-α to stimulate their own EGF/TGF-α receptors and those of their neighbors.[1,55] This is reminiscent of BALB/c 3T3 mouse cells in which PDGF-induced early gene expressions and competence-inducing protein syntheses make the cells responsive to EGF.[1] Up to this point, the liver cells have had the normal number of EGF/TGF-α receptors on their surfaces and have been sending EGF·EGF/TGF-α receptor complexes to the lysosomes for destruction. But now the liver cells shut this lysosomal pathway down and start vigorously endocytosing EGF·EGF/TGF-α (and TGF-α·EGF/TGF-α) receptor complexes and delivering them directly to their nuclei where free EGF and EGF complexed with a 180-kDa EGF receptor-like protein start appearing.[1] These complexes may be transported from the cell surface to the nuclear envelope and pore complexes in shuttle vesicles that are pulled along microtubule trackways by motor molecules.[1] The nuclear EGF and TGF-α must do something such as stimulate nuclear PKCs and thus stimulate gene transcription and mRNA transport through the nuclear pore complexes. Here is one place where the external Ca^{2+} concentration could be critically important, because extracellular Ca^{2+} deprivation is known to stop WI-38 human fibroblasts from endocytosing EGF·EGF/TGF-α receptor complexes.[1]

Until now the cell has been growing to an optimal size for initiating chromosome replication under the influence of the activated velcroceptors and 7-TM receptors of the "Portal Symphony" of hormones and the subsequent steady urgings of insulin and the IGFs and their velcroceptors. Then, the growth-dependent cyclic AMP surge somehow urged it to make the first of the cyclin-dependent PSTAIRE-motif STPK complexes of the cell cycle engine and with them the transcription factor(s) that coordinately stimulates the genes encoding the various proteins needed to replicate the chromosomes. Now, the cell must translate the new messages into chromosome parts and replication machinery. These last minutes and hours of the G_1 build-up are mainly for translation rather than transcription. Indeed, T51B rat liver cells stalled at the end of the G_1 build-up by external Ca^{2+} deprivation can be stimulated by Ca^{2+} to start replicating DNA (albeit abortively) in the presence of actinomycin D, a gene transcription inhibitor, but not in the presence of cycloheximide, a mRNA translation inhibitor.[1]

During the Ca^{2+}-independent, cyclic AMP-driven stage of the G_1 build-up, regenerating rat liver cells, as well as cultured BALB/c 3T3 cells and human foreskin fibroblasts, load their Ca^{2+} storage vesicles.[1] As we shall soon see, these reloaded Ca^{2+} storage vesicles are parts of a replication-starting mechanism.

The dramatic Ca^{2+}-dependence of this part of the normal cell cycle is demonstrated by the ability of the rather small (about 50%) reduction of the blood Ca^{2+} concentration by PTX or TPTX to selectively block regenerating rat liver cells at this late-G_1 point without affecting earlier events.[1] Thus, HPX in the hypocalcemic rat still starts the remnant cells growing, accumulating mRNA and ribosomal RNA. It still triggers a burst of ODC activity and both

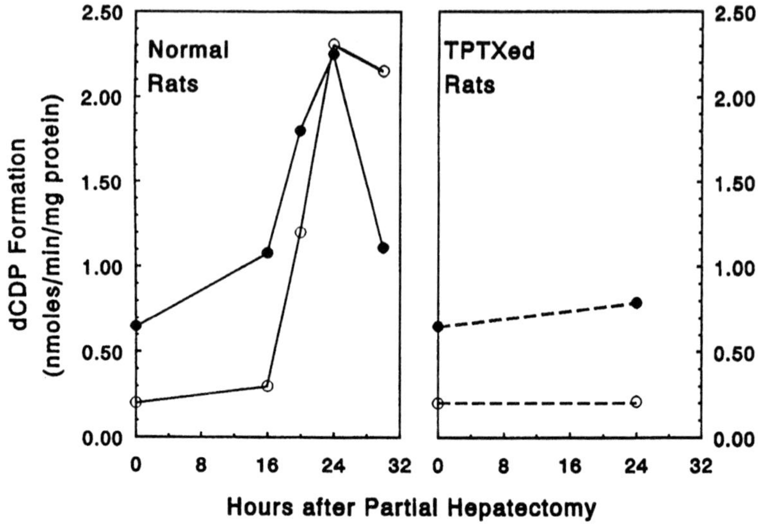

FIGURE 16 Regenerating liver cells in a hypocalcemic rat do not accumulate functional cytosolic ribonucleotide reductase M1 subunit (O) or the glycosylated M1 subunits (O) needed to form the nuclear envelope-associated ribonucleotide reductase holoenzymes that actually makes the deoxyribonucleotide precursors for DNA replication. Young (200-g) male Sprague-Dawley rats underwent TPTX, followed 3 d later by HPX to trigger proliferation of the remaining hepatocytes.

cyclic AMP surges.[1] And they accumulate the PKAs' RI and RII subunits.[1] However, they progressively lose their ability to make enough of the PKAs' C subunits to start the cyclin-dependent cycle engine because of the progressive drop in the $1\alpha,25(OH)_2D_3$ concentration in the blood.[79] They CANNOT: accumulate and plant the replication-relevant ribonucleotide reductase holoenzyme in the outer membranes of their nuclear envelopes (Figure 16); accumulate active thymidylate synthase, make deoxyribonucleotides, accumulate DNA polymerase-α, or expel Ca^{2+} from their reloaded Ca^{2+} storage vesicles.[1,125,126] The message is dramatically clear: external Ca^{2+} is somehow needed to make the various parts of the DNA-replicating machines and the next set of cyclin-dependent STPKs to activate them.

The lateness of this block in the G_1 build-up and the importance of external Ca^{2+} are seen dramatically in the response of remaining hepatocytes to HPX in rats that have undergone PTX or TPTX 1 d before the HPX.[1] One 30-min surge of blood Ca^{2+} produced at any time between 12 and 15 h after HPX by an i.p. injection of 2 mg of $CaCl_2$/100 g of rat enables the remnant's hepatocytes to assemble and operate their late-G_1-cyclin-dependent PKs and finish the G_1 build-up in the next 3 to 4 h and start replicating their chromosomes *on time*.

By contrast, continuous intravenous infusion of Ca^{2+} *only during the first 8 h* after HPX does not enable the cells to start replicating DNA.[1] It seems that there is a Ca^{2+}-dependent component that appears on the cell surface later in the G_1 build-up and functions only in the presence of Ca^{2+} concentration of 0.5 m*M* or higher. This Ca^{2+}-dependent surface sensor could be similar to that of the low-affinity Ca^{2+} receptors on the surface of keratinocytes (see Chapter 4), osteoclasts,[127] and parathyroid gland chief cells.[128] Or it could be, or could be like, the myoblast's surface (ecto-) protein kinase that is optimally active when the external Ca^{2+} concentration is 0.5 m*M* or higher.[1] If there is not enough external Ca^{2+} when this sensor/receptor appears or should become operational, the G_1 build-up of the unstable replication-related components stops, promptly starts dissipating, and the stalled late-G_1 cells can no longer be induced to replicate their chromosomes by injecting Ca^{2+} (or a plasma Ca^{2+}-elevating dose of PTH). The cells must start the G_1 build-up over again in order to replicate their chromosomes.[1]

Lowering the external Ca^{2+} concentration to a suboptimal level (e.g., from 1.5 to 0.02 m*M*) also stalls BALB/c 3T3 mouse cells, C3H10T1/2 mouse fibroblasts, T51B rat liver cells, and human WI-38 fibroblasts either at the beginning or at the end of the G_1 build-up.[1] Ca^{2+} deprivation does not stop quiescent primary rat thigh muscle cells or Swiss albino 3T3 cells from initiating prereplicative development when growth factors are added to the medium, because when quiescent they keep enough mobilizable calcium in their external stores for prereplicative activation.[1] But they too need external Ca^{2+} at the end of the G_1 build-up where they will stall unless enough Ca^{2+} has been added to the medium by the time they reach this critical point. The importance of Ca^{2+}-binding surface components at this stage is vividly demonstrated by Swiss albino 3T3 mouse fibroblasts (Figure 17). Adding serum to medium containing an optimal amount of Ca^{2+} stimulates quiescent Swiss albino 3T3 cells to start their prereplicative development. Removing the normal medium, chelating the residual Ca^{2+} in the culture and stripping calcium from cell surface receptors by washing the cells for just 5 min with EGTA-buffer mixture, and then returning them to their *original* high-Ca^{2+} medium, does not prevent them from initiating DNA replication until this brief Ca^{2+} stripping is done when they are only a couple of hours away from the scheduled time of initiation of DNA replication.[1] Then, this thorough, 5-min external Ca^{2+} stripping becomes dramatically effective.[1] The cells irreversibly lose key, late-appearing, late-operating, surface Ca^{2+}-binding components; giving them back the original Ca^{2+}-rich medium does *not* restore cell-associated Ca^{2+} as it does before or after this critical time. BALB/c 3T3 and T51B rat liver cells cannot linger long in the late-G_1 state in *confluent*, Ca^{2+}-deprived cultures. The G_1 build-up rapidly dissipates and the cells cannot immediately restart it if Ca^{2+} is added 2 h beyond the expected time of onset of DNA replication.[1] By contrast, when BALB/c 3T3 mouse cells or T51B rat liver cells are in *sparse* cultures with an ample supply of growth factors and EGTA is used to reduce the Ca^{2+} concentration in the serum component of the low-Ca^{2+} medium, some of them are blocked in a different late-G_1 state in which they can maintain pools of late

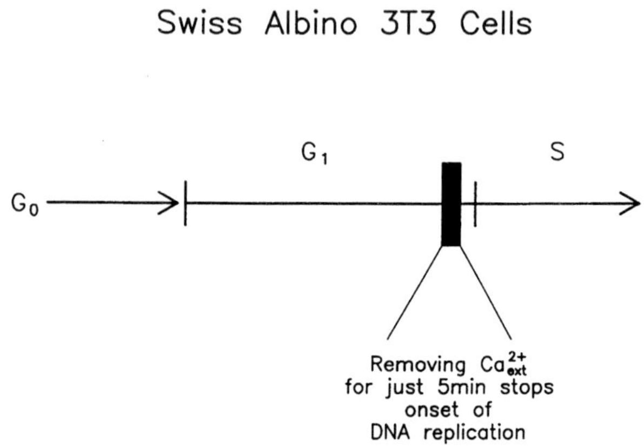

FIGURE 17 There is a brief, external Ca^{2+}-dependent stage shortly before the onset of DNA replication in Swiss Albino 3T3 mouse cells. This has been shown by stimulating quiescent G_0 cells to start cycling by adding more serum to their high-Ca^{2+} medium. Removing this normal medium, chelating the residual Ca^{2+} in the culture, and stripping calcium from the cell surfaces by washing them for 5 min with EGTA-buffer mixture, and then returning them to the original high-Ca^{2+} medium prevents them from initiating DNA replication only when it is done at the end of the G_1 build-up. See Whitfield[1] for the original reference to the experiments of Damluji and Riley.

mRNA transcripts which they start translating and thus initiate DNA replication 1 or 2 h after being given enough Ca^{2+}.[1]

Parathyroid removal causes the plasma Ca^{2+} concentration to drop promptly and the $1\alpha,25(OH)_2D_3$ concentration to drop more slowly (Figure 18).[79] These drops are associated with a waning responsiveness of liver cells to HPX. Feeding the rats a phosphorus-free diet for 5 d before TPTX prevents plasma Ca^{2+} and $1\alpha,25(OH)_2D_3$ from falling after TPTX.[79] Under these conditions, the liver cells no longer progressively lose their ability to make active DNA polymerase-α and replicate DNA after HPX (Figure 18). Keeping the plasma Ca^{2+} and $1\alpha,25(OH)_2D_3$ concentrations at their preoperative values maintains the full responsiveness of rat liver cells to HPX, despite the lack of PTH.

As we have seen, a single i.p. Ca^{2+} injection as late as 15 h after HPX prevents the remnant's cells from stalling in late G_1 in rats that have undergone TPTX 1 d before HPX. But with increasing time after HPX, something else happens alongside the simple external Ca^{2+} shortage to prevent the initiation of DNA replication. By the third day after TPTX, the hepatocytes can no longer be primed for DNA replication by an i.p. Ca^{2+} injection just before the scheduled time of onset of DNA replication. Now, an injection of $1\alpha,25(OH)_2D_3$ (e.g., 200 to 400 ng/200 g of body weight), preferably at the time of HPX, is essential. $1\alpha,25(OH)_2D_3$ does two things. It brings the plasma Ca^{2+} concentration back to the normal value between 12 and 15 h, when it is most needed. It

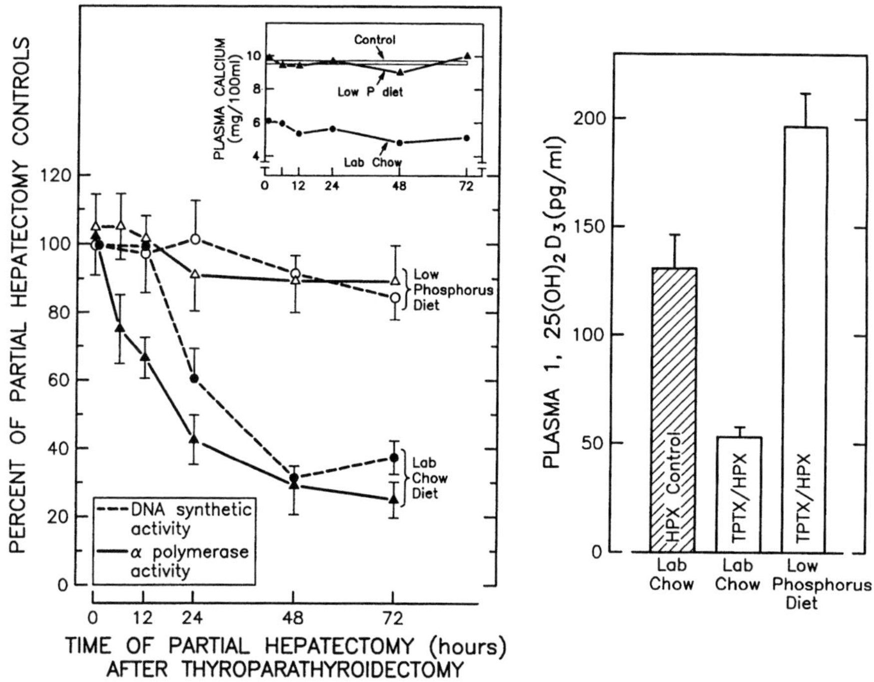

FIGURE 18 TPTX almost immediately reduced the plasma calcium concentration of 200-g male Sprague-Dawley rats given a standard diet, but it did not affect the plasma calcium concentration in TPTX rats given a low-phosphorus diet (inset, *left panel*). The plasma $1\alpha,25(OH)_2D_3$ concentration *(right panel)* fell more slowly, and by 72 h after TPTX it was about 35% of the normal value in rats given a normal diet, but it actually rose above the normal value in TPTX rats given the low-phosphorus diet.[79] The drops of the plasma Ca^{2+} and $1\alpha,25(OH)_2D_3$ concentration in the TPTX rats given the normal diet were associated with a progressive loss of the proliferative responsiveness of hepatocytes to HPX while the hepatocytes stayed fully responsive to HPX in TPTX rats when the Ca^{2+} and $1\alpha,25(OH)_2D_3$ concentrations were kept high by giving the animals the low-phosphorus diet.[79]

also enables the cell to accumulate the PKAs C subunits needed for the G_1 build-up. TPTX causes the synthesis of PKAs C subunits to drop by nearly 50% between 24 and 72 h without affecting the increased production of RI and RII subunits.[1] HPX at 72 h after TPTX is still followed by large surges of PKA-RI and RII subunit production, but there is no corresponding surge of the PKAs' C subunits.[1] $1\alpha,25(OH)_2D_3$ injection promptly stimulates the expression of C subunits, the level of which returns to the stimulated normal HPX value by 4 h.[1]

In summary, the steadily worsening hypocalcemia and $1\alpha,25(OH)_2D_3$ deficiency caused by PTX or TPTX hits the G_1 build-up at two different points. Its almost immediate post-HPX effect is to prevent the late, external Ca^{2+}-dependent coordinate accumulation of the parts of the replication mechanism.

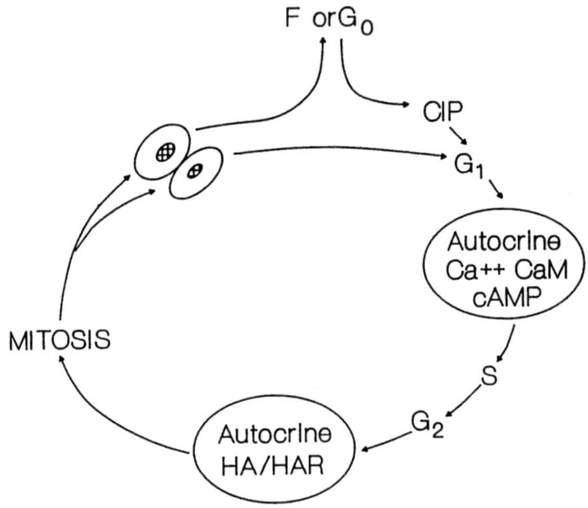

FIGURE 19 The principal autocrine-triggered stages of the cell cycle. The newborn cell decides to (1) stop cycling and function (**F**), or (2) enter a quiescent, $\mathbf{G_o}$, state, or immediately start the G_1 build-up. Functioning or G_0 cells can be induced to enter a competence-inducing phase (**CIP**) leading to the G_1 build-up. At the end of the G_1 build-up, CaM and cyclic AMP are released onto the cell surface where they stimulate a burst of PKs activity that is necessary for the initiation of DNA replication. During the $\mathbf{G_2}$ phase the cell secretes head activator (**HA**), a polypeptide autocrine/paracrine "cycle completion factor", and its receptor (**HAR**), which are needed for the initiation of mitosis.

Its second, longer-term effect is to block the earlier action of the G_1 cyclic AMP/PKA surge.

Part of the final starter signal is the crashing of the cyclic AMP surge (Figure 11). It seems that the G_1 cyclic AMP surge crashes because of the activation of cyclic nucleotide phosphodiesterase (PDE),[1] increased membrane-associated adenylyl cyclase-inhibiting $G_{i\alpha}$ protein,[80] and expulsion of the nucleotide into the medium.[81] Although the G_1 cyclic AMP surge, and the burst of PKAs activity it caused, were needed to produce and start the cell cycle engine, they must subside otherwise they will somehow block the subsequent operation of the engine and the initiation of DNA replication. We will see further on that exactly the same applies to the G_2 cyclic AMP surge, which also must crash if the Cdc2·cyclin B complexes that will prevail at that time are to be able to initiate mitosis.

The possibility of cyclic AMP being part of a late G_1 autocrine/paracrine loop (Figure 19) is raised by certain lines of evidence. Various cells display PKAs on their surfaces,[1,129] and T51B cells, an established rat liver line with hepatocyte characteristics, and neonatal rat hepatocytes start accumulating PKAs, mainly PKAI, on their surfaces soon after being aroused from confluence-induced quiescence by serum growth factors.[1] If the cells have enough external Ca^{2+} to retain their accumulated cyclic AMP[81] and generate the G_1 cyclic AMP surge and start replicating DNA, the surface PKA build-up stops and the PKAs R subunits separate from the PKAs C subunits and fall off into the medium, probably when they bind the cyclic AMP that is being released onto the cell surface.[1] When the medium contains only 0.02 mM Ca^{2+} instead of the normal 1.0 to 1.8 mM, the T51B cells do not generate the G_1 cyclic AMP surge or start making DNA, but they still load their surfaces with PKAs.[1] Indeed, this build-up does not stop, but keeps rising, though more slowly, after the initial surge.[1] Restimulating the confluent, Ca^{2+}-deprived cells with high-Ca^{2+} medium-serum triggers a further surge of surface PKA accumulation.[1] Now the cells can generate the G_1 cyclic AMP surge, initiate DNA replication, and lose the cyclic AMP-binding R subunits from their surfaces. Neonatal rat hepatocytes generate two cyclic AMP surges when stimulated by EGF and, like cells in a regenerating liver, start replicating DNA after the peaking of the second, G_1 cyclic AMP surge.[81] These cells dump a substantial part of their newly made cyclic AMP into the medium. But when the external Ca^{2+} concentration is only 0.02 mM, the EGF-stimulated cells still increase their cyclic AMP production, but now they dump *all* of the additional cyclic AMP into the medium and cannot generate the internal G_1 cyclic AMP surge they need to initiate DNA replication.[81] Thus, external Ca^{2+} restrains the release of cyclic AMP from stimulated liver cells and thus controls the activation of surface PKAs and at the same time enables the generation of the all-important G_1 internal cyclic AMP surge.

The importance of the surface PKAs in the final stages of the G_1 build-up has been indicated in several ways.[1] First, BALB/c 3T3 mouse cells, rat thymic lymphoblasts, and T51B rat liver cells stalled in late G_1 phase by external Ca^{2+} deprivation can be booted into the S phase by adding cyclic AMP which does not enter the cell. They, and Friend erythroleukemia cells, can also be stimulated to start replicating DNA by active PKAs' C subunits and by PKAII holoenzyme, but not by PKAI holoenzyme.[1] All of these agents, except PKAI holoenzyme, also stimulate the phosphorylation of a specific set of surface proteins.[1] Another indication of the importance of surface PKAs is the ability of the PKAs' C subunit inhibitor protein, which cannot enter the cell, to block the initiation of DNA replication that rapidly follows the addition of Ca^{2+} to cultures of T51B rat liver cells stalled in late G_1 phase by external Ca^{2+} deprivation.[130]

After accumulating in the cytosol, CaM has been shipped into the nucleus and the nuclear CaM content nearly triples.[46] Something now triggers the release of Ca^{2+} from the storage vesicles that were reloaded earlier in the G_1 build-up.[1] In C127 mouse mammary cells, the Ca^{2+} surge starts in the perinuclear region,[40] possibly from the emptying of Ins(1,4,5)P_3-responsive stores in the nuclear membrane.[46–48] Could this be the response to a Ras-mediated signal from IGF-1 velcroceptors which, as in Swiss albino 3T3 cells, selectively target the nuclear envelope?[102] Soon, [Ca^{2+}] starts oscillating throughout the cell: the cell has started singing its preinitiation Ca^{2+} song (Figure 7). It will continue singing for a few minutes into the S phase and then the oscillations will stop (Figure 7).[40] The singing cell may pump Ca^{2+} into the nucleus with the Ca^{2+}•CaM-activated ATPase pumps in the nuclear envelope, which causes a surge of nuclear Ca^{2+} and triggers the redistribution of CaM to the nuclear periphery.[46–48] Ca^{2+} is not released from the storage vesicles in the cells of liver remnants in TPTX rats that do not initiate DNA replication.[1] This signal could be from α_1-adrenergic receptors which generate the Ca^{2+}-mobilizing Ins(1,4,5)P_3 from PtdIns(4,5)P_2 by activating a PLC.[1,46] This is indicated by the fact that blocking α_1-receptors late in the G_1 build-up of hepatocytes (e.g., 11 h after HPX) does not affect the loading of nuclei with CaM, but prevents the association of Ca^{2+}•CaM with the nuclear skeleton (matrix) and the initiation of DNA replication.[1,46] However, it should be noted that adrenergic blockade might not directly affect the hepatocytes, but instead block the release of other late-G_1-promoting factors, such as GLPs and insulin, into the blood.[1]

The nuclear Ca^{2+} surge can affect several important things. One of these might be the Ca^{2+}•CaM-activated nuclear contractile system consisting of actin, caldesmon, α-fodrin, MLCK, and myosin.[46] This system seems to be associated with the nuclear envelope and involved in transporting replication-related proteins from the production sites into the nucleus. One of these proteins is a 68-kDa Ca^{2+}•CaM-binding protein which has been found to enter the nucleus of CHEF/18 hamster cells and bind to the matrix-associated replication complexes at the time of onset of DNA replication (Figure 20).[131] The oscillating Ca^{2+} helps to trigger replication by binding to the heavy chain of the Ca^{2+} signal-mediating annexin II (calpactin I) associated with the multisubunit replication factor-C (RF-C), a primer-recognition protein accessory factor of Pol-α that triggers the binding of the polymerase to the primer terminus (Figure 20).[132,134] The surge of Ca^{2+}•CaM complexes could also promote the assembly and activity of the replication fork complexes through the several CaM-binding proteins that have been found in affinity-purified Pol-α.[135,136] Indeed, it is the CaM-binding proteins in the matrix-associated replication complexes that are mainly responsible for the attachment of CaM to the nuclear skeleton (matrix) upon the onset of replication.[46]

The nucleus also contains CaMKII, which is stimulated by the Ca^{2+} and CaM surges.[46] This enzyme must affect a wide variety of activities by phosphorylating parts of the replication complexes and stimulating the expression of genes with CaMRE elements in their upstream regulatory regions. The Ca^{2+} oscillations and perhaps DGs generated by an α_1-receptor signal probably also stimulate PKCs, such as PKC-β, in rat liver nuclei. PKCs could do at least three very important things at this late stage: phosphorylate Pol-α and thereby increase the enzyme's affinity for DNA, stimulate the transport of mRNA through the nuclear pore complexes in the cytoplasm, and activate so-called TPA-responsive genes.[1] The possible importance of the PKCs is indicated by the ability of the potent, specifically PKCs-stimulating TPA to prevent T51B

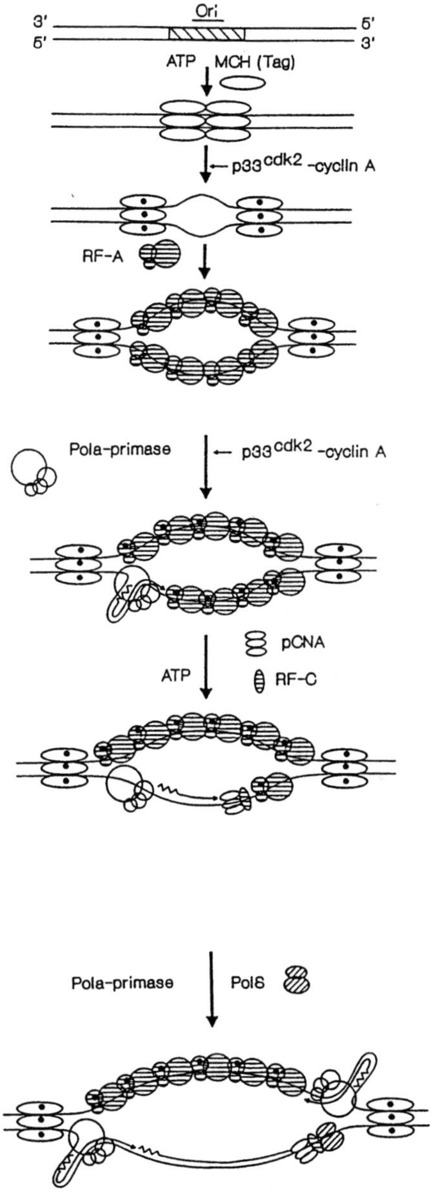

FIGURE 20 The initiation of DNA replication. The reader should consult the text for a discussion of the events and Abbreviations (found on page 217) for the keys to the abbreviations used in the diagram. It should be noted that the symbol (•) indicates the phospho-forms of the key MCH/Tag helicase and RF-A subunits that are phosphorylated by Cdk2·cyclin A.

rat liver cells from stalling at the end of their G_1 build-up in low (0.07 to 0.10 m*M*)-Ca^{2+} medium. But this could be deceptive because of TPA's nonphysiologically potent ability to prolongedly stimulate cellular PKCs. However, even with TPA external Ca^{2+} is not entirely dispensable, because TPA does not prevent late G_1 stalling when the external Ca^{2+} concentration is only 0.02 m*M*.[1]

Clearly Ca^{2+} and Ca^{2+}•CaM are key participants in the initiation of DNA replication. This has been proven in many ways. Expressing CaM antisense mRNA prevents mouse C127 mammary cells from initiating DNA replication.[40] Exposing CHEF/18 hamster embryo fibroblasts or T51B rat liver cells to Ca^{2+}•CaM blockers, such as the naphthalenesulfonamides W7 and W13, when they reach the end of their G_1 build-up (in normal, high-Ca^{2+} medium) "freezes" the cells on the threshold of chromosome replication (Figure 21).[1] Removing the inhibitor instantly lifts the block and the cells immediately start replicating DNA (Figure 21). Although the nuclear CaM content rises at this time, the total CaM content in the cells of the regenerating liver remnant, serum-stimulated BALB/c 3T3 mouse cells, and serum-stimulated T51B rat liver cells, which has doubled, now plunges down to the starting level.[1,78] It seems that the G_1 CaM surge crashes at least in part because the protein is released from the cells to serve as an autocrine/paracrine progression factor (Figure 19).[1] The first hint that CaM could stimulate something on the cell surface was our discovery that external Ca^{2+}•CaM can stimulate T51B cells, stalled in late G_1 phase by external Ca^{2+} deprivation, to start replicating DNA.[1] Then it was found that Ca^{2+}•CaM can also stimulate other cells, such as B16 mouse melanoma cells, K562 human erythroleukemia cells, NRK rat kidney cells, some peripheral blood lymphocytes, and *Planaria* cells, to initiate DNA replication.[1] Clearly, expelled CaM would stimulate something on the cell surface if there is enough external Ca^{2+} to produce Ca^{2+}•CaM complexes.[1] This was actually proven by Crocker et al.[137] when they showed that human K562 leukemia cells, like regenerating rat liver cells and other cells, accumulate CaM during the G_1 build-up, and then they dump it into the medium just before the onset of DNA replication. The autocrine/paracrine importance of this released CaM was demonstrated by the ability of the anti-CaM drug, W7, to prevent the K562 cells from initiating DNA replication even when it was bound to Agarose® beads which could not have got into the cells. Thus, CaM is released in a single pulse at the end of the G_1 build-up to do something important on the cell surface which is essential for the initiation of DNA replication. Since CaM can stimulate the initiation of DNA replication by T51B liver cells stalled in late G_1 phase by external Ca^{2+} deprivation (i.e., incubation in medium containing 0.02 m*M* Ca^{2+}) and since external Ca^{2+}•CaM can stimulate a protein kinase on the surface of T51B rat liver cells and myoblasts, it is likely that the important "something" that the released Ca^{2+}•CaM does on the cell surface is to stimulate the phosphorylation of some surface signaling receptor (e.g., an integrin).[1]

During the G_1 build-up, two proteins, Ran and RCC1, have been doing a very important job, probably in collaboration with PKAs — driving enhanced

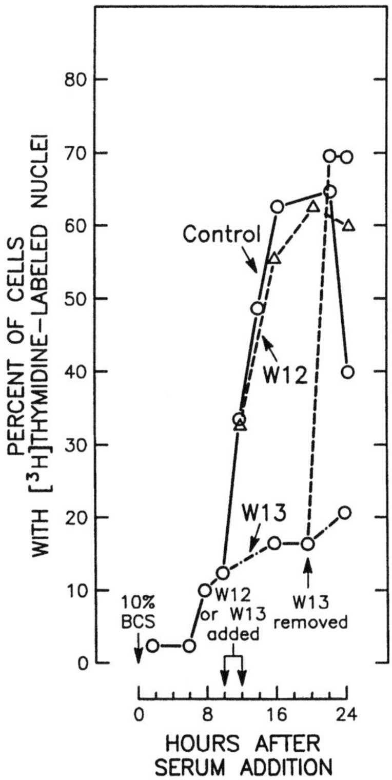

FIGURE 21 A selective Ca^{2+}•CaM blocker, the naphthalene sulfonamide W13, but not the much less active control naphthalene sulfonamide W12, traps T51B rat liver cells (in confluent monolayer cultures) at the end of the G_1 phase of their serum-triggered prereplicative development. W13 or the control W12 (10^{-5} mol/l) was added twice, at 10 and 12 h, after serum addition and then removed (by discarding the culture medium and replacing it with inhibitor-free medium) at 20 h. The almost immediate onset of DNA replication upon the removal of W13 reflects the importance of Ca^{2+}•CaM in the assembly and activation of replication complexes (replitases).

nucleocytoplasmic traffic.[1,138–140] The 45-kDa RCC1 resides in the nucleus where it activates a G protein, the 25-kDa Ran (*RA*s-related [though not closely] *N*uclear protein) by stimulating the exchange of GDP for GTP to produce active GTP•Ran complexes.[140] Moore and Blobel[140] have suggested that active cytoplasmic GTP•Ran complexes (produced from GDP•Ran complexes by the cytoplasmic B-2 factor) bind to [import-cargo protein•carrier] complexes moored to docking sites on the cytoplasmic surface of the NPCs (*N*uclear *P*ore *C*omplexes). This commits the [GTP•Ran•import — cargo•carrier] complexes to a PKAs-assisted[1] passage through the nuclear pore and delivery into the cytoplasm. The incoming complexes break down and the cargo is released when a GAP protein stimulates the GTP•Rans to autoinactivate to GDP•Ran complexes. The export arm of the cycle starts when RCC1 stimulates the exchange of GDP to GTP to convert the inactive GDP•Rans to active GTP•Rans. The GTP•Ran complexes then bind to export cargo protein•carrier complexes waiting at docking sites on the nuclear sides of the NPCs. This causes the [GTP•Ran•export cargo•carrier] complexes to leave the docking sites and move through the nuclear pores and into the cytoplasm where the GTP•Rans autoinactivate to release the export-cargos, carriers, and GDP•Rans. Cytoplasmic B-2 GDP/GTP exchangers then convert the Rans to active GTP•Rans, which head for the NPC docking sites and waiting [import-cargo•carrier] complexes. If RCC1 should disappear for some reason (e.g., the thermal inactivation of a thermolabile mutant RCC1 protein), mRNA export stops, mRNAs accumulate in the nucleus, and a host of nuclear functions, such as assembly of stable DNA replication complexes and on-going DNA replication, fail.[138,139] And we shall see later on that RCC1 inactivation can also trigger premature chromosome condensation and nuclear membrane breakdown.

In this last part of the G_1 build-up, the CksHs1/2-oligomerized Cdk2-cyclin C or D complexes are broken up and the next wave of cyclins, cyclins A and E take over the liberated Cdk2s.[2,138,139] Cyclin A has a NLS which enables it to be transported promptly by the Ran/RCC-1-operated mechanism into the nucleus to form oligomerized Cdk2•cyclin A PK complexes, which will start and drive chromosome replication.

C. The S Phase

At this point the deoxyribonucleotide triphosphate-producing enzymes have been made and at least one of them, the all-important enzyme ribonucleotide diphosphate reductase, has been strategically planted on the nuclear envelope to channel deoxyribonucleotide precursors into the nucleus with minimal access of exogenously supplied deoxyribonucleotides to the replication complexes inside the nucleus.[1,125,126,136] The various parts of the replication machinery have been made and shipped into the nucleus. The p53-GADD-45 safety check mechanism is now silenced by Cdk2•cyclin A PKs and it starts being dismantled and destroyed.

The unreplicated chromosomes are organized into loop domains, the ends of which are fastened by their replication origins to the nuclear matrix (scaffold) by the 170-kDa topoisomerase II (ScI) which in its structural mode controls overall loop structure and in its enzyme mode will relieve the supercoiling strains and cut the tangles that will arise when replication forks meet during replication.[133,142–146] The chromosomes are subdivided into replicons (stretches of DNA that are replicated under the control of a single replication origin), a strategy that enables the chromosomes to be replicated in minutes to hours instead of the days to weeks that would be needed if each chromosome was a single replicon like the much smaller bacterial chromosomes.[1,133,147] The replicons are organized into tandem clusters, all of which start replicating upon receipt of a signal from a leader region and finish replicating 15 min to 1 h later.[1,148] Each replicon cluster is, in turn, part of a battery of replicon clusters on the same and/or different chromosomes which fire synchronously or within a few minutes of each other at a specific point in the S-phase program.[1,148,149] The first batteries scheduled to fire contain the vital housekeeping genes and the tissue-specific genes that the cell and its daughters are likely to express, whereas the later-firing batteries contain the many other genes that the cell and its daughters would not express under normal circumstances and were accordingly silenced and buried in the heterochromatin.[1]

Most of our ideas about the initiation of DNA replication are based on the results of experiments with the SV40 virus's T antigen (Tag), a replication-initiating helicase that recognizes and attaches to replication origin sequences.[133,150–152] A plausible scenario for the events leading up to the firing of the first salvo of replicons might begin with the generation of a Tag-like Master of Ceremonies Helicase (MCH) in the cytoplasm (Figure 20). An important feature of the anatomy of this enzyme is a nuclear localization sequence (NLS) which can be recognized, for example, by the active, phosphoform of NBP-70, a 70-kDa NLS-binding protein. NBP-70 is a shuttle which carries NLS-tagged proteins into the nucleus with the help of the Ran/RCC1 mechanism.[140,153–156] There is a heightened level of the active phospho-NBP-70 because of the increased activities of PKAs and/or Ca^{2+}•CaM PKs in the last stages of the G_1 build-up. The shuttle or cargo-carrier binds the Tag-like helicase, and the Ran/RCC1-dependent mechanism drives the Tag-Carrier through a nuclear pore into the nucleus, dumps it, and then returns to the cytoplasm for another NLS-tagged molecule.[153,156] The inactive helicase molecules accumulating in the nucleus may aggregate, like SV40's Tag, into double hexamers which recognize and bind to available origins.[133,150–152] So at some point the leading replicons are ready for the starting signal with their origins encrusted with inactive Tag-like helicase (Figure 20).

The other proteins needed to replicate the chromosomes also have NLSs.[154] They too are being brought into the nucleus by an ATP-dependent, Ran/RCC1-dependent mechanism centered in the nuclear pore complexes.[140,153,154] For example, enough histones must be imported into the nucleus to build another set of nucleosomes. Although the histones are small, and thus could simply diffuse through the nuclear pores, they are in fact endowed with NLSs to enable fast, ATP-fueled, directed transport to the places where they are needed.[154] Indeed, the S-phase nucleus may transport as many as 100 to 500 histone molecules per minute.[154]

Then come the nuclear Cdk2•cyclin A complexes. Cyclin A and the substrate-binding integrin receptors are the keys to the mystery of why most normal cells must be attached to a solid substrate during the Ca^{2+}-dependent, Ca^{2+}-singing final stages of the G_1 build-up to finish the build-up and start replicating DNA.[157] When integrins such as the α_5 β_1 integrins are bound to substrate fibronectin, they activate a Src-family YPK (c-Fyn, c-Lyn, c-Yes) which phosphorylates, and thus activates, the pivotal 125-kDa FAK (*F*ocal *A*dhesion *K*inase) YPK.[157a] FAK phosphorylates, and thus velcroizes, the 80-kDa focal adhesion complex component, paxilin.[157a] c-Crk, a universal adaptor protein,[157b] binds with its SH2 domains to the phosphorylated Y residues of the velcroized paxilin and collects a second team of signalers with its SH3 domains. The integrin-activated focal adhesion complex has become a large, functional modular velcroceptor (like the modular velcroceptors to be discussed in Chapter 6), which organizes a team of signalers that trigger Ca^{2+} surges, bursts of PKCs activities, and activate "matrix-responsive" genes, among which is the gene encoding cyclin A.[157,157a] Thus, when the cells reach the end of the G_1 build-up, it is signals from the substrate-activated integrins that stimulate the expression of cyclin A. Cells containing a cyclin A gene coupled to a retroviral promoter continuously express protein independently from adhesion to a substrate and therefore can cycle in suspension.[157] But cells constitutively expressing a cyclin E or Cdk2 gene coupled to the same viral promoter still need to be attached to a solid substrate in order to make cyclin A and start replicating DNA.[157]

It is the Cdk2·cyclin A complexes that actually start DNA replication. They associate with the helicases on the leading origins and activate them by phosphorylating a key residue like the T^{134} of SV40's Tag (Figure 20, below).[151] The activated helicase then unwinds the origin DNA (Figure 20).[144,150] This attracts RP-A (replication protein-A), a complex of three proteins weighing 70, 32, and 11 kDa, which binds to single-strand DNA (Figure 20).[133,150–152] RP-A is activated when Cdk2·cyclin A phosphorylates a set of serine residues on RP-A's 32-kDa subunit (Figure 20).[151–152] The possibility of a cyclin-dependent PK being the replication trigger is indicated by the ability of purified Cdc2 to start the unwinding of origin DNA and DNA replication when added to a cyclin-containing extract of G_1-phase cells containing SV40 Tag and RP-A.[152]

The helicase·RP-A complex now binds Polα-primase, which is made up of a 180-kDa polymerase catalytic subunit, a 72-kDa subunit, and two 58- and 48-kDa subunits containing the primase catalytic site (Figure 20).[133,150] The 180-kDa subunit associates with the activated helicase and the primase subunits associate with RP-A's 70-kDa subunit (Figure 20).[133,150] The Pol-α primase makes the first RNA primer and the Ca^{2+}·annexin II-activated, 5-subunit RF-C accessory factor causes the Pol-α catalytic subunit to bind to the primer terminus and start making the first Okazaki fragment (Figure 20).[133,150] The helicase plows ahead, unwinding the helix as it goes (Figure 20). Pol-α follows along making the lagging strand. The first Okazaki fragment primes the assembly of the leading-strand-replicating complex.[150] First, the Ca^{2+}·annexin II-modulated RF-C binds to the 3'-end of the Okazaki fragment in an ATP-dependent manner but without requiring ATP hydrolysis (ATP is an allosteric factor).[150] RF-C then recruits PCNA (*P*roliferating *C*ell *N*uclear *A*ntigen) which, in turn, recruits Pol-δ and serves as a "sliding clamp" to prevent Pol-δ from slipping off the DNA rail while moving off in the opposite, 5′ to 3′, direction from Pol-α-primase and synthesizing the leading strand (Figure 20).[150] PCNA is responsible for Pol-δ's high processivity. At some point another Pol-α-primase makes the first Okazaki fragment on the opposite strand.[150] Thus is established a bidirectionally moving replication fork (Figure 20).

During the S phase, the E2F transcription factor has formed complexes with cyclin A, p107 (a relative of Rb), and Cdk2 (Figure 15).[103,105] It seems that E2F is used to direct the cyclin-dependent PK into contact with factors on specific chromosome regions. Also, it seems that Cdk2·cyclin A PKs have been blocking the expression of Cdc2·cyclin B PKs which will be needed to trigger mitosis. This block is lifted when chromosome replication has been completed and the time for mitosis has come.[105]

D. Mitosis

Cells must not start mitosis until they have done two things: finished replicating their DNA and reached an optimal size for dividing (Figure 1).[3] If they should initiate mitosis prematurely they risk producing genetically defective daughters, or dividing themselves to death. Therefore, cells have devised ways to avoid premature mitosis.

Nuclei in the process of replicating DNA somehow suppress initiation of mitosis. This suppressor signal was first seen by Rao and Johnson when they fused S-phase and G_2-phase cells and found that the G_2 nuclei did not enter mitosis until the S-phase nuclei had finished replicating their DNA.[158] As stated above, the Cdk2·cyclin A PKs operating in the S phase suppress the expression of the 34-kDa Cdc2 and cyclin B components of the MPFs (mitosis-promoting factors). But it also appears that the intensity of suppression is a function of the amount of single-stranded DNA, and above a certain threshold amount, Cdc2 is inactivated by phosphorylation. It seems that the monitor of the level of unreplicated DNA is RCC1, which besides being involved in nucleocytoplasmic transport, is an essential part of replication complexes, clearly an ideal place for sensing single-stranded DNA and assessing the amount of unreplicated DNA.[138–140] The RCC1 gene is located in the 11 to 13 region of the long (q) arm of the X chromosome in human cells.[138,159–170] RCC1's homolog in the fission yeast *Sc. pombe* is Pim-1, SRM1/PRP20 in the budding yeast *Sa. cerevisiae*, and BRJ1 in the fruit fly *Drosophila*.[163–166] If all is in order, RCC1 is ubiquitinated and thus marked for selective degradation.[171] The importance of ubiquitin at this point in the cycle is illustrated with *ts*85 mouse mammary carcinoma cells, which have a mutant temperature-sensitive ubiquitin-activating enzyme E1.[172] These cells can replicate their chromosomes at 39°C with their thermolabile E1 inactivated, but they cannot start the G_2 build-up.[172]

RCC1 is transported into, and then retained by, the nucleus where it binds to the chromatin, possibly selectively to the centromere (CEN) regions of the interphase chromatids, by its N-terminus, but neither its nuclear retention nor its function require the chromatin-binding domain.[167,168] We have already met, albeit briefly, RCC1 at the end of the G_1 build-up in its capacity as a player in nucleocytoplasmic transport and the assembly of replication complexes.[138–140,173]

RCC1 revealed itself with the BN2 mutation which produced the *ts*BN2 mutant of the BHK21 baby hamster cell. This mutation caused the hamster protein's S^{256} residue to be replaced by an F residue.[170] This single amino acid substitution made the protein thermolabile. So when *ts*BN2 cells are incubated at 39.5°C instead of 37°C, the RCC1 protein disappears.[165,168] If the cell is in the G_1 phase when the temperature rises, the disappearance of RCC1 prevents the transport of mRNA from the nucleus, the assembly of replication complexes, and the initiation of DNA replication, but there is no premature chromosome condensation.[140,165,168] However, things change when the cell is in the S phase, because now when the temperature is raised DNA replication stops and a striking cascade of events is triggered by the disappearance of RCC1 and ends 3 h or so later in premature mitosis.[165,168] It seems that RCC1 somehow suppresses the synthesis of a protein(s) needed for the transport into the nucleus of the Cdc25C, STY-protein phosphatase.[168,173,174] Eliminating the *ts*BN2 cell's thermolabile RCC1 derepresses the synthesis of the transport factor and a surge of Cdc25C into the nucleus where it dephosphorylates, and thus activates the mitosis-promoting Cdc2·cyclin complexes.[168,173] There is a large CaM surge which must be important (and might be the derepressor protein) because premature mitosis is prevented by a Ca^{2+}·CaM blocker, the naphthalene sulfonamide W7.[175] The cell stops making DNA, possibly because of a collapse of RCC1-dependent replication complexes.[138] It rounds up as microtubules are recruited from the interphase network to build a mitotic spindle.[165,168] The partly replicated chromosomes condense prematurely into the fragmented pattern typical of the premature chromosome condensation (PCC) that occurs when S-phase cells are fused with mitotic cells.[165,168] The nuclear envelope disintegrates and the cell is in mitosis. By 6 h the premature mitosis is over, but the cell has not divided.[165,168] The Cdc2·cyclin B complexes that triggered it have disappeared as their cyclin B components are destroyed.[165,168] The mitotic spindle disperses and the interphase network is re-established. The nuclear envelope is regenerated, but micronuclei now form from some individual chromosomes.[165,168] The cell is again in the G_1 phase, but without any RCC1 to help assemble replication complexes there is where it will stay.[165,168]

RCC1 has a partner who seems to be the real operator. As we have learned this partner is a 25-kDa member of a novel class of Ras-like G proteins.[138–140,163,176] In humans, this G protein is known as Ran or TC4 and it is very closely related (81% homologous) to the *Sc. pombe* fission yeast's Spi 1 protein.[163] It seems that RCC1's job is to keep Ran in an active, mitosis-suppressing GTP·Ran complex by promoting the exchange of GDP for GTP.[140,163,176]

The budding yeast *Sa. cerevisiae* has yet another way of restraining the oligomerization and activation of the mitosis-promoting Cdc2·cyclin B complexes. This is the RAD 9 mechanism involving a network of 6 proteins, the products of the RAD 9, 17, and 24 genes and the MEC 1, 2, and 3 genes.[177] This mechanism is not normally needed for proliferation. But it kicks into action when chromosome replication is defective or chromosomes have been damaged by UV- or X-radiation. Normally, when UV- or X-radiation produces lesions in, or breaks, chromosomes, the RAD 9 mechanism, like the mammalian p53 safety officer, somehow senses these lesions and temporarily delays the initiation of mitosis until the G_2 cell has had at least a chance to repair the damage and improve the prospects of producing viable progeny. But when any one of the RAD 9 genes has suffered a loss-of-function mutation the cell does not pause to repair damaged chromosomes and produces lethally stricken progeny. The RAD 9 mechanism also swings into action to stop the initiation of mitosis when chromosome replication has failed in some way. Again, the goal of the RAD 9 mechanism is to give the cell some time to repair damage and avoid mitotic catastrophe.

When the cell finishes replicating its chromosomes it generates another cAMP surge which peaks during the G_2 build-up to the initiation of mitosis and then crashes abruptly when the cell enters mitosis (see Figure 3 of Chapter 6). It may crash because the accumulated cAMP is destroyed by a burst of PDE activity or it is released from the cell like the cAMP accumulated before the onset of DNA replication. This G_2 surge has so far been seen in many different cells: blastema cells of a regenerating annelid worm *Owenia fusiformis*, Chinese hamster ovary (CHO) cells, fetal rat liver cells, Hela human cervical adenocarcinoma cells, human RPMI 8866 lymphoid cells (Chapter 6, Figure 4), rat kidney tubule cells, seminal vesicle cells, and *Tetrahymena pyriformis*.[78] What causes it? Because it has been seen in a wide variety of cells cycling in radically different environments, it, like its G_1 predecessor, must be caused by something common and endogenous. This something might be analogous to the wasp venom peptide mastoparan, substance P, or polyamine mast cell activator that can operate from within the cell in directly activating G_s protein and thus stimulating adenylyl cyclase independently from cell surface receptors.[178] It could be an adenylyl cyclase-activating deformation of the cell membrane that happens at this point in the cycle.[179]

What does the G_2 cAMP surge do? It seems certain that one of its jobs is to help prevent a premature onset of mitosis. (Cells certainly do have a mitotic *horror aborti*!) This is indicated by the fact that keeping the G_2 surge from crashing down with dibutyryl cAMP (which easily penetrates the cell membrane), an adenylyl cyclase stimulator, or a PDE inhibitor prevents cells such as the RPMI 8866 cells (Chapter 6, Figure 3) from initiating mitosis after replicating their chromosomes.[1] This means that the G_2 burst of PKAs activity must somehow stop the dephosphorylation of the key threonine and tyrosine

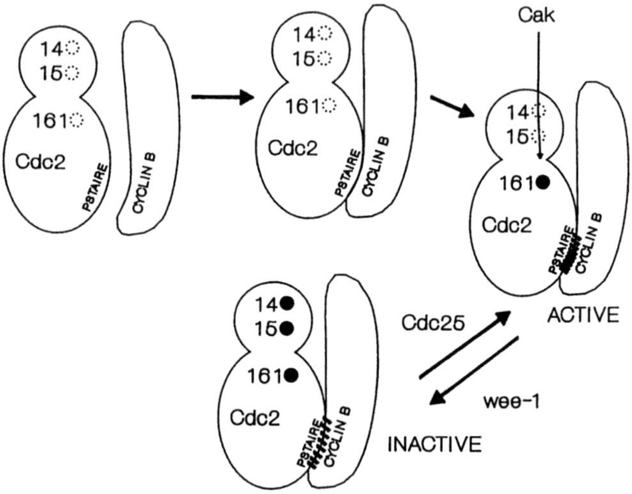

FIGURE 22 The assembly and activation or deactivation of Cdc2·cyclin B PKs. Cyclin B binds tightly to the PSTAIRE motif-bearing region of the Cdc2 catalytic subunit when the CaK kinase phosphorylates the T^{161} residue of Cdc2. The resulting complex is inactivated when the "wee 1" STYPK phosphorylates the Cdc2 T^{14} and Y^{15} residues. The complex is activated when the dual-function Cdc25 phosphatase dephosphorylates Cdc2 T^{14} and Y^{15} residues.

residues needed to activate the Cdc2·cyclin protein kinase complexes that trigger mitosis (Figure 22).[3] Indeed, this appears to be the case in starfish oocytes.[180] The cAMP surge probably also prevents premature nuclear envelope breakdown by phosphorylating certain S and T residues, which prevents the dissociation of lamins A and C from the inner surface of the envelope.[181] On the other hand, the G_2 cAMP surge, like its predecessor, the G_1 surge, might have other jobs to do.

The G_2 surge might indirectly (there is no reason to suspect a direct action) stimulate the phosphorylation of threonine 161 by CakPK which promotes the association of Cdc2 with cyclin B to form mitosis-promoting complexes primed for later activation (Figure 22).[85] Indeed, there are indications of a positive role for cyclic AMP in the initiation of prophase. Although artificially maintaining a high intracellular cyclic AMP concentration maintains mitotic arrest in pig oocytes, a *transient* cyclic AMP surge precedes, and *accelerates*, the MPFs-induced resumption of meiotic prophase ("germinal vesicle breakdown") induced by human chorionic gonadotropin (hCG) or luteinizing hormone (LH).[181a] It appears that this meiotic prophase-accelerating cyclic AMP surge is triggered by a soluble factor secreted by hCG or LH-stimulated follicular granulosa cells.[181a] A transient cyclic AMP surge also occurs at the time of resumption of meiotic prophase in rabbit and sheep oocytes.[181b,181c]

The first of the mainstream events leading to the generation of mitosis-promoting Cdc2-cyclin complexes is the phosphorylation by CakPK of the T^{161}

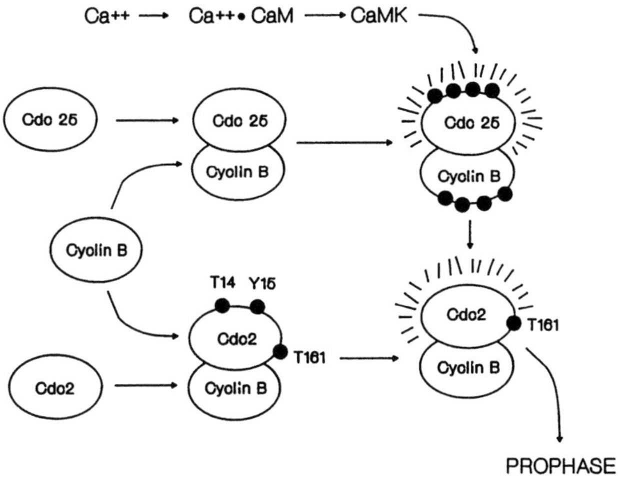

FIGURE 23 The activation of the mitosis-triggering Cdc2·cyclin B PKs by the preprophase Ca^{2+} spike (Figure 7) generates Ca^{2+}·CaM complexes which activate CaMPKs. The CaMPKs phosphorylate and thus activate (activity is symbolized by radiating lines) Cdc25·cyclin B STYPPases. These dual-function phosphatases then activate phospho-Cdc2·cyclin B PKs (activity is symbolized by radiating lines around the Cdc2 catalytic subunit) by dephosphorylating Cdc2 T^{14} and Y^{15} residues without affecting the phosphorylated T^{161} residue, which is needed for the formation of active Cdc2·cyclin B complexes. The symbol (•) indicates phosphorylation.

residue of cytoplasmic Cdc 2[3,85,182–185] (Figure 22). This enhances the binding of cyclin B to Cdc2's PSTAIRE motif-containing region to form a complete STPK (Figure 22). The cytoplasmic phospho-Cdc2-cyclin B PK is thus primed for phosphorylation of both its T^{14} and Y^{15} residues by the dual-function "wee-1" STYPK (Figures 22 and 23).[182,185] T^{14} and Y^{15} are part of Cdc2's catalytic domain and thus their phosphorylation inactivates the Cdc2-cyclin B PK. This inactivation is meant to delay the initiation of mitosis until the chromosomes have been replicated and both Cdc2-cyclin PKs and the cell size have reached optimal levels.[3]

When the G_2 cell's size and its load of inactive Cdc2-cyclin B complexes have reached their critical levels, the dual-function "wee-1" STYPK is shut off and the cytoplasmic inactive complexes are shipped into the nucleus where they are activated by a waiting, dual-function nuclear STY protein phosphatase(s) that unblocks the Cdc2 catalytic subunit's catalytic site by removing the phosphates attached to the T^{14} and Y^{15} residues (Figures 22 and 23).[3,85,180] This family of phosphatases is named after the *cdc*25 gene product of *Sc. pombe*. It includes the Cdc25Hu1, Hu2 or Cdc25A,B,C of human cells and the "string" enzyme of *Drosophila*.[3,186–191] It appears that there may be G_1 and S-phase Cdc25s, just as there are G_1/S and mitotic cyclins. The activation of the mitotic Cdc25s requires that there be no unreplicated DNA in order to prevent the premature activation of Cdc2-cyclin complexes and the initiation of mitosis while the Cdc25s responsible for triggering the G_1 build-up and DNA replication are unaffected by the presence of unreplicated DNA.

Members of the Src family of nonreceptor protein tyrosine kinases[12] may act *downstream* from Cdc2·cyclin in the initiation and progression of mitosis.[192] p60$^{c\text{-}src}$ is activated during mitosis in cells such as NIH3T3 fibroblasts probably in large part by having its Y^{527} residue dephosphorylated.[12,192] Normal p60$^{c\text{-}src}$ is kept under very tight control, because of its potent oncogenic power, either by prompt autophosphorylation or phosphorylation by another YPK, CSK, of its Y^{527} residue.[12,193] The C-terminus of the velcroized, phosphoenzyme bends around to shove its phosphorylated Y^{527} into its SH2 pocket and thus cover and block the catalytic domain.[12] It seems that either the activation of the dual-function Cdc25 phosphatase that switches on Cdc2·cyclin complexes also dephosphorylates and activates p60$^{c\text{-}src}$, or activated Cdc-cyclin complexes phosphorylate and they activate another protein-tyrosine phosphatase or inactivate the CSK YPK. Direct phosphorylation of threonine (T^{34}, T^{36}) and serine (S^{72}) residues, the so-called MSP (mitosis-specific phosphorylation) sites, by activated Cdc2·cyclin B complexes contributes to the activation of p60$^{c\text{-}src}$[192].

Is the stimulation of p60$^{c\text{-}src}$ activity one of the mainstream events that drive the cell into and through mitosis? Or is it just an incidental byproduct of the cascade of protein kinase and phosphatase activations that trigger mitosis? It seems intuitively unlikely that the activation of this potent oncogenic enzyme would be a purposeless side event! Indeed, microinjecting the constitutively activated oncogenic p60$^{v\text{-}src}$ enzyme into a frog *(Xenopus)* oocyte accelerates the resumption of meiotic prophase.[194] Results of experiments of Durkin et al.[195] proved that p60$^{c\text{-}src}$ *could* be a mainstream player in the initiation of mitosis. Serum-starved *ts*RSV-NRK rat kidney fibroblasts, carrying a mutant (LA23) Rous sarcoma virus with a v-*src* gene that codes for a thermolabile p60$^{v\text{-}src}$ YPK, stop cycling at 41°C, which inactivates the mutant p60$^{v\text{-}src}$. (These cells do not need exogenous serum factors to proliferate when driven by p60$^{v\text{-}src}$.) The quiescent, hot cells can be induced to start the G_1 build-up of a few key proteins by cooling them to 36°C, which reactivates the labile p60$^{v\text{-}src}$. If the viral kinase is inactivated again by heating the cells to 41°C when they are in the middle of the S phase, the cells will finish replicating their chromosomes, but they will not initiate mitosis. They are stalled indefinitely in a G_2 limbo. However, the stalled cells will start mitosis if p60$^{v\text{-}src}$ is reactivated by dropping the temperature back to 36°C or serum is added to the medium of the hot cultures.

p60$^{v\text{-}src}$ is a powerful agent that could artificially kick the G_2 build-up mechanism into action and p60$^{c\text{-}src}$ might not, in fact, be the scripted player in the events leading up to the initiation of mitosis. But the fact that serum can replace p60$^{v\text{-}src}$ in the stalled G_2 *ts*RSV-NRK cells suggests that there may be some exogenous G_2 promoter that operates by activating a velcroceptor YPK or somehow activates p60$^{c\text{-}src}$. Indeed, the results of experiments of Armelin and Armelin[196] and Swierenga et al.[197] suggest that fetal bovine serum and rat blood plasma contain a "cycle-completion" factor that cells need to complete the G_2 build-up and start mitosis. This factor may be autocrine HA (head activator), an 11-amino acid peptide (GPPGGSKVILF) that was originally found in *Hydra* where it stimulates the G_2/mitosis transition of head cells.[198] The peptide is secreted by a wide variety of proliferating cells and is in the blood of all mammals.[198] HA receptor (HAR) is expressed specifically, and the autocrine/paracrine HA peptide is made and secreted in a carrier complex, by G_2 and mitotic cells (Figure 19).[198] Inactivating HA by adding specific antibody to the culture medium prevents the HA-producing G_2 cells from completing the cycle and dividing.[198]

There is good reason to believe that Ras proteins may also be involved in triggering mitosis. First, there are the results of experiments of Durkin and Whitfield[199,200] on *ts*KSV-NRK cells. Serum-starved *ts*KSV-NRK cells, like *ts*RSV-NRK cells, stop cycling at 41°C, which inactivates the thermolabile mutant viral K-Ras protein that keeps them cycling without help from serum factors at 36°C. The quiescent, hot *ts*KSV-NRK cells start cycling again when cooled to 36°C to reactivate the K-Ras protein. If the viral protein is inactivated again by heating the cells to 41°C when they are in the middle of the S phase they will finish replicating their chromosomes and stay indefinitely in a G_2 limbo, exactly like *ts*RSV-NRK cells that have had their thermolabile $p60^{v\text{-}src}$ inactivated in the middle of the S phase. But there is an important difference. Serum cannot stimulate the hot, stalled G_2 cells — Ras activity is needed (the normal K-*ras* gene seems to be suppressed by the mutant viral gene). The only way to induce them to enter mitosis is to drop the temperature back to 36°C[199] and this entry into mitosis can be prevented by microinjecting an antibody to the reactivated K-Ras protein.[200] Both Ras and $p60^{c\text{-}src}$ are also key mediators of the triggering of meiotic prophase I (maturation) in *Xenopus* oocytes by activated insulin velcroceptors.[201–203] Thus, maybe an activated receptor such as the HAR receptor would activate the Src YPK and stimulate the accumulation of active GTP·Ras protein complexes in somatic cells. The active GTP·Ras complexes would activate Raf-1 and trigger a PKs cascade that stimulates the accumulation of mitosis-specific proteins.

Is Ca^{2+} a player in the main stream of events leading to mitosis? As mentioned above, the premature chromosome condensation and nuclear envelope breakdown triggered by eliminating the *ts*BN2 hamster cell's thermolabile RCC1 protein is preceded by a CaM surge and it is stopped by inhibiting Ca^{2+}·CaM activity.[175] Indeed, Ca^{2+} and Ca^{2+}-binding proteins have a key, hence evolutionarily conserved, role in the initiation of mitosis.[40] Thus, although there is no nuclear envelope breakdown or chromosome condensation in the mitosis of the yeast *Sa. cerevisiae*, a Ca^{2+} transient coincides with the initiation of this intranuclear mitosis which requires active *cdc*31 gene product, a Ca^{2+}-binding CaM-like protein.[204–207] A role for CaM in the initiation of animal cell mitosis is also indicated by the fact that inhibiting Ca^{2+}·CaM PK activity by microinjecting anti-Ca^{2+}·CaM antibody prevents or significantly delays nuclear envelope breakdown in sea urchin eggs.[40,208] Moreover, microinjecting Ca^{2+} or the Ca^{2+} stores-emptying Ins(1,4,5)P_3 stimulates premature chromosome condensation and nuclear envelope breakdown in sea urchin embryos.[209] Caffeine also releases Ca^{2+} from internal stores,[209a,209b] and it stimulates mammalian cells blocked in S phase to induce prematurely a true mitosis and thereafter undergo multiple rounds of chromosome condensation and decondensation.[210] Indeed, it seems that there is a Ca^{2+} surge from internal stores and a single Ca^{2+} spike before normally scheduled nuclear envelope breakdown (Figure 7) and that nuclear envelope breakdown is a Ca^{2+}-dependent process which can be prevented by intracellular Ca^{2+}-chelating[40,209,211–216] compounds. For example,[55] the cells of two-cell sand dollar embryos have a 40- to 45-s "window" 6 min before the scheduled time of nuclear membrane breakdown during which lowering the intracellular Ca^{2+} concentration prevents this breakdown.[215] Dropping the internal Ca^{2+} concentration before or after this critical time has no effect. If the Ca^{2+} concentration is reduced below the critical level at this time, the cell will not immediately initiate prophase when Ca^{2+} is microinjected. Indeed, it must wait for a whole cycle before it can try again to initiate prophase. This means that something critical that needs a Ca^{2+} surge appears for only a short time. If there is not enough Ca^{2+} during that time, it will vanish and the cell will have to go through another cycle to make it appear again. It is interesting that something critical, Ca^{2+}-dependent, and ephemeral also appeared just before the end of the G_1 build-up when the replication-triggering Cdk2·cyclin A PKs were being produced. Indeed, if there had not been enough Ca^{2+} around at that time, this ephemeral something would have vanished and the cell would have had to start the G_1 build-up over again to generate it and try again to initiate DNA replication.

From the start of the S phase onwards, cyclin A has been in the nucleus and the cell has started to make more Cdc2. Cyclin Bs, such as cyclin B1, appear and start accumulating in the cytoplasm.[2,3,217] There they bind to the PSTAIRE domains of Cdc2 catalytic subunits which have had their T^{161} residues phosphorylated by CaK kinase (Figures 22 and 23). These complexes are inactivated when the "wee-1" STY protein kinase phosphorylates their T^{14} and Y^{15} residues (Figure 22). It appears that nuclear Cdc2·cyclin A complexes are first to be activated to start prophase while something happens, maybe the Ca^{2+}·CaM surge resulting from the premitotic Ca^{2+} spike, that moves the inactive cytoplasmic Cdc2·cyclin B complexes through the nuclear pore complexes. There they meet one of the Cdc25 dual-function protein phosphatases which activates them by dephosphorylating their T^{14} and Y^{15} residues[2,3,218] (Figures 22 and 23). The now-activated Cdc2-cyclin B complexes join forces with the Cdc2·cyclin A complexes to start other parts of the prophase-metaphase program.

The nuclear Cdc2·cyclin A complexes are activated by mitotic Cdc25 (when there is no restraining signal from unreplicated DNA free ends). These complexes start chromosome condensation, while Cdc2·cyclin B complexes are still building up and waiting in the cytoplasm, by superphosphorylating the H1 histones associated with the internucleosomal ("linker") DNA and are also responsible for phosphorylating the H3 histones in the nucleosomal cores.[2,3,219,220] The crashing of the G_2 cyclic AMP surge and the abrupt drop in PKA activity has terminated the phosphorylation of specific domains of the H1 histones that would have prevented chromatid condensation.[221] The Cdc2-cyclin B complexes now phosphorylate a different set of domains in the H1 histone tails with conserved (-K[S/T]PXK-) or (-K[S/T]PK-) motifs.[221] This H1 histone kinase activity was the first of the "MPF" enzyme activates to be identified.[222] This H1 histone superphosphorylation it produces is somehow involved in condensing the replicated chromatids.[220–222] But how can this be? The additional negative charges would weaken the H1 histones' association with the linker DNA. In fact, H1 histone phosphorylation is commonly associated with chromosome *decondensation* and enhanced access of transcription factors to their target genes.[221] Instead of condensing and suppressing gene activity mitotic H1 histone superphosphorylation decondenses chromatin to admit key condensing factors such as topoisomerase II.[223] This topoisomerase is essential for chromatid condensation by stacking chromatin loop domains along an axial scaffold of which the enzyme is a key structural part.[223,224] Chromatid condensation also requires that Cdc2·cyclin B complexes stimulate the deacetylation of the H3 and H4 nucleosomal core histones.[220]

As the chromatids condense, DNA-binding proteins (the CENPs) and motor proteins, such as CEPN-E and cytoplasmic dynein, begin collecting on their heterochromatin-rich centromeres made up of long stretches of highly repeated CEN base sequences organized into several microtubule-binding segments separated by linker segments.[224a-228] Eventually, mature kinetochore motors will be assembled on the centromeres to drag the chromosomes toward the spindle poles at anaphase.

Before the nuclear membrane breaks down, the centrosomes are being pushed apart by activated microtubule plus-end-directed kinesin motors attached to microtubules.[229] While the centrosomes are separating, their microtubules are still long, relatively stable (turnover time about 10 to 20 min), and still connected to cortical actin in order to orient the spindle axis.[229] While this is going on, the triply phosphorylated cytoplasmic Cdc2·cyclin B complexes are still inactive because the Cdc25 PPase is locked in the nucleus.[1,3,217,218]

When the Cdc2·cyclin B complexes are kicked into the nucleus and activated by the waiting mitotic Cdc25 phosphatases, they take over from Cdc2·cyclin A complexes to complete chromosome condensation (cyclin A does not associate with highly condensed chromosomes). They also home onto different targets. They trigger the production of, and/or activate, a clathrin-like vesicularization factor that breaks the nuclear envelope into vesicles and they trigger the dismantling of the nuclear pore complexes.[229] The Cdc2·cyclin B PKs also phosphorylate, and thus directly depolymerize, or prime the depolymerization of, the lamina network that lines the envelope and connects with the cytoplasmic systems of intermediate filaments through the nuclear pores.[230–235] The mammalian nuclear lamina is a mesh of polymers of lamins A and C (the product of alternative splicing of the lamin A gene transcript) and lamin B which attaches the lamina both to the nuclear membrane (via a farnesyl [C-19 isoprenoid] covalently linked to a cysteine residue of the lamin molecule) and the cytoplasmic network of intermediate filaments. The lamins have the central α-helical rod domain characteristic of all intermediate filament proteins as well as a NLS for transportation into the nucleus where they polymerize into the lamina. Cdc2·cyclin B PKs phosphorylate S and T residues in GTPLSPTR and LSPSPSSR regions flanking, respectively, the N- and C-terminal sides of the α-helical domain of a lamin such as chicken lamin B_2.[232] Each lamin filament is an assembly of eight protofilaments, each of which consists of lamin tetramers linked end to end.[230,233] The phosphorylation of the lamin filaments by Cdc2·cyclin complexes either directly causes the filaments to break up into protofilaments, or it primes their break up.[232–234] It must not be assumed that lamina disassembly causes nuclear envelope breakdown. Lamina disassembly is indeed necessary, but not sufficient, for nuclear envelope breakdown.[232] Other things, such as CaM PKII activated by the prophase Ca^{2+}·CaM transient, MAP kinase and S6 kinase II, activated perhaps by the autocrine HA factor as well as the clathrin-like vesiculation factor, are all players in this breakdown. It is important, and not a little surprising, that the lamina may not in fact vanish upon nuclear envelope breakdown. In some mitotic cells the lamina persists as a well-defined shell around the nuclear region until it finally disperses at anaphase.[235]

The Ca^{2+}·CaM-, Cdc2·cyclin B-triggered nuclear envelope breakdown releases the Cdc25 PSTYPase(s) into the cytoplasm.[218] This dual-function phosphatase activates cytoplasmic Cdc2·cyclin B complexes by removing phosphates from T^{14} and Y^{15} residues that were blocking the Cdc2·kinase's catalytic domain. The activated Cdc2·cyclin B PKs directly or indirectly phosphorylate a cytoplasmic factor which proceeds to increase the "catastrophe" rates (the rate of transition from the growing to shrinking modes) of astral and other cytoplasmic microtubules.[236] These microtubules start shrinking and may eventually vanish.[236] But the "catastrophe" rate of microtubules growing from the centrosomes that are associated with, or near, the condensing chromosomes is not stimulated. Consequently, tubulin pools and microtubule growth are redirected toward spindle formation.[236] This selective orientation of microtubules towards the chromosomes may be due in part to a type I ST PPase in the chromosomes which dephosphorylates and inactivates the "catastrophe"-enhancing factor. Thus, the steady-state length of microtubules are functions of the relative levels of Cdc2·cyclin B STPK and phosphatase activities. Another factor, the 145-kDa STOP (*S*table *M*icrotubules *O*nly *P*rotein) protein, a Ca^{2+}·CaM target, associates specifically with, and stabilizes, kinetochore-attached microtubules.[237]

The plus ends of the microtubules growing from the centrosomes probe the cytoplasmic space for the chromosomes' centromeres and associated kinetochores. When they find one, as many as 30 to 40 of them may latch onto the motor's several microtubule attachment sites where they are stabilized by the CaM-binding/responsive STOP protein.[237] Now they must line up along the

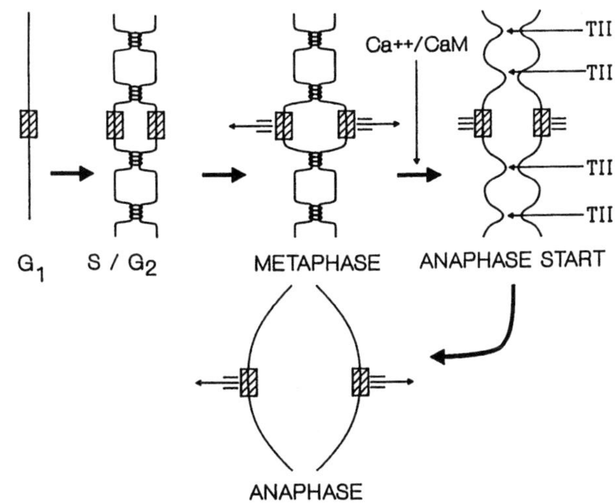

FIGURE 24 At the end of the S phase the replicated chromosomes are still joined by residual tangles where the replication forks had met. At the end of metaphase the still-joined chromosomes are being pulled and stretched in opposite directions by their kinetochore motors. They are released by a Ca^{2+} spike and the resulting Ca^{2+}·CaM complexes that somehow activate topoisomerase II (TII) which cuts the tangles and thus releases the chromosomes for their anaphase dash to the spindle poles. The same "anaphase start" signal stimulates the protease activity that also destroys the Cdc2·cyclin B complexes.

mid-zone of the spindle.[238,239] One of the players in this alignment, or *congression,* is a centromere-binding protein, CENP-E.[224a,240,241] This is a microtubule plus-end-directed motor protein belonging to the kinesin superfamily[224a] which the cell began accumulating during the S phase. It now joins the many proteins collecting on the centromere and kinetochore, such as the other CENPs and the microtubule minus-end-directed cytoplasmic dynein motor, as well as CaM, CaMPKII, α-CKI (α-casein kinase I), CKII (casein kinase II), and STOP protein.[242,243] There it helps drive the congression of the chromosomes at the spindle mid-zone.

When the metaphase congression is complete each duplicated chromosome is poised quivering in place in a state of dynamic equilibrium resulting from a tug-of-war between its two kinetochores which are in their stable poleward orientation and thus pulling the sister chromatids toward opposite poles (Figure 24). The sister chromatids are held together by the many tangles or catenated regions where the DNA strand replication forks met during the S phase (Figure 24).[244] Without these many entangled, catenated regions the replicated chromatids could not be aligned, held on the spindle, and then divided equally among the daughter cells.[244]

Something monitors metaphase congression by sensing a phosphoepitope on unattached kinetochores and prevents the initiation of anaphase until all of the chromosomes have taken their place on the spindle (Figure 1).[224a] When congression is complete, the kinetochore phosphoepitopes have been eliminated, and *all* of the chromosomes are lined up on the spindle mid-zone, something gives the "Anaphase START" signal (Figure 24).

A lot of things happen now. There is another Ca^{2+} spike (Figure 7), due to a Cdc2·cyclin B-induced discharge of Ca^{2+} from storage vesicles associated with the spindle and the opening of membrane Ca^{2+} channels, which generate Ca^{2+}·CaM complexes.[40,211,212,214,244–267] Ca^{2+}·CaM must be an important part of the "Anaphase START" signal because reducing or stopping CaM production with antisense CaM-RNA prevents metaphase cells (eg., C127 mouse cells) from initiating anaphase.[40,268] One key event is the release of the 135- to 150-kDa INCENPs (internal centromere proteins) which have been holding the sister centromeres together.[269] The other key event is the disentangling or decatenation of the sister chromatids by the chromatid knot-cutting topoisomerase II in the chromosome scaffolds (Figure 24).[244,270] The disentangled chromosomes can now be pulled to opposite spindle poles (anaphase A) by their kinetochore motors (Figure 24). There is no consensus about how the kinetochores move their chromosomes. It has been regarded as an ATP-independent process in which a "Pac Man"-like kinetochore chews up, and thus ratchets its way along, the kinetochore microtubules toward the poles.[212,230,271–273] The energy for this would come from energy stored in the microtubules by GTP hydrolysis during the polymerization of GTP-tubulin complexes.[273] But others have found cytoplasmic dynein in kinetochores.[226,227] This is an ATP-fueled motor which moves toward the minus ends of microtubules and could thus pull the chromosomes toward the minus ends of the kinetochore microtubules in the centrosome region. The Pac Man-like chewing up of the kinetochore microtubules by the kinetochore motors is probably primed by the Ca^{2+}·CaM-triggered release of the stabilizing 145-kDa STOP protein from the microtubules.[237]

The "Anaphase START" signal must also trigger the destruction of the Cdc2·cyclin complexes, otherwise the cell will be locked in metaphase just like a mature egg in meiotic metaphase II with its Cdc2·cyclin complexes protected by the Ca^{2+}-inactivated cMos STPK.[3,274–276] In *Xenopus* frog oocytes, the maturation signal triggers the expression of a STPK c-Mos which directly or indirectly phosphorylates and stabilizes cyclin B and activates inactive phosphorylated Cdc2·cyclin B complexes by stimulating the Cdc25 STYPPase.[3,277–280] This triggers meiotic prophase I and the completion of the cycle with the pre-anaphase destruction of both cyclin B in the Cdc2·cyclin B complexes and c-Mos. Cdc2·cyclin B complexes are formed anew in meiosis II, but now c-Mos persists as the cytostatic factor CSF (or a key part of the CSF), which holds the egg cell in metaphase II by maintaining an active pool of Cdc2·cyclin B complexes.[3,279–282] Fertilization triggers a single spreading wave of Ca^{2+} release from internal stores in frog eggs or sustained Ca^{2+} oscillations in mammalian eggs.[1,209b] The surging Ca^{2+} independently activates the proteolysis

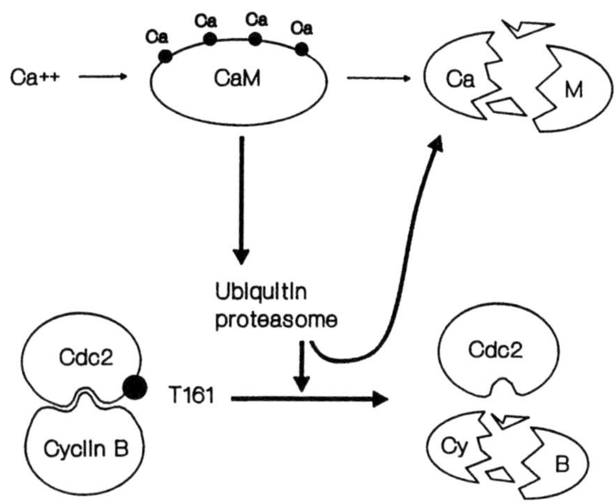

FIGURE 25 The Ca^{2+} spike that triggers anaphase destroys the Cdc2·cyclin B PK complexes which, if they were to persist, would lock the cell in metaphase like an unfertilized egg. Ca^{2+}·CaM complexes generated by the Ca^{2+} spike trigger the ubiquitination of the cyclin B components of the Cdc2·cyclin B PKs. This marks the cyclin Bs for proteolysis by ubiquitin proteasomes. Inactive Cdc2 catalytic subunits are then released, perhaps for recycling. The Ca^{2+}·CaM complexes may also be destroyed by the ubiquitin proteasome. It should be noted that the (•) symbol indicates the phosphorylated T^{161} residue of Cdc2.

by the Ca^{2+}-activated calpain of both the cyclin B in the metaphase-maintaining Cdc2·cyclin B complexes and c-Mos.[279,282,283] Thus, this burst of proteolytic activity releases the Cdc2·cyclin B-imposed metaphase block, the cell starts anaphase, and it completes meiosis II. In somatic cells the anaphase Ca^{2+} spike and the Cdc2·cyclin complexes themselves switch on ubiquitin ligase which pins ubiquitin onto lysine residues near the cyclins' N-terminal "destruction box" domains.[274–276] This marks the ubiquitinated cyclins for destruction by the 26-S ubiquitin conjugate-degrading proteasome (Figure 25).[274,275,284] At the same time a ST-PPase snips the phosphate off T^{161} that activated Cdc2. If this dephosphorylation were not to happen, the catalytic subunit would stay attached to cyclin B, remain active, and prevent, or delay, chromosome decondensation and exit from mitosis.[244a] The now inactivated, liberated Cdc2 may be saved and recycled if one or both daughter cells should immedidately decide to start a cycle and another parade of cyclins.

Another batch of microtubules growing from the poles have met and mingled in the mid-zone of the spindle. After chromosome congression, the CENP-E kinesin motor molecules move from the kinetochores to the interdigitating microtubules in the spindle mid-zone.[228,240,241] There they cause the microtubules to slide along one another and in doing so push the poles apart and elongate the cell in the process known as "anaphase B". At the same time these interzonal microtubules are lengthened by adding to their ends the tubulin dimers that are being snipped off the kinetochore microtubules by the poleward chewing kinetochore Pac Men.[245,271]

So far we have been focusing on the chromosomes and mitotic spindle, but we are now nearing the actual division of the cell, or cytokinesis, in which actin and myosin are the principal players. The cell lost its functional, interphase shape when it entered mitosis as the microtubules making up the interphase cell's backbone of girders linking centrosome to cortex were disassembled and their components diverted to spindle formation. This left the cortical mesh of actin filaments and the network of actomyosin stress fibers without support. In the interphase cell, the actin-binding caldesmon (a Ca^{2+}·CaM-binding protein) had been interlinking the stress fibers' F-actin and myosin.[219,230,231] Caldesmon thus locked or latched together the actomyosin fibers of the interphase cell. The Cdc2·cyclin B activity that destabilized the interphase microtubular girders also phosphorylated caldesmon.[219,230,231] The phospho-caldesmon then separated from the actomyosin fibers and this unlatched the fibers, which then relaxed to complete the cellular rounding up.[231] Phospho-caldesmon also separated from the cortical actin filaments. This made the filaments more accessible to myosin II and enabled the formation of a cortical band of actomyosin filaments. These filaments now generate a surge of cortical tension at the equator of the elongating cell.[230] This pulls more actin, myosin, and other proteins such as α-actinin and filamin from other regions into an equatorial contractile ring of actomyosin filaments attached to, and running parallel with, the plasma membrane.[230] The rising Ca^{2+} level during anaphase and the accumulation of CaM in the equatorial region produces Ca^{2+}·CaM complexes which activate myosin light-chain kinase (MLCK).[236] MLCK then phosphorylates (a N-terminal S residue) and thus activates the myosin motor heads. The actin and myosin filaments now start sliding along each other to contract the ring. As the ring contracts, it squeezes the elongating cell into two daughters and dissociates while doing so. It must be noted that CaM has concentrated in the mid-zone of the spindle, not in the overlying cleavage furrow itself, but MLCK may have been activated there by Ca^{2+}CaM before moving up into the furrow. Another thing to consider is that certain residues of the light chains in the myosin molecules' motor heads were phosphorylated and inactivated by the Cdc2·cyclin B complexes.[230] But this stopped when the complexes were dephosphorylated and destroyed by the "Anaphase START" signal. The dephosphorylation of these residues and the phosphorylation of other residues by Ca^{2+}CaM-activated MLCK activates the myosin motors. Interestingly, an activated H-Ras protein significantly speeds up cytokinesis in rat fibroblasts.[285] This acceleration might be due to the Ras protein increasing the Ca^{2+} surges that drive cytokinesis or it could be due to the downregulation of PKCs activity commonly seen in Ras-transformed cells if these enzymes are other MLCK inhibitors.[285]

Mitosis now ends. The daughter cells are in telophase and the chromosomes have finally arrived at the poles of the separating daughter cells. The potent Cdc2·cyclin A and Cdc2·cyclin B kinase complexes have gone. The chromosomes' superphosphorylated H1 histones were dephosphorylated during anaphase. Now the chromosomes decondense when Ca^{2+}-activated calpain degrades the topoisomerase II that stacked their loop domains along scaffold complexes during prophase.[1,244a,286] Dephosphorylated-recycled and new lamins repolymerize and reconnect recycled and new nuclear envelope vesicles to the decondensing chromosomes. The vesicles fuse via an ATP-dependent and GTP-modulated process into a new nuclear envelope with pore complexes. Soon a new functioning nucleus appears. The cytoplasmic network of intermediate filaments reconnects through the nuclear pores to the new lamin mesh. A new network of interphase microtubules extends from the centrosome to the cell cortex and elements of the new Golgi apparatus are driven along these microtubules by dynein motors to an assembly point near the centrosome.[287] During this time, the Rb protein, which had been hyperphosphorylated by Cdk2·cyclin A or E or Cdk4·cyclin D to lift a block to the coordinate expression of replication enzymes at the end of the G_1 build-up, is partially dephosphorylated to the active state (Figure 15).[103,104] The dephosphorylation of Rb and the restoration of its nuclear affinity started with the activation of protein phosphatase I after the destruction of cyclin B at the metaphase-anaphase transition. The dephosphorylation will now continue to completion in the newborn cells.[288,289]

E. Stromal Cell Activation and Termination of the Proliferative Response in the Regenerating Liver Remnant

The other regenerating liver cells, the stromal cells, do not seem to respond immediately to the chorus of signals that stimulates hepatocyte proliferation and they do not start replicating their chromosomes until 30 h or longer after the hepatocytes.[1] This delay suggests that the stromal cells might be proliferatively activated by a signal from a paracrine growth factor(s) such as ECM-bound bFGF which is phosphorylated and mobilized by the hepatocytes' ecto-PKAs activated by the cAMP released in later G_1 stages, and/or from TGF-α from the hepatocytes, and/or from EGF, which peaks in the portal blood 12 to 18 h after partial hepatectomy.[1] The production of such growth factors could be the reason, or part of the reason, why partial hepatecomy enables intraportally injected colon and mammary cancer cells to colonize the liver.[290,291] However, some stromal cells, the sinusoidal cells, but not hepatocytes, start making TGF-β1, around 4 h after partial hepatectomy.[1,55] It is likely that the sinusoidal cells release the factor in a latent TGF-β1-LAP (latency-associated protein) complex which consists of two disulfide-bridged, mature TGF-β1 molecules, each of which is noncovalently linked to a 39-kDa, mannose-containing, TG1-β1 precursor molecule.[292] This complex probably binds to the hepatocytes' mannose-6-phosphate/IGF-2 receptors after which the active TGF-β1 homodimers are released to activate types I and II TGF-β receptors and thus inhibit proliferation by preventing Rb inactivation.[292,293] It appears that when the flow of TGF-β1 from the sinosoidal cells peaks at around eight times the basal level at 72 h after HPX it stops further hepatocyte proliferation by preventing the cells from completing the G_1 build-up.[1,55] Since injecting TGF-β1 into HPX rats prevents the initiation of DNA replication in the liver remnant,[55] the TGF-β1 released from the TGF-β1-LAP secreted by the sinusoidal cells is likely to be the principal terminator of hepatocyte proliferation following hepatectomy and other proliferogenic injuries. TGF-β1 may stop the newborn hepatocytes from proliferating further by suppressing the expression of the key proliferation-determining c-*myc* gene and by preventing the phosphorylation and inactivation of the Rb suppressor protein near the end of the G_1 build-up. This is supported by Whitson and Itakura[294] using WB-5334 rat liver epithelial cells. Thus, when quiescent WB cells with underphosphorylated Rb protein were stimulated by fresh medium containing 10% fetal bovine serum, Rb protein was phosphorylated and DNA replication began between 8 and 12 h and peaked between 18 and 20 h. When the stimulatory medium contained 200 p*M* TGF-β1, Rb remained underphosphorylated and the cells did not start replicating DNA. The likelihood of Rb phosphorylation being part of TGF-β's mode of action is supported further by the fact that with increasing passage WB cells became resistant to TGF-β1 which then could no longer inhibit Rb phosphorylation or stop the initiation of DNA replication.

To proliferate in response to the HPX signal, the hepatocytes in the liver remnant had to stop making the powerful 30-kDa proliferation-suppressing protein, prohibitin.[295,296] If the lost liver cells have now been replaced the newborn hepatocytes will start making prohibitin to lock themselves safely into proliferative quiescence.

As we have seen, partial hepatectomy triggers the release of a host of hormones and local factors that collectively and redundantly signal the remaining hepatocytes to express their c-*fos,* c-*myc, and* c-*max* genes, whose products are believed to be essential for the initiation of the prereplicative cascade. However, hepatocytes can also be induced to proliferate in the intact rat by liver mitogens such as the synthetic sex steroid, cyproterone, ethylene dibromide, and lead nitrate.[297–301] In this case, there is probably *no* triggering of a portal symphony of factors and no stimulation of c-*fos* and c-*myc* expression![297,301] As an example of this hyperplastic response, the liver can be induced to double its size within 4 to 8 d by giving the rat one injection of lead nitrate or 7 doses of cyproterone (100 ng/kg/d by gavage for the first 3 d; 130 ng/kg/d by gavage for the next 4 d).[298–301] If mitogen treatment is stopped, cells start randomly self-destructing which returns the liver to its normal size. Similarly, a burst of proliferation induced by EGF in the noncycling fraction of a primary neonatal rat liver culture is followed by a wave of cell suicide which reduces the cell number to the initial value.[302] Hepatocytes kill themselves with apoptogenic, autocrine TGF-β1, or with apoptogenic paracrine TGF-β1 collected from other cells with their mannose-6-phosphate/IGF-2 receptors.[293,303,304] TGF-β1 triggers the apoptogenic mechanism which includes the Ca^{2+} activation of hydrolytic enzymes and membrane-stabilizing/strengthening TGase to keep the hydrolytic enzymes inside the dead shell and prevent them from leaking out and damaging neighbors before the dead shell is swept away by macrophages.[305,306] However, in these cells there seems to be *no* measurable DNA fragmentation by Ca^{2+}-activated endonuclease.[293]

The regeneration of a partially resected rat liver is not followed by random cell suicide. It replaces lost cells and restores the liver to its original functional mass (Figure 9). By contrast, mitogen-induced hyperplasia of an intact liver produces an excess of cells, which the liver senses and remedies when the mitogen is withdrawn. We do not know how the rat liver senses and responds to a surfeit of cells with induction of TGF-β1 production and Ca^{2+}-dependent killing. As we shall see, because of this homeostatic cellular self-sacrifice, the mitogen-stimulated intact liver, unlike the partially resected liver, does not produce preneoplastic foci when the tissue is also treated with a carcinogen then a tumor promoter.[297]

F. Hepatoma Cells

The proliferation of normal hepatocytes is triggered by factors such as HGF released by sinusoid and more remote cells in response to partial hepatectomy. These factors start the cells growing to a size and metabolic configuration at which it uses cyclic AMP and its PKAs to start the cell cycle engine with its parade of PSTAIRE-motif PK catalytic subunits and stage-specific cyclins to initiate the expression and accumulation of the various parts of the chromosome-replicating machinery. Ca^{2+} is a key player in all of this. As we have seen, normal hepatocytes need a lot (1.0 to 1.5 m*M*) of external Ca^{2+}, especially in the last stages of the G_1 build-up.[1] Their cycling is stopped by TGF-β issuing from sinusoidal cells, and they dismantle their cell cycle engines.

Hepatoma cells proliferate spontaneously. They do not need signals from blood-borne hepatocyte growth factor, hormones, and eicosanoids to proliferate. They, like other neoplastic cells such as chemically and virally transformed human and murine fibroblasts,[1] no longer need a lot of external Ca^{2+} to initiate and later complete the G_1 build-up (Figure 26). The first evidence that hepatoma cells, like other chemically and virally transformed human and murine cells,[1] drop their dependence on external Ca^{2+} at some point during neoplastic transformation was reported by Swierenga et al.[307] Dropping the Ca^{2+} concentration from the usual 1.25 m*M* to 0.02 m*M* in the medium (15% horse serum, 2.5% fetal bovine serum, 82.5% Ham's F-10 medium) of 3-d-old primary cultures of livers from 6-d-old Buffalo-strain or Sprague-Dawley rats dramatically reduced

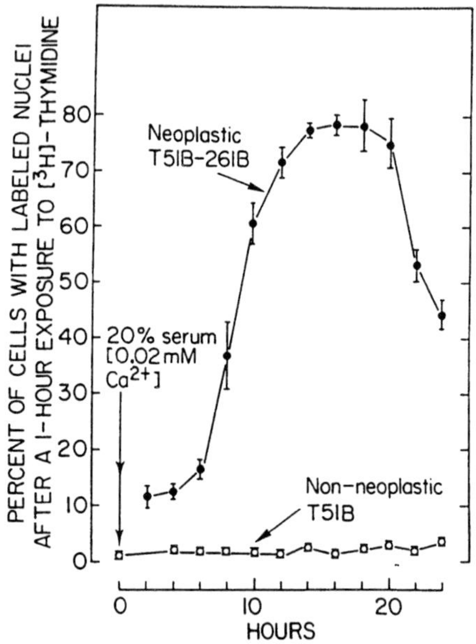

FIGURE 26 Confluent tumorigenic T51B-261B rat liver cells, but not their nontumorigenic T51B counterparts, can be stimulated to initiate DNA replication and transit the S phase in Ca^{2+}-deficient Eagle's basal medium.

the DNA-synthetic activities of *both* hepatocytes and stromal cells.[307] This severe external Ca^{2+} deprivation did not affect the DNA synthetic activity of the neoplastic cells in the hepatocyte islands in 3-d primary cultures of a Morris hepatoma (5123 tc) in a Buffalo-strain rat, but it still dramatically dropped the proliferation of the normal stromal cells surrounding the neoplastic hepatocyte islands.[307] Raising the Ca^{2+} concentration of the medium to 1.25 m*M* did not affect the neoplastic hepatocytes, but it strongly stimulated the declining proliferative activity of the normal stromal cells. Reducing the Ca^{2+} concentration of the medium did not affect the flow of cells into the S phase and mitosis of a cultured line of neoplastic cells derived from a Morris 7795 hepatoma.[307] The loss of external Ca^{2+} dependence seems to be an early happening in the carcinogenic process. Thus, cells from preneoplastic or premalignant rat liver nodules initiated by hexachlorodicyclohexane were found by Swierenga[308] to be already able to proliferate in medium containing only 0.02 m*M* Ca^{2+}. Boynton et al.,[309] San et al.,[310] Swierenga et al.,[311] and Williams[312] reported that a dozen or so chemically transformed tumorigenic hepatocytes could replicate DNA and form colonies in medium containing 0.01 to 0.04 m*M* Ca^{2+} in which nontumorigenic hepatocytes could not proliferate.

The ability of most, but not all, tumorigenic liver and other cells to proliferate and form colonies in Ca^{2+}-deficient medium is commonly accompanied by an ability to form colonies while suspended in soft agar. An adhesion-dependent signal is required near the end of the G_1 build-up just when the cells are most dependent on external Ca^{2+}.[313,314] The reason for this dependence on adhesion may be at least partly due to the dependence of PtdIns(4,5)P_2 synthesis on adhesion to a solid substrate and, perhaps more importantly, on the signals from integrin receptors when they bind to ECM proteins.[315] Thus, Ca^{2+} receptors or other receptors on suspended cells can be activated and stimulate PLCs, but the cells do not have enough PtdIns(4,5)P_2 for the PLCs to break down into PKCs-stimulating DGs and the internal Ca^{2+} stores-emptying Ins(1,4,5)P_3 to trigger the cascade of events leading to the G_1 finale, specifically the expression of cyclin A which is needed to produce the Cdk2·cyclin A PKs that trigger DNA replication.[157,157a] The signal from integrins, particularly $\alpha_5\beta_1$ integrins, triggers the phosphorylation of Y residues of FAK, a 125-kDa YPK, which itself becomes an active velcroceptor whose hooked signaler team stimulates "matrix-responsive" genes such as the *cyclin* A gene.[157,157a] Either tumorigenic cells no longer need a signal from integrin receptors to make Cdk2·cyclin A PKs, or they generate the same signal without adhesion, because, for example, FAK is phosphorylated and activated independently from integrins in anchorage-independent cells by a constitutively active, oncogenic Src-family YPK.[157a] On the other hand, the anchorage dependence of transformed cells can be restored, and their growth properties normalized, by increasing the expression of $\alpha_5\beta_1$ integrins.[157a]

An example of a transformed cell producing a signaler that replaces both external Ca^{2+} and substrate-triggered integrin signals is an avian or mammalian fibroblast transformed by the ASV's p60$^{v\text{-}src}$ YPK.[316–319] In these transformed cells, the constitutively expressed, hyperactive p60$^{v\text{-}src}$ does what activated integrin receptors do in normal cells — phosphorylates the FAK YPK.[315] In other words, the hyperactive viral YPK eliminates the need for adhesion and the cells can initiate DNA replication and form colonies in soft agar. These ASV-transformed cells can also proliferate in low-Ca^{2+} medium, but they still need Ca^{2+}·CaM near the end of the G_1 build-up.[316,318,319] Thus, NRK rat kidney fibroblasts infected with a mutant ASV producing a thermolabile p60$^{v\text{-}src}$ YPK cannot proliferate in medium containing only 0.02 m*M* Ca^{2+} when the viral enzyme is inactivated by raising the temperature to 40°C, but reactivating the enzyme by dropping the temperature back to 36°C stimulates the cells to proliferate indefinitely despite the severe external Ca^{2+} shortage. While the p60$^{v\text{-}src}$-driven cells can proliferate in Ca^{2+}-deficient medium, they still need Ca^{2+}·CaM and become *extremely* sensitive to inhibition by the naphthalene sulfonamide Ca^{2+}·CaM blocker, W7, between 4 and 2 h before the expected time of initiation of DNA synthesis, but they are completely unaffected before or after this time.[319]

As we have seen, Ras proteins are adaptor-coupled parts of velcroceptor teams that stimulate the cell to grow toward the synthesis and assembly of the cell cycle engine by stimulating protein synthesis and the movement of gene transcription factors to their target genes in the nucleus or activating PKs that phosphorylate and activate transcription factors already in place on the regulatory elements of target early cell cycle genes. A very common happening along the road to malignancy is the appearance of a mutant, persistently active Ras protein. Activated Ras proteins, such as an oncogenic, viral K-Ras protein, enable the cell to form colonies in soft agar, avoid apoptosis, and proliferate rapidly and indefinitely in very-low-Ca^{2+} medium.[1] Thus, the presence of a persistently active mutant Ras protein anchored to the inner surface of the plasma membrane eliminates the cell's need for external Ca^{2+} and signals from adhesion-activated cyclin A-stimulating integrin receptors to proliferate.

It is noteworthy that a persistently active mutant Ras protein (which has lost its GTPase activity and thus cannot convert itself from the active GTP·Ras to the inactive GDP·Ras form) has provided us with a dramatic example of "proto-cancer".[1] As we have seen, activation of the budding yeast's RAS 2 (by a CDC25-induced exchange of Ras-bound GDP for GTP) is the growth-triggered event that stimulates adenylyl cyclase and the expression of the cyclins needed to form the Cdc2/28·CLN PKs that START the G_1 build-up to chromosome replication and division.[1–3] If a normal yeast cell has not got enough nutrients to grow to a critical size, its RAS proteins will not be activated and it will not START the G_1 build-up. But a budding yeast with mutant GTPase-less, permanently activated RAS proteins is no longer responsive to its growth status, and it will suicidally start its cell cycle engine and try to proliferate even when it cannot grow in a nutritionally inadequate medium.[1,3]

Are Ca^{2+} and anchorage independence really linked? This question has been answered by Boynton et al.,[320] who exposed cultures of immortalized, but non-neoplastic, T51B rat liver cells to the carcinogens AFB or MNNG and in some cases promoted the postinitiation transformation process with saccharin. Then they determined whether cells in individual variant clones emerging from these treatments which could proliferate in medium containing only 0.02 m*M* Ca^{2+} were also able to proliferate while suspended in soft agar and produce tumors in BALB/c nu/nu athymic nude mice. They also determined whether cells of variant clones that were able to proliferate in soft agar were also able to proliferate in Ca^{2+}-deficient medium and produce tumors in nude mice. The cells from 10 of 11 colonies that formed in Ca^{2+}-deficient medium did *not* form colonies in soft agar or produce tumors. Cells of the remaining clone, T51B-261B, were strongly tumorigenic and also proliferated in soft agar (Figure 26). By contrast, the cells of 6 of 9 colonies that grew in soft agar did not form a significant number of colonies in Ca^{2+}-deficient medium, and the cells of 5 of these 6 anchorage-independent clones did *not* produce tumors. The cells of 2 of the remaining 3 anchorage-independent clones were able to proliferate in Ca^{2+}-deficient medium, but they were *not* tumorigenic. Finally, the cells of the last of the 9 anchorage-independent clones were both tumorigenic and able to proliferate in Ca^{2+}-deficient medium. Clearly, Ca^{2+} and anchorage independence are not tightly linked. The same message was sent by Grisham et al.,[321] Tsao and Grisham,[322] and Tsao et al.,[323] who determined the abilities of carcinogen-treated rat liver cells to proliferate in Ca^{2+}-deficient medium, form colonies in soft agar, and produce tumors. Clearly, the ability to proliferate in Ca^{2+}-deficient medium and form colonies in soft agar are unrelated, independently expressed indicators of *impending* tumorigenicity. Thus, Ca^{2+} and anchorage independence are "paraneoplastic" properties which are not essential or sufficient for tumorigenicity, but are commonly co-expressed by neoplastic cells. But the immortalized rat liver cells in these studies almost certainly had already started making autocrine growth factors, and already had very much reduced extracellular Ca^{2+} requirements compared to hepatocytes in the rat. Thus, the primary reduction of the extracellular Ca^{2+} requirement is tightly linked to the essential "immortalization" step of neoplastic transformation, but it is the second reduction of the Ca^{2+} requirement to very low levels that is paraneoplastic.

As expected from the importance of cyclic AMP and the PKAs in driving the cycles of normal cells, there is an increased responsiveness of adenylyl cyclase and an increased cyclic AMP level in the early stages of chemically induced carcinogenesis.[1,92–94] The hepatoma cell's cyclic AMP-dependent controls also seem to break down, as indicated by the loss of the 47-kDa suppressor protein which binds to the CRE of cyclic AMP-responsive genes such as the overexpressed immediate-early c-*fos* gene in precancerous liver and liver tumors.[99,324,325] Unfortunately there are no reports of the marked shift in the balance of the expression of the RI and RII PKA subunits in favor of the proliferation-pushing RI, that we will see in Chapter 5, is a striking feature of colon and other cancer cells.[92–94]

Hepatoma cells do not need exogenous hormones or HGF from sinusoidal cells or distant organs to be switched into the proliferative mode or a fast-cycling, nonsenescing mode. Unlike the proliferatively inactive hepatocytes in a normal liver, hepatoma cells constitutively express the c-*fos* and c-*myc* cell cycle genes, make and secrete the TGF-β-inactivatable Rb cycle-suppressor protein.[103–105,326,327] They are consequently permanently competent to start the G_1 build-up and able to respond immediately to the secreted TGF-α. Since EGF reduces the Ca^{2+} dependence of proliferation as dramatically as neoplastic transformation,[1] the stimulation of EGF receptors by large amounts of self-made autocrine TGF-α could explain the greatly reduced external Ca^{2+} requirements of hepatoma cells.[1] An autocrine, bFGF-like factor may be responsible for the anchorage-independent proliferation of its producer hepatoma cells.[55] It may also stimulate the angiogenesis needed for sustained tumor growth as well as the proliferation of the tumor's stromal cells.[55]

Hepatoma cells, like many other tumor cells, overproduce CaM, which could be regarded as an autocrine growth factor because *external* Ca^{2+}-CaM stimulates the proliferation of rat liver cells and other cells stalled in late G_1 phase by external Ca^{2+} deprivation.[1] Indeed, human K562 myeloid leukemia cells release CaM onto their surfaces where it activates a process required for the initiation of DNA replication and proliferation.[137] Since normal cells adjust the production of CaM to avoid an excess of CaM over its target proteins, the overproduction of CaM by hepatoma cells must greatly affect Ca^{2+} signaling and upset the regulation of a variety of cell functions ranging from cytoskeleton activities to PK and phosphatase activities.[1,328,329] Indeed, constitutive CaM overexpression stimulates the proliferation of mouse fibroblasts.[40,328] Hepatoma cells also make another Ca^{2+}-binding signaling protein, the oncotrophoblastic 11.7-kDa oncomodulin, which was discovered and named in our laboratory.[1] While we know almost everything about oncomodulin's structure we know nothing about its function in placenta, where it is normally made, and in cancer cells.[1] However, we know that under oxidizing conditions it can dimerize into a calmodulin-like dumbell-shaped dimer which binds to, and activates, all calmodulin targets as effectively as calmodulin itself, but in its monomeric form it can inhibit glutathione reductase, an enzyme that is unaffected by CaM.[1,330] It is a member of the parvalbumin family of Ca^{2+}-binding proteins.[1] But it is a parvalbumin with a punch! Parvalbumins are usually Ca^{2+}-binding/Ca^{2+}-buffering, but not -signaling, proteins that do not bind to target proteins and thus simply blot up and terminate Ca^{2+} signals.[1,329] Excessive levels of the Ca^{2+}-buffering parvalbumins slow the progression through the G_1 and mitotic phases of the cell cycle.[331] However, oncomodulin does bind to target proteins and is thus a *signaling parvalbumin* which can stimulate DNA replication.[1] Oncomodulin is normally made and functions in placental cytotrophoblast cells, but not in proliferating or quiescent cells of the embryo proper or adult.[1] However, the oncomodulin gene is switched on, and oncomodulin appears, in the mutant clones emerging during carcinogenesis in various tissues including colon, liver, and skin.[1] The inhibition of glutathione reductase by the reappearing oncomodulin could promote proliferation and carcinogenesis by reducing the cellular pool of reduced glutathione (GSH) which would result in increased accumulations of peroxides and peroxide-induced arachidonic acid derivatives, which are normally restrained by the GSH-dependent activity of glutathione peroxidase. Clearly, the seemingly universal changes in the production of these two Ca^{2+}-binding proteins (and perhaps similar changes in other Ca^{2+}-binding proteins in fewer kinds of cancer cells), and the resulting disturbances of Ca^{2+}-dependent functions must be important events in carcinogenesis.

The malignant transformation of liver cells, like that of the cells to be discussed in the following chapters, appears to be associated with the inactivation loss or change of function of their p53 (or TP53) genes.[327] As we have learned, the p53 protein operates a late-G_1 checkpoint mechanism which involves the stimulation by the protein of the expression of so-called GADD and other genes whose products prevent the initiation of DNA replication until the chromosomes have been safety-checked. Without this safety-check mechanism damaged genes and chromosomes can be replicated, which enables the appearance of clones of malignant cells.

The appearance of invasive, malignant cells that can scatter throughout the body also requires that they no longer be tightly bound to their neighbors through adherens junction molecules, the E-cadherins.[335–338] Epithelial cells, but not fibroblasts and other mesenchymal cells, express their E-cadherin genes and display E-cadherins on their surfaces.[335] The cadherins are attached by their cytoplasmic domains to catenins, which are somehow connected to the microfilament network. The N-termini of the cadherins on the surface of an epithelial cell are bonded to the adjacent cell's cadherin N-termini by Ca^{2+}. Thus, the cadherins are *Ca^{2+} sensors* which, depending on the external Ca^{2+} concentration, determine cell and tissue morphogenesis and restrain cell motility and proliferation. Various types of malignant cells, including hepatocellular carcinoma cells, do not express E-cadherins or express nonfunctional mutant cadherins.[335,338] The human gene for E-cadherin on chromosome 16q 22.1 can be regarded as one of the genes that suppress tumor progression, because the loss of heterozygosity of the q (long) arm of chromosome 16 has been found in hepatocellular carcinomas.[337]

The growth of foci of transforming liver cells is a function of the rate of cell proliferation and the frequency of cell suicide by Ca^{2+}-mediated apoptosis, a homeostatic mechanism which we have seen is designed to eliminate excess liver tissue. PKCs-activating phorbol esters promote the proliferation and survival of premalignant cells and the ultimate emergence of cancer cells in two ways. While phorbol esters do not directly stimulate normal hepatocytes to proliferate they do directly stimulate carcinogen-initiated hepatocytes to proliferate and at the same time they enhance tumor growth by inhibiting the triggering of tissue-homeostatic apoptotic cell death.[339–341]

REFERENCES

1. **Whitfield, J. F.**, *Calcium, Cell Cycles, and Cancer*, lst ed., CRC Press, Boca Raton, FL, 1990.
2. **Hunter, T., Ed.**, *Regulation of the Eukaryotic Cell Cycle*, Ciba Foundation Symp. 170, John Wiley & Sons, Chichester, 1992.
3. **Murray, A. and Hunt, T.**, *The Cell Cycle*, W. H. Freeman, New York, 1993.
4. **Altman, M. and Trachsel, H.**, Regulation of translation initiation and modulation of cellular physiology, *Trends Biochem. Sci.*, 18, 42a, 1993.
5. **Frederickson, R. M. and Sonenberg, N.**, Signal transduction and regulation of translation initiation, *Semin. Cell Biol.*, 3, 107, 1992.
6. **Koromilas, A. E., Lazaris-Karatzas, A., and Sonenberg, N.**, mRNAs containing extensive secondary structure in their 5′ non-coding region translate efficiently in cells overexpressing initiation factor eIF-4E, *EMBO J.*, 2, 4153, 1992.
7. **Morley, S. J. and Thomas, G.**, Intracellular messengers and the control of protein synthesis, in *Intracellular Messengers*, Taylor, C. W., Ed., Pergamon Press, Oxford, 1993, 447.
8. **Rhoads, R. E.**, Regulation of eukaryotic protein synthesis by initiation factors, *J. Biol. Chem.*, 268, 3017, 1993.

9. **Rhoads, R. E., Joshi-Barve, S., and Rinker-Schaeffer, C.**, Mechanism of action and regulation of protein synthesis initiation factor 4E: effects on mRNA discrimination, cellular growth rate, and oncogenesis, *Prog. Nucleic Acid Res. Mol. Biol.,* 46, 183, 1993.
10. **Smith, M. R., Jaramillo, M., Liu, Y.-L., Dever, T. E., Merrick, W. C., Kung, H.-F., and Sonenberg, N.**, Translation initiation factors induce DNA synthesis and transform NIH 3T3 cells, *New Biol.,* 2, 648, 1990.
11. **Smith, M. R., Jaramillo, M., Tuazon, P., Traugh, J. A., Liu, Y.-L., Sonenberg, N., and Kung, H.-F.**, Modulation of the mitogenic activity of eukaryotic translation initiation factor-4E by protein kinase C, *New Biol.,* 3, 601, 1991.
11a. **De Boni, U.,** The interphase nucleus as a dynamic structure, *Int. Rev. Cytol.,* 150, 149, 1994.
12. **Brickell, P. M.**, The $p60^{c\text{-}src}$ family of protein-tyrosine kinases: structure, regulation, and function, *Crit. Rev. Oncogenesis*, 3, 401, 1992.
13. **Fante, W. J., Johnson, D. E., and Williams, L. T.**, Signalling by tyrosine kinases, *Annu. Rev. Biochem.*, 62, 453, 1993.
13a. **Varticovski, L., Harrison-Findik, D., Keeler, M. L., and Susa, M.,** Role of PI3-kinase in mitogenesis, *Biochim. Biophys. Acta*, 1226, 1, 1994.
14. **Birge, R. B. and Hanafusa, H.**, Closing in on SH2 specificity, *Science*, 262, 1522, 1993.
15. **Mayer, B. J. and Baltimore, D.**, Signalling through SH2 and SH3 domains, *Trends Cell Biol.*, 3, 8, 1993.
16. **Pawson, T. and Gish, G. D.**, SH2 and SH3 domains: from structure to function, *Cell*, 71, 359, 1992.
16a. **Bar-Sagi, D.**, The Sos (son of sevenless) proteins, *Trends Endocrinol. Metab.*, 5, 165, 1994.
17. **Bruder, J. T., Heidecker, G., and Rapp, U. R.**, Serum-, TPA-, and Ras-induced expression from AP-1/Ets-driven promoters requires Raf-1 kinase, *Genes Dev.,* 6, 545, 1992.
17a. **Hall, A.,** A biochemical function for Ras — at last, *Science*, 264, 1413, 1994.
18. **deVries, A. M. M., Burgering, B. M. T., Leevers, S. J., Marshall, C. J., and Bos, J. L.**, Involvement of $p21^{ras}$ in activation of extracellular signal-regulated kinase 2, *Nature*, 357, 602, 1992.
19. **Douville, E., Duncan, P., Abraham, N., and Bell, J. C.,** Dual specific kinases — a new family of signal tranducers, *Cancer Metastasis Rev.,* 13, 1, 1994.
19a. **Pelech, S. L. and Sanghera, J. S.**, MAP kinases: charting the regulatory pathways, *Science*, 257, 1385, 1992.
19b. **Robbins, D. J., Zhen, E., Cheng, M., Xu, S., Ebert, D., and Cobb, M. J.**, MAP kinases ERK1 and ERK2: pleiotropic enzymes in a ubiquitous signaling network, *Adv. Cancer Res.*, 63, 93, 1994.
20. **Karin, M. and Smeal, T.**, Control of transcription factor by signal transduction pathways: the beginning of the end, *Trends Biochem. Sci.*, 17, 418, 1992.
21. **Seth, A., Gonzalez, F. A., Gupta, S., Raden, D. L., and Davis, R. J.**, Signal transduction within the nucleus by mitogen-activated protein kinase, *J. Biol. Chem.*, 267, 24796, 1992.

22. **Bissonnette, R. P., Shi, Y., Mahboubi, A., Glynn, J. M., and Green, D. R.**, c-myc and apoptosis, in *Apoptosis II*, Tomei, L.D. and Cope, F.O., Eds., Cold Spring Harbor Laboratory Press, Cold Spring Harbor, NY, 1994, 327.

22a. **Evan, G. I., Wyllie, A. H., Gilbert, C. S., Littlewood, T. D., Land, H., Brooks, M., Waters, C. M., Penn, L. Z., and Hancock, D. C.**, Induction of apoptosis in fibroblasts by c-myc protein, *Cell*, 69, 119, 1992.

23. **Martin, T. F. J.**, Receptor regulation of phosphoinositase C, in *Intracellular Messengers*, Taylor, C. W., Ed., Pergamon Press, Oxford, 1993, 63.

24. **Goldshmidt-Clermont, P. J., Kim, J. W., Machesky, L. M., Rhee, S. G., and Pollard, T. D.**, Regulation of phospholipase C-γ1 by profilin and tyrosine phosphorylation, *Science*, 251, 1231, 1991.

25. **Bray, D.**, *Cell Movements*, Garland Publishing, New York, 1992.

26. **Asaoka, Y., Nakamura, S., Yoshida, K., and Nishizuka, Y.**, Protein kinase C, calcium and phospholipid degradation, *Trends Biochem. Sci.*, 17, 414, 1992.

27. **Cook, S. J. and Wakelham, M. J. O.**, Phospholipases C and D in mitogenic signal transduction, *Rev. Physiol. Biochem. Pharmacol.*, 119, 13, 1992.

28. **Liscovitch, M.**, Cross-talk among multiple signal-activated phospholipases, *Trends Biochem. Sci.*, 17, 393, 1992.

29. **Stabel, S. and Parker, P. J.**, Protein kinase C, in *Intracellular Messenger*, Taylor, C.W., Ed., Pergamon Press, Oxford, 1993, 167.

30. **Chakravarthy, B. R., Whitfield, J. F., and Durkin, J. P.**, Inactive membrane protein kinase Cs — a possible target for receptor signalling, *Biochem J.*, 304, 809, 1994.

31. **Blank, V., Kourilsky, P., and Israël, A.**, NF-κB and related proteins: Rel/dosal homologies meet ankyrin-like repeats, *Trends Biochem. Sci.*, 17, 135, 1992.

32. **Exton, J. H., Taylor, S. J., Blank, J. S., and Bocckino, S. B.**, Regulation of phosphoinositide and phosphatidylcholine phospholipases by G proteins, *Ciba Found. Symp.*, 164, 36, 1992.

33. **Smith, W. L., Marrnett, L. J., and DeWitt, D. L.**, Prostaglandin and thromboxane biosynthesis, in *Intracellular Messengers*, Taylor, C. W., Ed., Pergamon Press, Oxford, 1993, 101.

34. **Putney, J. W. and Bird, G. S. J.**, The inositol phosphate-calcium signalling system in non-excitable cells, *Endocr. Rev.*, 14, 610, 1993.

35. **Rossier, M. F., Bird, G. S. J., and Putney, J. W.**, Subcellular distribution of the calcium-storing inositol 1,4,5-trisphosphate-sensitive organelle in rat liver, *Biochem. J.*, 274, 643, 1991.

36. **Rossier, M. F. and Putney, J. W.**, The identity of the calcium-storing, inositol 1,4,5-triphosphate-sensitive organelle in non-muscle cells: calciosome, endoplasmic reticulum or both?, *Trends Neurosci.*, 14, 310, 1991.

37. **Taylor, C. W. and Richardson, A.**, Structure and function of inositol trisphosphate receptors, in *Intracellular Messengers*, Taylor, C. W., Ed., Pergamon Press, Oxford, 1993, 199.

38. **Parekh, A. B., Terlau, H., and Stühmer, W.**, Depletion of $InsP_3$ stores activates a Ca^{2+} and K^+ current by means of a phosphatase and a diffusible messenger, *Nature*, 364, 814, 1993.

39. **Randriamampita, C. and Tsien, R. Y.**, Emptying of intracellular Ca^{2+} stores releases a novel small messenger that stimulates Ca^{2+} influx, *Nature*, 364, 809, 1993.

40. **Lu, K. P. and Means, A. R.**, Regulation of the cell cycle by calcium and calmodulin, *Endocr. Rev.*, 14, 40, 1993.
41. **Tucker, R. W. and Fay, F. S.**, Distribution of intracellular free calcium in quiescent BALB/c. 3T3 cells stimulated by platelet-derived growth factor, *Eur. J. Cell Biol.*, 51, 120, 1990.
42. **Short, A. D., Bian, J., Ghosh, T. K., Waldson, R. T., Rybak, S. L., and Gill, D. L.**, Intracellular Ca^{2+} pool content is linked to control of cell growth, *Proc. Natl. Acad. Sci. U.S.A.*, 90, 4986, 1993, 427.
43. **Aderem, A.**, Signal transduction and the actin cytoskeleton: the roles of MARCKS and profilin, *Trends Biochem. Sci.*, 17, 438, 1992.
44. **Cheek, T. R.**, Calcium signalling and the triggering of secretion in adrenal chromaffin cells, in *Intracellular Messengers*, Taylor, C. W., Ed., Pergamon Press, Oxford, 1993.
45. **Naim, A. C. and Aderem**, A., Calmodulin and protein kinase C cross-talk: the MARCKS protein is an actin filament and plasma membrane cross-linking protein regulated by protein kinase C phosphorylation and by calmodulin, *Ciba Found. Symp.*, 164, 145, 1992.
46. **Bachs, O., Agell, N., and Carafoli, E.**, Calcium and calmodulin function in the cell nucleus, *Biochim. Biophys. Acta*, 1113, 259, 1992.
47. **Nicotera, P., Bellomo, G., and Orrenius, S.**, Calcium-mediated mechanisms in chemically induced cell death, *Annu. Rev. Pharmacol. Toxicol.*, 32, 449, 1992.
48. **Nicotera, P., McConkey, D. J., Jones, D. P., and Orrenius, S.**, ATP stimulates Ca^{2+} uptake and increases the free Ca^{2+} concentration in isolated liver nuclei, *Proc. Natl. Acad. Sci. U.S.A.*, 86, 453, 1989.
49. **Lee, K. A. W. and Masson, N.**, Transcriptional regulation by CREB and its relatives, *Biochim. Biophys. Acta*, 1174, 221, 1993.
50. **Kreis, T. and Vale, R., Eds.**, *Guidebook to the Cytoskeletal and Motor Proteins*, Oxford University Press, Oxford, 1993.
51. **Kumar, R. V., Wolfman, A., Panniers, R., and Henshaw, E. C.**, Mechanism of inhibition of polypeptide chain initiation in calcium-depleted Ehrlich ascites tumor cells, *J. Cell Biol.*, 108, 2107, 1989.
52. **Rinker-Schaeffer, C. W., Austin, V., Zimmer, S., and Rhoads, R. E.**, *ras* transformation of cloned rat embryo fibroblasts results in increased rates of protein synthesis and phosphorylation of eukaryotic initiation factor 4E, *J. Biol. Chem.*, 267, 10659, 1992.
53. **Whitfield, J. F.**, Calcium switches, cell cycles, differentiation, and death, in *Calcium, Vitamin D, and the Prevention of Colon Cancer*, Lipkin, M., Newmark, H. L., and Kelloff, G., Eds., CRC Press, Boca Raton, FL, 1991, 31.
54. **Whitfield, J. F.**, Calcium: driver of cell cycles, trigger of differentiation, and killer of cells, in *Cellular and Molecular Targets for Chemoprevention*, Steele, V. E., Stoner, G. D., Boone, C. W., and Kelloff, G. J., Eds., CRC Press, Boca Raton, FL, 1992, 257.
55. **Whitfield, J. F.**, Calcium signals and cancer, *Crit. Rev. Oncogenesis*, 3, 55, 1992.
56. **Diliberto, P. A., Bernacki, S. H., and Herman, B.**, Interrelationships of platelet-derived growth factor isoform-induced changes in c-fos expression, intracellular free calcium and mitogenesis, *J. Cell. Biochem.*, 44, 39, 1990.

57. **Mohammadi, M., Dionne, C. A., Li, W., Li, N., Spivak, T., Honegger, A. M., Jaye, M., and Schlessinger, J.**, Point mutation in FGF receptor eliminates phosphatidylinositol hydrolysis without affecting mitogenesis, *Nature*, 358, 681, 1992.
58. **Peters, K. G., Marie, J., Wilson, E., Ives, H. E., Escobedo, J., DelRosario, M., Mirda, D., and Williams, L. T.**, Point mutation of an FGF receptor abolishes phosphatidylinositol turnover and Ca^{2+} flux but not mitogenesis, *Nature*, 358, 678, 1992.
58a. **Rubin, J. S., Bottaro, D. P., and Aaronson, S. A.**, Hepatocyte growth factor/scatter factor and its receptor, the c-*met* proto-oncogene product, *Biochim. Biophys. Acta*, 1155, 357, 1993.
59. **Deser-Wiest, L. and Desser, H.**, The proliferation activity of the rat thymocytes after partial hepatectomy, *Exp. Pathol.*, 18, 167, 1990.
60. **Abou-Samra, A.-B., Jüppner, H., Force, T., Freeman, M. W., Kong, X.-F., Schipani, E., Usena, P., Richards, J., Bonventre, J. V., Potts, J. T., Kronenberg, H. H., and Segre, G. V.**, Expression of a common receptor for parathyroid hormone and parathyroid hormone-related peptide from rat osteoblast-like cells: a single receptor stimulates intracellular accumulation of both cAMP and inositol triphosphate and increases intracellular free calcium, *Proc. Natl. Acad. Sci. U.S.A.*, 89, 2732, 1992.
61. **Birnbaumer, M., Seibold, A., Gilbert, S., Ishido, M., Barberis, C., Antaramian, A., Brabet, P., and Rosenthal, W.**, Molecular cloning of the receptor for human antidiuretic hormone, *Nature*, 357, 333, 1992.
62. **Georgoussi, Z., Taylor, S. J., Bocckino, S. B., and Exton, J. H.**, Hepatic vasopressin receptor: a key receptor of phosphoinositide metabolism, in *Biological Signal Transduction*, Ross, E. M. and Wirtz, K. W. A., Eds., CRC Press, Boca Raton, FL, 1991, 73.
63. **Jouishomme, H., Whitfield, J. F., Chakravathy, B., Durkin, J. P., Gagnon, L., Isaacs, R. J., MacLean, S., Neugebauer, W., Willick, G., and Rixon, R. H.**, The protein kinase-C activation domain of the parathyroid hormone, *Endocrinology*, 130, 53, 1992.
64. **Lolait, S. J., O'Carroll, A.-M., McBride, O. W., Koning, M., Morel, A., and Brownstein, M. J.**, Cloning and characterization of a vasopressin V2 receptor and possible link to diabetes insipidus, *Nature*, 357, 336, 1992.
65. **Buckley, A. R., Crowe, P. D., and Russel, D. H.**, Rapid activation of protein kinase C in isolated rat liver nuclei by prolactin, a known hepatic mitogen, *Proc. Natl. Acad. Sci. U.S.A.*, 85, 8649, 1988.
66. **Buckley, A. R., Putman, C. W., Evans, R. E., Laird, H. E., Shah, G. N., Montgomeny, D. W., and Russell, D. H.**, Hepatic protein kinase C: translation stimulated by prolactin and partial hepatectomy, *Life Sci.*, 41, 2827, 1987.
67. **Villalba, M., Zabala, M. T., Martinez-Serrano, A., de la Calina, R., Satrustegu, J., and Gavera-Ruiz, J. P.**, Prolactin increases cytosolic free calcium concentration in hepatocytes of lactating rats, *Endocrinology*, 129, 2857, 1991.
68. **Johnson, R. M., Napier, M. A., Cronin, M., and King, K.**, Growth hormone stimulates the formation of 1–2 diacylglycerol in rat hepatocytes, *Endocrinology*, 127, 2099, 1990.

69. **Kelly, P. A., Djiane, J., Postel-Vinay, M.-C., and Edery, M.**, The prolactin/growth hormone receptor family, *Endocr. Rev.*, 12, 235, 1991.
70. **Baffy, G., Yang, L., Michalopoulos, G. K., and Williamson, J. R.**, Hepatocyte growth factor induces calcium mobilization and inositol phosphate production in rat hepatocytes, *J. Cell. Physiol.*, 153, 332, 1992.
71. **Kaneko, A., Hayashi, N., Tsubouchi, H., Tanoka, Y., Ito, T., Saski, Y., Fusamoto, H., Daikuhara, Y., and Kamada, T.**, Intracellular calcium as a second messenger for human hepatocyte growth factor in hepatocytes, *Hepatology*, 15, 1173, 1992.
72. **Maher, J. J.**, Cell-specific expression of hematocyte growth factor in liver. Upregulation in sinusoidal endothelial cells after carbon tetrachloride, *J. Clin. Invest.*, 91, 2244, 1993.
73. **Matsumoto, K. and Nakamura, T.**, Hepatocyte growth factor: molecular structure, roles in liver regeneration, and other biological functions, *Crit. Rev. Oncogenesis*, 3, 27, 1992.
74. **Osada, S., Saji, S., Nakamura, T., and Nozawa, Y.**, Cytosolic calcium oscillations induced by hepatocyte growth factor (HGF) in single fura-2-loaded cultured hepatocytes: effects of extracellular calcium and protein kinase C, *Biochim. Biophys. Acta*, 1135, 229, 1992.
75. **Putney, J. W. and Bird, G. S.**, The inositol-phosphate-calcium signalling system in non-excitable cells, *Endocr. Rev.*, 14, 610, 1993.
76. **Meyer, T., Hanson, P. I., Stryer, L., and Schulman, H.**, Calmodulin trapping by calcium-calmodulin-dependent protein kinase, *Science*, 256, 1199, 1992.
77. **Fewtrell, C.**, Ca^{2+} oscillations in excitable cells, *Annu. Rev. Physiol.*, 55, 427, 1993.
78. **Boynton, A. L. and Whitfield, J. F.**, The role of cyclic AMP in cell proliferation: a critical assessment of the evidence, *Adv. Cyclic Nucleotide Res.*, 15, 193, 1983.
79. **Rixon, R. H., Isaacs, R. J., and Whitfield, J. F.**, The control of DNA polymerase-α activity in regenerating rat liver by calcium and $1\alpha,25(OH)_2D_3$, *J. Cell. Physiol.*, 139, 354, 1989.
80. **Diehl, A. M., Yang, S. Q., Wolfgang, D., and Wand, G.**, Differential expression of guanine nucleotide-binding proteins enhances cAMP synthesis in regenerating rat liver, *J. Clin. Invest.*, 89, 1706, 1992.

80a. **Vintarmyr, O. K., Torunn, R. O., Bruland, T., Houge, G., and Doskeland, S. O.**, Elevated cAMP gives short-term inhibition and long-term stimulation of hepatocyte DNA replication. Roles of the cAMP-dependent protein kinase subunits, *J. Cell. Physiol.*, 156, 166, 1993.

81. **Armato, U., Ribecco, M., Guerriero, C., Testolin, L., and Whitfield, J. F.**, Extracellular calcium modulates prereplicative cyclic AMP surges in EGF-stimulated primary neonatal rat hepatocytes, *J. Cell. Physiol.*, 161, 55, 1994.
82. **Sheffield, L. G.**, Oligonucleotides antisense to catalytic subunit of cyclic AMP-dependent protein kinase inhibit mouse mammary epithelial cell DNA synthesis, *Exp. Cell Res.*, 192, 307, 1991.
83. **Gadbois, D. M., Grissman, H. A., Tobey, R. A., and Bradbury, E. M.**, Multiple kinase arrest points in the G_1 phase of non-transformed mammalian cells are absent in transformed cells, *Proc. Natl. Acad. Sci. U.S.A.*, 89, 8626, 1992.

84. **Scott, J. D.**, Cyclic nucleotide-dependent protein kinases, in *Intracellular Messengers*, Taylor, C. W., Ed., Pergamon Press, Oxford, 1993, 137.
84a. **Koch, C. and Nasmyth, K.**, Cell cycle regulated transcription in yeast, *Curr. Opinion Cell Biol.*, 6, 451, 1994.
85. **Draetta, G.**, cdc 2 activation: the interplay of cyclin binding and thr161 phosphorylation, *Trends Cell Biol.*, 3, 287, 1993.
86. **Johnson, L. H.**, Cell cycle control of gene expression in yeast, *Trends Cell Biol.*, 2, 353, 1992.
87. **Parge, H. E., Arvai, A. S., Murtari, D. J., Reed, S. I., and Tainer, J. A.**, Human CksHs2 atomic structure: a role for its hexameric assembly in cell cycle control, *Science*, 262, 387, 1993.
88. **Marini, N. J. and Reed, S. I.**, Direct induction of G_1-specific transcripts following reactivation of the Cdc28 kinase in the absence of de novo protein synthesis, *Genes Dev.*, 6, 557, 1992.
89. **Heideman, W., Casperson, G. F., and Bourne, H. R.**, Adenyl cyclase in yeast: antibodies and mutations identify a regulatory domain, *J. Cell. Biochem.*, 42, 229, 1990.
90. **Krupinsky, J., Coussen, F., Bakalyar, H. A., Tang, W.-J., Feinstein, P. G., Orth, K., Slaughter, C., Reed, R. R., and Gilman, A. G.**, Adenylcyclase-amino acid sequence: possible channel- or transporter-like structure, *Science*, 244, 1558, 1989.
91. **Young, D., Riggs, M., Field, J., Vojtek, A., Broek, D., and Wigler, M.**, The adenylyl cyclase gene from *Schizosaccharomyces pombe, Proc. Natl. Acad. Sci. U.S.A.*, 86, 7989, 1989.
92. **Cho-Chung, Y. S.**, Role of cyclic AMP receptor proteins in growth, differentiation, and suppression of malignancy: new approaches to therapy, *Cancer Res.*, 50, 7093, 1990.
93. **Cho-Chung, Y. S. and Clair, T.**, The regulatory subunit of cAMP-dependent protein kinase as a target for chemotherapy of cancer and other cellular dysfunctional-related diseases, *Pharmacol. Ther.*, 60, 265, 1994.
94. **Cho-Chung, Y. S., Clain, T., Tortora, G., and Yokozaki, H.**, Role of site-selective cAMP analogues in the control and reversal of malignancy, *Pharmacol. Ther.*, 50, 123, 1991.
95. **Borrelli, E., Montmayeur, J.-P., Faulkes, N. S., and Sassone-Corsi, P.**, Signal transduction and gene control: the cyclic AMP pathway, *Crit. Rev. Oncogenesis*, 3, 321, 1992.
96. **Delmas, V., Molina, C. A., Lalli, E., de Groot, R., Foulkes, N. S., Masquilier, D., and Sassone-Corsi, P.**, Complexity and versatility of the transcriptional response to cAMP, *Rev. Physiol. Biochem. Pharmacol.*, 124, 1, 1994.
97. **Kwast-Welfeld, J., Walker, P. R., Whitfield, J. F., and Sikorska, M.**, Evidence that an additional conserved element with the consensus C/G AG A/C is essential for maximal responsiveness of the cyclic AMP enhancer, *Cell. Mol. Biol. Res.*, 39, 231, 1993.
98. **Tokeda, J., Mackawa, T., Sudo, T., Seino, Y., Imura, H., Sato, N., Tanaka, C., and Ishii, S.**, Expression of the CRE-BP1 transcriptional regulator binding to the cyclic AMP response element in central nervous system, regenerating liver, and human tumours, *Oncogene*, 6, 1009, 1991.

99. **Kwast-Welfeld, J., de Belle, I., Walker, P. R., Whitfield, J. F., and Sikorska, M.**, Changes in cyclic AMP-responsive element binding proteins in rat hepatomas, *Cancer Res.*, 51, 528, 1991.
100. **Calabretta, B. and Gewirtz, A. M.**, Functional requirements of c-myb during normal and leukemic hematopoiesis, *Crit. Rev. Oncogenesis*, 2, 187, 1991.
101. **Katsaros, D., Clair, T., Tortora, S., Ally, S., Robins, R. K., and Cho-Chung, Y. S.**, Transcriptional activation of the RII cAMP receptor protein and c-Ras (H) proto-oncogene in the growth inhibition of human colon cancer cell line treated with 8-Cl-cAMP, *Proc. Annu. Meet. Am. Assoc. Cancer Res.*, 2a, A231, 1988.
102. **Divecha, N., Banfic, H., and Irvine, R. F.**, The polyphosphoinositide cycle exists in the nuclei of Swiss 3T3 cells under the control of a receptor (for IGF-1) in the plasma membrane, and stimulation of the increased nuclear diacylglycerol and apparently induces translocation of protein kinase C to the nucleus, *EMBO J.*, 10, 3207, 1991.
103. **Zacksenhaus, E., Bremner, R., Jiang, Z., Gill, R. M., Muncaster, M., Sopta, M., Phillips, R. A., and Gallie, B. L.**, Unravelling the function of the retinoblastoma gene, *Adv. Cancer Res.*, 61, 115, 1993.
104. **Hamel, P. A., Gallie, B. L., and Phillips, R. A.**, The retinoblastoma protein and cell cycle regulation, *Trends Genet.*, 8, 180, 1992.
105. **Müller, R., Mumberg, D., and Lucibello, F. C.**, Signals and genes in the control of cell-cycle progression, *Biochim. Biophys. Acta*, 1155, 151, 1993.
106. **Mercer, W. E.**, Cell cycle regulator and p53 tumor suppressor protein, *Crit. Rev. Eukaryotic Gene Expression*, 2, 251, 1992.
107. **Montenarh, M.**, Biochemical, immunological, and functional aspects of the growth-suppression/oncoprotein p53, *Crit. Rev. Oncogenesis*, 3, 233, 1992.
108. **Bischoff, J. R., Friedman, P. N., Marshak, D. R., Prives, C., and Beach, D.**, Human p53 is phosphorylated by p60-cdc 2 and cyclin B-cdc 2, *Proc. Natl. Acad. Sci. U.S.A.*, B87, 4766, 1990.
109. **Thompson, N. L., Mead, J. E., Braun, L., Goyette, M., Shanks, P. R., and Fausto, N.**, Sequential protooncogene expression during rat liver regeneration, *Cancer Res.*, 46, 3111, 1986.
110. **Quartin, R. S., Finlay, C. A., and Levin, A. J.**, The p53 gene and gene product, in *Nuclear Processes and Oncogenes*, Sharp, P. A., Ed., Academic Press, New York, 1992, 87.
111. **Addison, C., Jenkins, J. R., and Stürzbecher, H.-W.**, The p53 nuclear localisation signal is structurally linked to a p34 cdc 2 kinase motif, *Oncogene*, 5, 423, 1990.
112. **Shaulsky, G., Goldfinger, N., Ben-Ze'en, A., and Rotter, V.**, Nuclear accumulation of p53 protein is mediated by several nuclear localization signals and plays a role in tumorigenesis, *Mol. Cell. Biol.*, 10, 6565, 1990.
113. **Shaulsky, G., Ben-Zeev, A., and Rotter, V.**, Subcellular distribution of the p53 protein during the cell cycle of Balb/c 3T3 cells, *Oncogene*, 5, 1707, 1990.
114. **Stürzbecher, H.-W., Maimets, T., Chumakor, P., Brain, R., Addison, C., Simanis, V., Rudge, K., Philp, R., Grimaldi U., Count, W., and Jenkins, J. R.**, p53 interacts with $p53^{cdc2}$ in mammalian cells: implications for cell cycle control and oncogenesis, *Oncogene*, 5, 795, 1990.

115. **Hermann, C. P. E., Kraiss, S., and Montenarh, U.**, Association of casein kinase II with immunopurified p53, *Oncogene*, 6, 877, 1991.
116. **Meek, D. W., Simon, S., Kikkawa, U., and Eckhart, W.**, The p53 tumor suppressor proteins is phosphorylated at serine 389 by casein kinase II, *EMBO J.*, 9, 3253, 1990.
116a. **de Murcia, G. and Ménissier de Murcia, J.**, Poly(ADP-ribose) polymerase: a molecular nick-sensor, *Trends Biochem. Sci.*, 19, 172, 1994.
117. **Fornace, A. J.**, Mammalian genes induced by radiation: activation of genes associated with growth control, *Annu. Rev. Genet.*, 26, 507, 1992.
118. **Kastan, M. B., Zhan, Q., El-Deiry, W. S., Carrier, F., Jacks, T., Walsh, W. V., Plunkett, B. S., Vagelstein, B., and Fornace, A. T.**, A mammalian cell cycle checkpoint pathway utilizing p53 and GADD45 is defective in ataxia telangiectasia, *Cell*, 71, 587, 1992.
119. **Marx, J.**, How p53 suppresses cell growth, *Science*, 262, 1644, 1993.
120. **Yamamoto, M., Yoshida, M., Ono, K., Fujita, T., Ontani-Fujita, N., Sakai, T., and Nikaido, T.**, Effect of tumor suppressors on cell cycle-regulatory genes: RB suppresses $p34^{cdc2}$ expression and normal p53 suppresses cyclin A expression, *Exp. Cell Res.*, 210, 94, 1994.
121. **Kastan, M. B., Onyckwere, O., Sidransky, D., Vogelstein, B., and Craig, R.**, Participation of p53 protein in the cellular response to DNA damage, *Cancer Res.*, 51, 6304, 1991.
122. **Anderson, C. W.**, DNA damage and the DNA-activated protein kinase, *Trends Biochem. Sci.*, 18, 433, 1993.
123. **Bakalkin, G., Yakovleva, T., Selivanova, G., Magnusson, K. P., Szekely, L., Kiseleva, E., Klein, G., Terenius, L., and Wiman, K. G.**, p53 binds single-stranded DNA ends and catalyzes DNA renaturation and strand transfer, *Proc. Natl. Acad. Sci. U.S.A.*, 91, 413, 1994.
124. **Yonish-Rouach, E., Resnitzki, D., Lotem, J., Sacks, L., Kinchi, A., and Oren, M.**, Wild-type p53 induces apoptosis of myeloid leukemia cells that is inhibited by interleukin-6, *Nature*, 352, 345, 1991.
125. **Sikorska, M., Brewer, L. M., Youdale, T., Richards, R., Whitfield, J. F., Houghten, R. A., and Walker, P. R.**, Evidence that mammalian ribonucleotide reductase is a nuclear membrane-associated glycoprotein, *Biochem. Cell Biol.*, 68, 880, 1990.
126. **Sikorska, M., Kwast-Welfeld, J., Whitfield, J. F., Youdale, T., Richards, R., and Walker, P. R.**, M1 subunit of rat liver ribonucleotide reductase is modified by ubiquitination, *Biochem. Cell Biol.*, 70, 215, 1992.
127. **Baron, R., Ravesloot, J.-H., Neff, L., Chakravarthy, M., Chattenyee, D., Lomri, A., and Home, W.**, Cellular and molecular biology of the osteoclast, in *Cellular and Molecular Biology of Bone*, Noda, M., Ed., Academic Press, San Diego, 1993, 445.
128. **Brown, E. M., Gamba, G., Riccardi, D., Lombardi, M., Butters, R., Kifor, O., Sun, A., Hedlger, M. A., Lytton, J., and Hebert, S. C.**, Cloning and characterization of an extracellular Ca^{2+}-sensing receptor from bovine parathyroid, *Nature*, 366, 575, 1993.
129. **Kübler, D., Pyerin, W., Bill, O., Hotz, A., Sonka, J., and Kinzel, V.**, Evidence for ecto-protein kinase activity that phosphorylates kemptide in a cyclic AMP-dependent mode, *J. Biol. Chem.*, 164, 14549, 1989.

130. **Boynton, A. L., Whitfield, J. F., MacManus, J. P., Armato, U., Tsang, B. K., and Jones, A.**, Involvement of cAMP and cAMP-dependent protein kinase in the initiation of DNA synthesis by rat liver cells, *Exp. Cell Res.*, 135, 199, 1981.
131. **Subramanyan, C., Honn, S. C., Reed, W. C., and Reddy, P. V.**, Nuclear localization of 68 kDa calmodulin-binding protein is associated with the onset of DNA replication, *J. Cell. Physiol.*, 144, 423, 1990.
132. **Jindal, H. K., Chaney, W. G., Anderson, C. W., Davis, R. G., and Vishwanatha, J. K.**, The protein-tyrosine kinase substrate, calpactin I heavy chain (p36), is part of the primer recognition protein complex that interacts with DNA polymerase α, *J. Biol. Chem.*, 266, 5169, 1991.
133. **Kornberg, A. and Baker, T.**, *DNA Replication*, W. H. Freeman, New York, 1992.
134. **Dedman, J. R. and Kaetzel, M. A.**, Annexins, in *Guidebook to the Cytoskeletal and Motor Proteins*, Kreis, T. and Vale, R., Eds., Oxford University Press, Oxford, 1993, 26.
135. **Hammond, R. A., Foster, K. A., Berchtold, M. W., Gassman, M., Holmes, A. M., Hubscher, V., and Brown, N. C.**, Calcium-dependent calmodulin binding proteins associated with mammalian DNA polymerase-alpha, *Biochem. J.*, 951, 315, 1988.
136. **Reddy, G. P. V. and Fager, R. S.**, Replitase: a complex integrating dNTP synthesis and DNA replication, *Crit. Rev. Eukaryotic Gene Expression*, 3, 255, 1993.
137. **Croker, G., Dawson, R. A., Barton, C. H., and MacNeil, S.**, An extracellular role for calmodulin-like activity in cell proliferation, *Biochem. J.*, 553, 877, 1988.
138. **Dasso, M., Nishitani, H., Kombluth, S., Nishimoto, T., and Newport, J. W.**, RCC1, a regulator of mitosis, is essential for DNA replication, *Mol. Cell. Biol.*, 12, 3337, 1992.
139. **Dasso, M., Smythe, C., Milarski, K., Kombluth, S., and Newport, J. W.**, DNA replication and progression through the cell cycle, *Ciba Found. Symp.*, 170, 161, 1992.
140. **Moore, M.S. and Blobel, G.**, A G protein involved in nucleocytoplasmic transport: the role of Ran, *Trends Biochem. Sci.*, 19, 211, 1994.
141. **Dou, Q.-P., Levin, A. H., Zhao, S., and Pardee, A. B.**, Cyclin E and cyclin A as candidates for the restriction point protein, *Cancer Res.*, 53, 1493, 1993.
142. **Gasser, S. M. and Laemmli, U. K.**, A glimpse of chromosomal order, *Trends Genet.*, 3, 16, 1987.
143. **Neilson, W. G., Pierta, K. J., Barrack, E. R., and Coffey, D. S.**, The role of the nuclear matrix in the organization and function of DNA, *Annu. Rev. Biophys. Chem.*, 15, 457, 1986.
144. **Razin, S. V.**, DNA interaction with the nuclear matrix and spatial organization of replication and transcription, *BioEssays*, 6, 19, 1987.
145. **Van Holde, K. E.**, *Chromatin*, Springer-Verlag, New York, 1988.
146. **Wolfe, A.**, *Chromatin, Structure and Function*, Academic Press, London, 1992.
147. **Huberman, J. A.**, Eukaryotic DNA replication: a complex picture partially clarified, *Cell*, 48, 7, 1987.

148. **Hand, K.**, Eukaryotic chromosome replication and its regulation, in *Cell Biology: A Comprehensive Treatise*, Vol. 2, Prescott, D. M. and Goldstein, L., Eds., Academic Press, New York, 1979, 389.
149. **Prescott, D. M.**, *Reproduction of Eukaryotic Cells*, Academic Press, New York, 1976.
150. **Diffley, J. F. X.**, Early event in eukaryotic DNA replication, *Trends Cell. Biol.*, 2, 298, 1992.
151. **Fotedar, R. and Roberts, J.**, Cell cycle regulated association of proteins within the replication initiation complex, in *Regulation of Eukaryotic DNA Replication, Proceedings of the Second McGill University Conference on Regulation of DNA Replication*, Abstract 89, McGill University, Montreal, 1992.
152. **Stillman, B., Bell, S. P., Dutta, A., and Marahrens, Y.**, DNA replication and the cell cycle, *Ciba Found. Symp.*, 170, 147, 1992.
153. **Dingwall, C. and Laskey, R.**, The nuclear membrane, *Science*, 258, 942, 1992.
154. **Silver, P. A.**, How proteins enter the nucleus, *Cell*, 64, 489, 1991.
155. **Shi, Y. and Thomas, J. O.**, The transport of proteins into the nucleus requires the 70-kilodalton heat shock protein or its cytosolic cognate, *Mol. Cell. Biol.*, 12, 2186, 1992.
156. **Stochaj, U. and Silver, P.A.**, A conserved phosphoprotein that specifically binds nuclear localization sequences is involved in nuclear import, *J. Cell Biol.*, 117, 473, 1992.
157. **Guadagno, T.M., Ohtsubo, M., Roberts, J.M., and Assoian, R.K.**, A link between cyclin A expression and adhesion-dependent cell cycle progression, *Science*, 262, 1372, 1993.
157a. **Giancotti, F. G. and Mainiero, F.**, Integrin-mediated adhesion and signaling in tumorigenesis, *Biochim. Biophys. Acta*, 1198, 47, 1994.
157b. **Cooper, G. M.**, *Oncogenesis*, Jones and Bartlett Publishers, Boston, 1990.
158. **Rao, P. and Johnson, R.**, Mammalian cell fusion: studies on the regulation of DNA synthesis and mitosis, *Nature*, 225, 159, 1970.
159. **Brown, C.J., Sekguchi, T., Nishimoto, T., and Willard, H.F.**, Regional localization of CCG1 gene which complements hamster cell cycle mutation BN462 to XQ11-XQ13, *Somatic Cell Mol. Genet.*, 15, 93, 1989.
160. **Clark, K. L., Ohtsubo, M., Nishimoto, T., Goebl, M., and Sprague, L. G. F., Jr.**, The yeast SRM1 protein and human RCC1 protein share analogous functions, *Cell Regul.*, 2, 781, 1991.
161. **Fleishmann, M., Clark, M.W., Forrester, W., Wickens, M., Nishimito, T., and Aebi, M.**, Analysis of yeast PRP20 mutations and functional complementation by the human homologue RCC1, a protein involved in the control of chromosome condensation, *Mol. Gen. Genet.*, 227, 417, 1991.
162. **Furuno, N., Nakagawa, K., Eguchi, U., Ohtsubo, M., Nishimoto, T., and Soeda, E.**, Complete nucleotide sequence of the human RCC1 gene involved in coupling between DNA replication and mitosis, *Genomics*, 11, 459, 1991.
163. **Matsumoto, T. and Beach, D.**, The *spi* 1 GTPase interacts with RCC1 in cell cycle dependency, *Cold Spring Harbor Symp. Quant. Biol.*, 56, 385, 1991.
164. **Nishitani, H., Kobayashi, H., Ohtsubo, M., and Nishimoto, T.**, Cloning of *Xenopus* RCC1 complementary DNA, a homolog of the RCC1 gene complementation of TSBN2 mutation and identification of the product, *J. Biochem. (Tokyo)*, 107, 228, 1990.

165. **Nishitani, H., Ohtsubo, M., Yamashita, K., Iida, H., Pines, J., Yasudo, H., Shibata, Y., Hunter, T., and Nishimoto, T.**, Loss of RCC1, a nuclear DNA-binding protein uncouples the completion of DNA replication from the activation of CD C2 protein, *EMBO J.*, 10, 1555, 1991.
166. **Ohtsubo, M., Yoshiba, T., Seino, H., Nishitani, H., Clark, K. L., Sprague, G. F., Jr., Frasch, M., and Nishimoto, T.**, The homologous genes of *Drosophila* and *Saccharomyces cerevisiae, EMBO J.*, 10, 1265, 1991.
167. **Seino, H., Hisamoto, N., Uzawa, S., Sckiguchi, T., and Nishimoto, T.**, DNA-binding domain of RCC1 protein is not essential for coupling with DNA replication, *J. Cell. Sci.*, 102, 393, 1992.
168. **Seino, H., Nishitani, H., Seki, T., Hisamoto, N., Tazunoki, T., Shiraki, N., Ohtsubo, M., Yamashita, K., Sekiguchi, T., and Nishimoto, T.**, RCC1 is a nuclear protein required for coupling activation of cdc 2 kinase with DNA synthesis and for start of the cell cycle, *Cold Spring Harbor Symp. Quant. Biol.*, 56, 367, 1991.
169. **Sekiguchi, T., Nohiro, Y., Nakamura, Y., Hisamoto, N., and Nishimoto, T.**, The human CCG1 gene essential for progression of the G-1 phase encodes a 210-kilodalton nuclear DNA-binding protein, *Mol. Cell. Biol.*, 11, 3317, 1991.
170. **Uchida, S., Sekiguchi, T., Nishitani, H., Miyauchi, K., Ohtsubo, M., and Nishimoto, T.**, Premature chromosome condensation is induced by a point mutation in the hamster RCC1 gene, *Mol. Cell. Biol.*, 10, 577, 1990.
171. **Nishimoto, T., Kai, R., Sekiguchi, T., Ohtsubo, M., Sasaki, H., Furuno, N., Uchida, S., Miyauchi, K., and Nishitani, N.**, Molecular cloning of human genes that complement the temperature-sensitive cell-cycle mutants of BHK-21 cells, in *Cell Cycle Control in Eukaryotes*, Beach, D., Basilico, C., and Newport, J., Eds, Cold Spring Harbor Laboratory, Cold Spring Harbor, NY, 169, 1988.
172. **Yasuda, H., Matsumoto, Y., Mita, S., Marunouchi, T., and Yamada, M.**, A mouse temperature-sensitive mutant defective in H1 histone phosphorylation is defective in deoxyribonucleic acid synthesis and chromosome condensation, *Biochemistry*, 20, 4414, 1981.
173. **Bishoff, F. R., Maier, G., Tilz, G., and Ponsting, H.**, A 47-kDa human nuclear protein recognized by the antikinetochrone autoimmune sera is homologous with the protein encoded by RCC1, a gene implicated in onset of chromosome condensation, *Proc. Natl. Acad. Sci. U.S.A.*, 87, 8617, 1990.
174. **Seki, T., Yamashita, K., Nishitani, H., Takagi, T., Russel, P., and Nishimoto, T.**, Chromosome condensation caused by a loss of RCC1 function requires the cdc 25 protein that is located in the cytoplasm, *Mol. Biol. Cell*, 3, 1373, 1992.
175. **Nishimoto, T., Ajiro, K., Hirata, M., Yamashita, K., and Sekiguchi, M.**, The induction of chromosome condensation in *ts*BN2, a temperature-sensitive mutant BHK21, is inhibited by the calmodulin antagonist, W-7, *Exp. Cell Res.*, 156, 351, 1985.
176. **Bischoff, F. R. and Ponstingl, H.**, Catalysis of guanine nucleotide exchange on Ran by the mitotic regulator RCC1, *Nature*, 354, 80, 1991.
177. **Brown, M., Garvik, B., Hartwell, L., Kadyk, L., Seekey, T., and Weinert, T.**, Fidelity of mitotic chromosome transmission, *Cold Spring Harbor Symp. Quant. Biol.*, 56, 359, 1991.

178. **Spiegel, A. M., Shenker, A., and Weinstein, L. S.**, Receptor-effector coupling by G proteins: implications for normal and abnormal signal transduction, *Endocr. Rev.*, 13, 536, 1992.
179. **Watson, P. A.**, Function follows form: generation of intracellular signals by cell deformation, *FASEB J.*, 5, 2013, 1991.
180. **Meijer, L. and Arion, D.**, Negative control of cdc2 kinase activation by cAMP in starfish oocytes, *Cold Spring Harbor Symp. Quant. Biol.*, 56, 591, 1991.
181. **Lamb, N. J. C., Cavadore, J,-C., Labbe, J.-C., Maurer, R., and Fernandez, A.**, Inhibition of cAMP-dependent protein kinase plays a key role in the induction of mitosis and nuclear envelope breakdown in mammalian cells, *Embo J.*, 10, 1523, 1991.

181a. **Mattioli, M., Galeati, G., Barboni, B., and Seren, E.**, Concentration of cyclic AMP during the maturation of pig oocytes *in vivo* and *in vitro*, *J. Reprod. Fertil.*, 100, 403, 1994.

181b. **Moor, R. M. and Heslop, J. P.**, Cyclic AMP in mammalian follicle cells and oocytes during maturation, *J. Exp. Zool.*, 216, 205, 1981.

181c. **Yoshimura, Y., Nakamura, Y., Oda, T., Ando, M., Ubukata, Y., Karube, M., Koyama, N., and Yamada, H.**, Induction of meiotic maturation of follicle-enclosed oocytes of rabbit by a transient increase followed by an abrupt decrease in cyclic AMP concentration, *J. Reprod. Fertil.*, 95, 803, 1992.

182. **Draetta, G., Ed.**, Cyclin-dependent kinase, *Semin. Cell Biol.*, 2, 193, 1991.
183. **Ducommun, B.**, From growth to cell cycle control, *Semin. Cell Biol.*, 2, 233, 1991.
184. **Ducommun, B., Brambila, P., Felix, M.-A., Franza, B. R., Jr., Karsenti, E., and Draetta, G.**, cdc 2 phosphorylation is required for its interaction with cyclin, *EMBO J.*, 10, 3311, 1991.
185. **Hunt, T.**, Cyclins and their partners: from a simple idea to complicated reality, *Semin. Cell Biol.*, 2, 213, 1991.
186. **Lehrer, C. F.**, Pulling the *string*: cell cycle regulation during *Drosophila* development, *Semin. Cell Biol.*, 2, 223, 1991.
187. **Nagata, A., Igarashi, M., Jinno, S., Suto, K., and Okayama, H.**, An additional homolog of the fission yeast cdc 25^{+} gene occurs in humans and is highly expressed in some cancer cells, *New Biol.*, 3, 959, 1991.
188. **Nishimoto, T., Uzawa, S., and Schlegel, R.**, Mitotic checkpoints, *Curr. Biol. Cell Biol.*, 4, 174, 1992.
189. **Russell, P., Moreno, S., and Reed, S. I.**, Conservation of mitotic controls in fission and building yeasts, *Cell*, 57, 295, 1989.
190. **Russel, P. and Nurse, P.**, cdc 25^{+} function as a mitotic inducer in the mitotic control of fission yeast, *Cell*, 45, 145, 1986.
191. **Sadhu, K., Reed, S. T., Richardson, H., and Russell, P.**, Human homolog of fission yeast cdc 2S mitotic inducer is predominantly expressed in G_2, *Proc. Natl. Acad. Sci. U.S.A.*, 87, 5139, 1990.
192. **Shalloway, D., Bagrodia, S., Chakalaparampil, I., Shenoy, S., Lin, P.-H., and Taylor, S. J.**, c-Src and mitosis, *Ciba Found. Symp.*, 170, 248, 1992.
193. **Nada, S., Okada, M., MacAuley, A., Cooper, J. A., and Nakagawa, H.**, Cloning of a complementary DNA for a protein-tyrosine kinase that specifically phosphorylates a negative regulatory site for $p60^{c\text{-}src}$, *Nature*, 351, 69, 1991.

194. **Spivak, J. G., Erikson, R. L., and Maller, J. L.**, Microinjection of pp60$^{v\text{-}src}$ into Xenopus oocytes increases phosphorylation of ribosomal S6 and accelerates the rate of progesterone-induced meiotic maturation, *Mol. Cell. Biol.*, 4, 1631, 1984.
195. **Durkin, J. P., Youdale, T., and Whitfield, J. F.**, *ts*LA23-NRR cells need pp60$^{v\text{-}src}$ protein-tyrosine kinase activity in G_2 phase to initiate mitosis in serum-free medium, *Cell. Signal.*, 3, 93, 1991.
196. **Armelin, H. A. and Armelin, M. C. S.**, Regulation of fibroblast growth in culture, *Biochem. Biophys. Res. Commun.*, 62, 260, 1975.
197. **Swierenga, S. H. H., Whitfield, J. F., and Boynton, A. L.**, Age-related and carcinogen-induced alterations of the extracellular growth factor requirements for cell proliferation in vitro, *J. Cell. Physiol.*, 94, 171, 1978.
198. **Schaller, H. C., Druffel-Augustin, S., and Dübel, S.**, Head activator acts as an autocrine growth factor for NH15-CA2 cell in the G_2/mitosis transition, *EMBO J.*, 8, 3311, 1989.
199. **Durkin, J. P. and Whitfield, J. F.**, The viral Ki-*ras* gene must be expressed in the G_2 phase if *ts* kirsten sarcoma virus-infected NRK cells are to proliferate in serum-free medium, *Mol. Cell. Biol.*, 7, 444, 1987.
200. **Durkin, J. P. and Whitfield, J. F.**, Characterization of the mitogenic signal from an oncogenic *ras* protein, *Anticancer Res.*, 9, 1313, 1989.
201. **Deshpande, A. K. and Kung, H. F.**, Insulin induction of *Xenopus laevis* oocyte maturation is inhibited by monoclonal antibody against p21 *ras* or *src* oncogenes, *Science*, 235, 1492, 1987.
202. **Korn, L. J., Siebel, C. W., McCormick, F., and Roth, R. A.**, *Ras* p21 as a potential mediator of insulin action in *Xenopus* oocytes, *Science*, 236, 840, 1987.
203. **Bischmeier, C. B., Broek, D., and Wigler, M.**, RAS proteins can induce meiosis in *Xenopus* oocytes, *Cell*, 43, 615, 1985.
204. **Anand, S. and Prasad, R.**, Status of calcium flux in the cell cycle of *S. cerevisiae, Biochem. Int.*, 14, 983, 1987.
205. **Baum, P., Furlong, R., and Byers, B.**, Yeast gene required for spindle pole body duplication: homology of its product with Ca^{2+}-binding proteins, *Proc. Natl. Acad. Sci. U.S.A.*, 83, 5512, 1986.
206. **Byers, B., Baum, P., Goebe, M. G., Goetsch, L., Johnson, S. L., and Yochem, J.**, Regulation of spindle pole duplication in *Saccharomyces*: the transition from G1 to S phase, in *Cell Cycle Control in Eukaryotes*, Beach, D., Basilico, C., and Newport, J., Eds., Cold Spring Harbor Laboratory, Cold Spring Harbory, NY, 1988, 201.
207. **Saavedra-Molina, A., Villalobos, R., and Borbolla, M.**, Calcium uptake during the cell cycle of *Saccharomyces cerevisiae*, *FEBS Lett.*, 160, 195, 1983.
208. **Baitinger, C., Alderton, J., Poenie, M., Schulman, H., and Steinhardt, R. A.**, Multifunctional Ca^{2+}/calmodulin-dependent protein kinase is necessary for nuclear envelope breakdown, *J. Cell Biol.*, 111, 1763, 1990.
209. **Twigg, J., Patel, R., and Whitaker, M.**, Translational control of $InsP_3$-induced chromatin condensation during the early cell cycles of sea urchin embryos, *Nature*, 332, 366, 1988.

209a. **Hille, B.**, *Ionic Channels of Excitable Membranes, 2nd Ed.*, Sinauer Associates, Sunderland, MA, 1992.

209b. **Swann, K. and Ozil, J.-P.**, Dynamics of the calcium signal that triggers mammalian egg activation, *Int. Rev. Cytol.,* 152, 183, 1994.

210. **Schlegel, R. and Craig, R. W.**, Mitosis: normal control mechanisms and consequences of aberrant regulation in mammalian cells, in *Perspectives on Cellular Regulation: From Bacteria to Cancer,* Campisi, J., Cunningham, D. D., Inouye, M., and Riley, M., Eds., Wiley-Liss, New York, 1991, 235.

211. **Hepler, P. K.**, Calcium transients during mitosis: observations in flux, *J. Cell Biol.*, 109, 2567, 1989.

212. **McIntosh, J. R. and Hering, G. E.**, Spindle fiber action and chromosome movement, *Annu. Rev. Cell Biol.*, 7, 403, 1991.

213. **Poenie, M. and Steinhardt, R. A.**, The dynamics of $[Ca^{2+}]_i$ during mitosis, in *Calcium and Cell Function*, Vol. 7, Cheung, W. Y., Ed., Academic Press, San Diego, 1987, 133.

214. **Poenie, M., Alderton, J., Tsien, R. Y., and Steinhardt, R. A.**, Changes in free calcium levels with stages of the cell division cycle, *Nature*, 315, 147, 1985.

215. **Silver, R. B.**, Nuclear envelope breakdown and mitosis in sand dollar embryos inhibited by microinjection of calcium buffers in a calcium-reversible fashion, and by antagonists of intracellular Ca^{2+} channels, *Dev. Biol.*, 131, 11, 1989.

216. **Steinhardt, R. A. and Alderton, J.**, Intracellular free calcium rise triggers nuclear envelope breakdown in the sea urchin embryo, *Nature*, 315, 147, 1988.

217. **Pines, J. and Hunter, T.**, Human cyclins A and B1 are differentially located in the cell and undergo cell cycle-dependent nuclear transport, *J. Cell Biol.*, 115, 1, 1991.

218. **Girard, F., Strausfeld, U., Cavadore, J.-C., Russell, P., Fernandez, A., and Lamb, N. J. C.**, cdc 25 is a nuclear protein expressed constitutively throughout the cell cycle in nontransformed mammalian cells, *J. Cell Biol.*, 118, 785, 1992.

219. **Nigg, E. A.**, The substrates of the cdc 2 kinase, *Semin. Cell Biol.,* 2, 261, 1991.

220. **Th'ng, J. P. H., Guo, X.-W., and Bradbury, E. M.**, Histone modifications associated with mitotic chromosome condensation, in *Molecular and Cellular Approaches to the Control of Proliferation*, Academic Press, San Diego, 1992, 381.

221. **Roth, S. Y. and Allis, C. D.**, Chromatin condensation: does histone H1 dephosphorylation play a role?, *Trends Biochem. Sci.*, 17, 93, 1992.

222. **Bradbury, E. M., Maclean, N., and Mathews, H. R.**, *DNA, Chromatin and Chromosomes*, Blackwell Scientific, Oxford, 1981.

223. **Earnshaw, W. C. and Heck, M. M. S.**, Cell biology of topoisomerase II, in *Cancer Cells 6/Eukaryotic DNA Replication*, Cold Spring Harbor Laboratory Press, Cold Spring Harbor, NY, 1988, 279.

224. **Van Holde, K. E.**, *Chromatin*, Springer-Verlag, New York, 1988.

224a. **Goldstein, L. S. B.**, With apologies to Scheherazade: tails of 1001 kinesin motors, *Annu. Rev. Genet.*, 27, 319, 1993.

225. **Brinkley, B. R.**, Chromosomes, kinetochores and the microtubule connection, *BioEssays*, 13, 675, 1991.

226. **Pfarr, C. M., Cour, M., Grisson, P. M., Hays, T. S., Porter, M. E., and McIntosh, J. R.**, Cytoplasmic dynein is localized to kinetochores during mitosis, *Nature*, 345, 263, 1990.

227. **Steuer, E. R., Wordman, L., Schroer, T. A., and Sheetz, M. P.**, Localization of cytoplasmic dynein to mitotic spindles and kinetochores, *Nature*, 345, 266, 1990.

228. **Yen, T. J., Compton, D. A., Wise, D., Zinkowski, R. P., Brinkley, B. R., Earnshaw, W. C., and Cleveland, D. W.**, CENP-E, a novel human centromere-associated protein required for progression from metaphase to anaphase, *EMBO J.*, 10, 1245, 1991.
229. **Newport, J. W. and Spann, T.**, Disassembly of the nucleus in mitotic extracts: membrane vesicularization, lamin disassembly, and chromosome condensation are independent processes, *Cell*, 48, 219, 1987.
230. **Bray, D.**, *Cell Movements*, Garland Publishing, New York, 1992.
231. **Yamashiro, S. and Matsumura, F.**, Mitosis-specific phosphorylation of caldesmon: possible mechanism of cell rounding during mitosis, *BioEssays*, 13, 636, 1991.
232. **Nigg, E. A.**, Assembly and cell cycle dynamics of the nuclear lamina, *Semin. Cell Biol.*, 3, 245, 1992.
233. **Lüscher, B., Brizuela, L., Beach, D., and Eisenman, R. N.**, A role for the $p34^{cdc}$ kinase and phosphatases in the regulation of phosphorylation and disassembly of lamin B_2 during the cell cycle, *EMBO J.*, 10, 865, 1991.
234. **Peter, M., Heitlinger, E., Häner, M., Aebi, U., and Nigg, E. A.**, Disassembly of *in vitro* formed lamin head-to-tail polymers by CDC 2 kinase, *EMBO J.*, 10, 1535, 1991.
235. **Paddy, M. R., Agard, D. A., and Sedat, J. W.**, An extended view of nuclear lamin structure, function and dynamics, *Semin. Cell Biol.*, 3, 255, 1992.
236. **Kassenti, E.**, Mitotic spindle morphogenesis in animal cells, *Semin. Cell Biol.*, 2, 251, 1991.
237. **Margolis, R. S., Rauch, C. T., Pirollet, F., and Job, D.**, Specific association of SROP protein with microtubules *in vitro* and with stable microtubules in mitotic spindles of cultured cells, *EMBO J.*, 9, 4095, 1990.
238. **Maniotis, A. and Schliwa, M.**, Microsurgical removal centrosomes blocks cell reproduction and centriole generation in BSC-1 cells, *Cell*, 67, 495, 1991.
239. **Sakai, H. and Ohta, K.**, Centrosome signalling at mitosis, *Cell. Signal.*, 3, 267, 1991.
240. **Niglow, C., Lombillo, V. A., Kuriyama, R., and McIntosh, J. R.**, A plus-end-directed motor enzyme that moves antiparallel microtubules *in vitro* localizes to the interzone of mitotic spindles, *Nature*, 359, 543, 1992.
241. **Yen, T. J., Li, G., Schaar, B. T., Szilak, I., and Cleveland, D. W.**, CENP-E is a putative kinetochore motor that accumulates just before mitosis, *Nature*, 359, 536, 1992.
242. **Brockman, J. L., Gross, S. D., Sussman, M. R., and Anderson, R. A.**, Cell cycle-dependent localization of casein kinase I to mitotic spindles, *Proc. Natl. Acad. Sci. U.S.A.*, 89, 454, 1992.
243. **Ohta, Y., Ohba, T., and Miyamoto, E.**, Ca^{2+}/calmodulin-dependent protein kinase II: localization in the interphase nucleus and the mitotic apparatus of mammalian cells, *Proc. Natl. Acad. Sci. U.S.A.*, 87, 534, 1990.
244. **Murray, A. W. and Szostak, J. W.**, Chromosome segregation in mitosis and meiosis, *Annu. Rev. Cell Biol.*, 1, 289, 1985.
244a. **Earnshaw, W. C. and MacKay, A. M.**, Role of nonhistone proteins in the chromosomal events of mitosis, *FASEB J.*, 8, 947, 1994.
245. **Cande, W. Z.**, Reactivation of mitosis *in vitro, Trends Biochem. Sci.*, 11, 447, 1986.

246. **Cande, W. A. and McDonald, K. L.**, *In vitro* reactivation of anaphase spindle elongation using isolated diatom spindles, *Nature*, 316, 168, 1985.
247. **Harris, P.**, The role of membranes in the organization of the mitotic apparatus, *Exp. Cell Res.*, 94, 409, 1975.
248. **Harris, P.**, Triggers, trigger waves, and mitosis: a new model, in *Cell Cycle Regulation*, Jeter, J. R., Cameron, I. L., Padilla, G. M., and Zimmerman, A. M., Eds., Academic Press, New York, 1978, 76.
249. **Hepler, P. K. and Wolniak, S. M.**, Membranous compartments and ionic transients in the mitotic apparatus, *Mod. Cell Biol.*, 2, 93, 1983.
250. **Hepler, P. K., Wolniak, S. M., and Wick, S. M.**, A role for calcium and membranes in mitosis, *Eur. J. Cell Biol.*, 22, C934, 1980.
251. **Izant, J. G.**, The role of calcium ions during mitosis: calcium participates in the anaphase trigger, *Chromosoma*, 88, 1, 1983.
252. **Keith, C. H.**, Effect of microinjected calcium-calmodulin on mitosis PtK2, *Cell Motility Cytoskeleton*, 7, 1, 1987.
253. **Keith, C. H., Ratan, R., Maxfield, F. R., Bayers, A., and Shelanski, M. L.**, Local cytoplasmic calcium gradients in living mitotic cells, *Nature*, 316, 848, 1985.
254. **Keirhart, D. P.**, Studies on the *in vitro* sensitivity of spindle microtubules to calcium ions: evidence for a vesicular calcium-sequestering system, *J. Cell Biol.*, 88, 604, 1981.
255. **Kinoshita, S. and Yazaki, I.**, The behavior and localization of intracellular relaxing system during cleavage in the sea urchin egg, *Exp. Cell Res.*, 47, 449, 1967.
256. **Mazia, D., Petzelt, C., Williams, R. O., and Mega, B. T.**, A Ca^{2+}-activated ATPase in the mitotic apparatus of the sea urchin egg, *Exp. Cell Res.*, 70, 325, 1972.
257. **Paweltz, N.**, Membranes in the mitotic apparatus of mammalian cells, in *Cell Reproduction*, Dirksen, E.R., Prescott, D.M., and Fox, C.F., Eds, Academic Press, New York, 1978, 76.
258. **Picard, A., Cavadere, J.-C., Lory, P., Bernego, J.-C., Ojedo, C., and Dorée, M.**, Microinjection of a conserved peptide sequence of $p34^{Cdc-2}$ induces a Ca^{2+} transient in oocytes, *Science*, 247, 327, 1990.
259. **Ratan, R. R., Maxfield, F. R., and Shelanski, M. L.**, Long-lasting and rapid calcium changes during mitosis, *J. Cell Biol.*, 107, 993, 1988.
260. **Ratan, R. R. and Shelano, M. L.**, Calcium and the regulation of mitotic events, *Trends Biochem.*, 11, 456, 1986.
261. **Ratan, R. R., Shelanski, M. L., and Maxfield, F. R.**, Transition from metaphase to anaphase is accompanied by local changes in cytoplasmic free calcium in PtK2 kidney epithelial cells, *Proc. Natl. Acad. Sci. U.S.A.*, 83, 5136, 1986.
262. **Schatten, G. and Schatten, H.**, Calcium sequestration by the mitotic apparatus in mammalian cell cultures and sea urchin zygotes, studies with the fluorescent probe chlorotetracycline, *Eur. J. Cell Biol.*, 22, C935, 1980.
263. **Silver, R. B., Cole, R. D., and Cande, W. Z.**, Isolation of mitotic apparatus containing vesicles with calcium sequestration activity, *Cell*, 19, 505, 1980.
264. **Silver, R. B., Cole, R. D., and Cande, W. Z.**, Localization of calcium sequestering vesicles and identification of a creatine phospho-kinase activity within the mitotic apparatus, *Eur. J. Cell Biol.*, 22, C937, 1980.

265. **Tombes, R. M. and Borisy, G. G.**, Intracellular free calcium and mitosis in mammalian cells: anaphase onset is calcium modulated, but is not triggered by a brief transient, *J. Cell Biol.*, 109, 627, 1989.
266. **Tsien, R. Y. and Poenie, M.**, Fluorescence ratio imaging: a new window into intracellular ionic signalling, *Trends Biochem. Sci.*, 11, 450, 1986.
267. **Wick, S. M. and Hepler, P. K.**, Localization of Ca^{2+} containing precipitates during mitosis, *J. Cell Biol.*, 86, 500, 1980.
268. **Rasmussen, C. D. and Means, A. R.**, Calmodulin is required for cell cycle progression during G1 and mitosis, *EMBO J.*, 8, 73, 1989.
269. **Pluta, A. F., Cooke, C. A., and Earnshaw, W. C.**, Structure of the human centromere at metaphase, *Trends Biochem. Sci.*, 15, 181, 1990.
270. **Holm, C., Steams, T., and Botstein, D.**, Topoisomerase II must act at mitosis to prevent nondisjunction and chromosome breakage, *Mol. Cell. Biol.*, 9, 159, 1989.
271. **Gorbsky, G. J., Sammak, P. J., and Borisy, G. G.**, Chromosomes move poleward along stationary microtubules that coordinately disassemble from their kinetochore ends, *J. Cell Biol.*, 104, 9, 1987.
272. **Hyams, J. S. and Brinkley, B. R., Eds.**, *Mitosis, Models and Mechanisms*, Academic Press, London, 1989.
273. **Mitchison, T. J.**, Microtubule dynamics and kinetochore function in mitosis, *Annu. Rev. Cell Biol.*, 4, 527, 1988.
274. **Glotzor, M., Murray, A. W., and Kirschner, M. W.**, Cyclin is degraded by the ubiquitin pathway, *Nature*, 349, 132, 1991.
275. **Hunt, T.**, Cyclins and their partners: from a simple idea to complicated reality, *Semin. Cell Biol.*, 2, 213, 1991.
276. **Lorca, T., Fesquet, D., Zindy, F., Le Bouffant, F., Cerruti, M., Brechot, C., Devauchelle, G., and Dorée, M.**, An akadaic acid-sensitive phosphatase negatively controls the degradation pathway in amphibian eggs, *Mol. Cell. Biol.*, 11, 1171, 1991.
277. **O'Keefe, S. J., Kiessling, A. A., and Cooper, G. M.**, The c-*mos* gene product is required for cyclin B accumulation during meiosis of mouse eggs, *Proc. Natl. Acad. Sci. U.S.A.*, 88, 7869, 1991.
278. **Roy, L. M., Singh, B., Gautier, J., Arlinghaus, R. B., Nordeen, S. K., and Maller, J. L.**, The cyclin B2 component of MPF is a substrate for the c-mos^{xe} proto-oncogene product, *Cell*, 61, 825, 1990.
279. **Shalloway, D. and Shenoy, S.**, Oncoprotein kinases in mitosis, *Adv. Cancer Res.*, 57, 185, 1991.
280. **Vande Woude, G. F., Buccione, R., Daar, I., Eppig, J. J., Oskarsson, M., Paules, R., Sagata, N., and Yew, N.**, *mos* proto-oncogene function, *Ciba Found. Symp.*, 150, 147, 1990.
281. **Sagata, N., Watanabe, N., Vande Woude, G. F., and Ikawa, Y.**, The c-*mos* proto-oncogene product is a cytostatic factor responsible for meiotic arrest in vertebrate eggs, *Nature*, 342, 512, 1989.
282. **Watanabe, N., Hunt, T., Ikawa, Y., and Sagata, N.**, Independent inactivation of MPF and cytostatic factor (Mos) upon fertilization of *Xenopus* eggs, *Nature*, 352, 247, 1991.
283. **Watanabe, N., Vande Woude, G. F., Ikawa, Y., and Sagata, N.**, Specific proteolysis of the c-*mos* proto-oncogene product by calpain on fertilization of *Xenopus* eggs, *Nature*, 342, 505, 1989.

284. **Finley, D.**, Ubiquitination, *Annu. Rev. Cell Biol.*, 7, 21, 1991.
285. **Ng, G., Boylan, J., Zimmer, S. G., and Sislen, J. E.**, Cytokinesis is more rapid in Ha-T24-*ras* transfected rat embryo fibroblast than in non-transfected control cells, *Cell Motility Cytoskeleton,* 21, 159, 1992.
286. **Earnshaw, W. C., and Heck, M. M. S.**, DNA topoisomerase decondensation during G1, in *Cell Cycle Control in Eukaryotes*, Beach, D., Basilico, C., and Newport, J., Eds, Cold Spring Harbor, Cold Spring Harbor Laboratory, NY, 1988, 176.
287. **Corthésy-Theulaz, I., Pauloin, A., and Pfeffer, S.**, Cytoplasmic dynein participates in the centrosomal localization of the Golgi complex, *J. Cell Biol.*, 118, 1333, 1992.
288. **Alberts, A. S., Thomburn, A. M., Shenolikar, S., Mumby, M. C., and Feramisco, J. R.**, Regulation of cell cycle progression and nuclear affinity of the retinoblastoma protein phosphatases, *Proc. Natl. Acad. Sci. U.S.A.*, 90, 388, 1993.
289. **Ladlow, J. W., Glendening, C. L., Livingston, D. M., and DeCaprio, J. A.**, Specific enzymatic dephosphorylation of the retinoblastoma protein, *Mol. Cell. Biol.*, 13, 367, 1993.
290. **Fisher, B. and Fisher, E. R.**, Experimental studies of factors influencing hepatic metastases. III. Effect of surgical trauma with special reference to liver injury, *Ann. Surg.*, 150, 731, 1959.
291. **van Dale, P. and Galand, P.**, Effect of partial hepatectomy on experimental liver invasion by intraportally injected colon carcinoma cells in rats, *Invasion Metastasis*, 8, 217, 1988.
292. **Laiho, M. and Keski-Oja, J.**, Transforming growth factors-β as regulators of cellular growth and phenotype, *Crit. Rev. Oncogenesis*, 3, 1, 1992.
293. **Oberhammer, F. A., Pavelka, M., Sharma, S., Tiefenbacher, R., Purchio, A. F., Bursch, W., and Schulte-Hermann, R.**, Induction of apoptosis in cultured hepatocytes and in segressing liver by transforming growth factor β1, *Proc. Natl. Acad. Sci. U.S.A.*, 89, 5408, 1992.
294. **Whitson, R. H. and Itakura, K.**, TGF-β1 inhibits DNA synthesis and phosphorylation of the retinoblastoma gene product in a rat liver epithelial cell line, *J. Cell. Biochem.*, 48, 305, 1992.
295. **McClung, J. K., Danner, D. B., Stewart, D. A., Smith, J. R., Schneider, E. L., Lumpkin, C. K., Dell'Orco, R. T., and Nuell, M. J.**, Isolation of a cDNA hybrid that selects antiproliferative mRNA from rat liver, *Biochem. Biophys. Res. Commun.*, 164, 1316, 1989.
296. **Nuell, M. J., Steward, D. A., Walker, L., Friedman, V., Wood, C. M., Owens, G. A., Smith, J. R., Schneider, E. L., Dell'Orco, R., Lumpkin, C. K., Danner, D. B., and McClung, J. K.**, Prohibitin, an evolutionarily conserved intracellular protein that blocks DNA synthesis in normal fibroblasts and Hela cells, *Mol. Cell. Biol.*, 11, 1372, 1981.
297. **Coni, P., Pichiri-Coni, G., Ledda-Columbano, G. M., Rao, P. M., Rajalakshmi, S., Sarma, D. S. R., and Columbano, A.**, Liver hyperplasia is not necessarily associated with increased expression of c-*fos* and c-*myc* mRNA, *Carcinogenesis*, 11, 835, 1990.
298. **Fesus, L. and Tomazy, V.**, Searching for the function of tissue transglutaminase: its possible involvement in the biochemical pathway of programmed cell death, *Adv. Exp. Med. Biol.*, 231, 119, 1988.

299. **Fesus, L., Tomazy, V., and Autouri, F.**, Apoptotic hepatocytes become insoluble in detergent and chaotropic agents as a result of transglutaminase action, *FEBS Lett.*, 245, 150, 1989.
300. **Fesus, L., Tomazy, V., and Falus, A.**, Induction and activation of tissue transglutaminase during programmed cell death, *FEBS Lett.*, 224, 104, 1987.
301. **Ledda-Columbano, G. M. and Columbano, A.**, Apoptosis and hepatocarcinogenesis, in *Apoptosis: The Molecular Basis of Cell Death*, Tomei, L. D. and Cope, F. O., Eds., Cold Spring Harbor Laboratory Press, Cold Spring Harbor, NY, 1991, 101.
302. **Armato, U., Romano, F., Andseis, P. G., Paccagnella, L., and Marchesini, C.**, Growth stimulation and apoptosis induced in cultures of neonatal rat liver cells by repeated exposures to epidermal growth factor/urogastrone with or without associated pancreatic hormones, *Cell Tissue Res.*, 245, 471, 1986.
303. **Oberhammer, F. A., Bursch, W., Parzefall, W., Breit, P., Erber, E., Stadler, M., and Schulte-Hermann, R.**, Effect of transforming growth factor β on cell death of cultured rat hepatocytes, *Cancer Res.*, 51, 2478, 1991.
304. **Oberhammer, F. A., Sedivy, R., Jirtle, R., and Schulte-Hermann, R.**, Role of pro and mature TGFβ-1 in cell death (apoptosis) of hepatocytes, *Naunyn-Schmiedeberg's Arch. Pharmacol. Suppl.*, 343, R24, 1991.
305. **Arends, M. J. and Wyllie, A. H.**, Apoptosis: mechanism and roles in pathology, *Int. Rev. Exp. Pathol.*, 32, 223, 1991.
306. **Goldstein, P., Ojcius, D. M., and Young, D. E.**, Cell death mechanisms and the immune system, *Immunol. Rev.*, 121, 30, 1991.
307. **Swierenga, S. H. H., Whitfield, J. F., and Morris, H. P.**, The reduced extracellular calcium requirement for proliferation by neoplastic hepatocytes, *In Vitro*, 14, 527, 1978.
308. **Swierenga, S. H. H.**, Use low calcium medium in carcinogenicity testing: studies with rat liver cells, in *In Vitro Models for Cancer Research*, Vol. 2, Webber, M. M. and Sekely, L. I., Eds., CRC Press, Boca Raton, FL, 1985, 61.
309. **Boynton, A. L., Swierenga, S. H. H., and Whitfield, J. F.**, The calcium-independence of neoplastic cell proliferation: a promising tool for carcinogen detection, in *Short-Term Tests for Chemical Carcinogens*, Stich, H. F. and San, R. H. C., Eds, Springer-Verlag, New York, 1981, 62.
310. **San, R. H. C., Shimada, T., Melansky, C. T., Kreiser, D. M., Laspia, M. F., Rice, J. M., and Williams, G. M.**, Growth characteristics and enzyme activities in a survey of transformation markers in adult rat liver epithelial-like cell cultures, *Cancer Res.*, 39, 4441, 1979.
311. **Swierenga, S. H. H., Whitfield, J. F., and Karasaki, S.**, Loss of proliferation calcium dependence: simple *in vitro* indicator of tumorigenicity, *Proc. Natl. Acad. Sci. U.S.A.*, 75, 6069, 1978.
312. **Williams, G. M.**, Liver culture indicators for the detection of chemical carcinogens, in *Short-Term Tests for Chemical Carcinogenesis*, Stich, H.F. and San, R.H.C., Eds., Springer-Verlag, New York, 1981, 277.
313. **Guadagno, T. M. and Assoian, R. K.**, G_0/S control of anchorage-independent growth in the fibroblast cell cycle, *J. Cell Biol.*, 115, 1419, 1991.
314. **Matsuhisa, T. and Mori, Y.**, An anchorage-dependent locus in the cell cycle for the growth of 3T3 cells, *Exp. Cell Res.*, 135, 393, 1981.
315. **Schwartz, M. A.**, Signalling by integrins: implications for tumorigenesis, *Cancer Res.*, 53, 1503, 1993.

316. **Durkin, J. P., Boynton, A. L., and Whitfield, J. F.**, The *src* gene product ($pp60^{src}$) of avian sarcoma virus rapidly induces DNA synthesis and proliferation of calcium-deprived rat cells, *Biochem. Biophys. Res. Commun.*, 103, 233, 1981.
317. **Durkin, J. P. and Whitfield, J. P.**, Partial characterization of the mitogenic action of $pp60^{v\text{-}src}$, the oncogenic protein product of the *src* gene of avian sarcoma virus, *J. Cell. Physiol.*, 120, 135, 1984.
318. **Durkin, J. P. and Whitfield, J. P.**, Transforming NRK cells with avian sarcoma virus reduces the extracellular Ca^{2+} requirement without affecting the calcicalmodulin requirement for the G_1/S transition, *Exp. Cell Res.*, 157, 544, 1985.
319. **Durkin, J. P. and Whitfield, J. F.**, The role of calmodulin in the proliferation of transformed and phenotypically normal *ts*ASV-infected rat cells, *J. Cell. Physiol.*, 115, 313, 1983.
320. **Boynton, A. L., Kleine, L. P., and Whitfield, J. F.**, Relation between colony formation in calcium-deficient medium, colony formation in soft agar, and tumor formation by T51B rat liver cells, *Cancer Lett.*, 21, 293, 1984.
321. **Grisham, J. W., Smith, J. D., and Tsao, M.-S.**, Colony forming ability in calcium-poor medium *in vitro* and tumorigenicity *in vivo* not coupled in clones of transformed rat hepato-epithelium cells, *Cancer Res.*, 44, 2831, 1984.
322. **Tsao, M.-S. and Grisham, J. W.**, Phenotypic modulation during tumorigenesis by clones of transformed rat liver epithelial cells, *Cancer Res.*, 47, 1282, 1987.
323. **Tsao, M.-S., Grisham, J. W., and Nelson, K. G.**, Clonal analysis of tumorigenicity and paratumorigenic phenotypes in rat liver epithelial cells chemically transformed *in vitro*, *Cancer Res.*, 45, 5139, 1985.
324. **Corral, M., Defer, N., Paris, B., Raymonjean, M., Corcos, D., Tichonicky, L., Kruh, J., Glaise, D., Kneip, B., and Guguen-Guillouzo, C.**, Isolation and characterization of complementary DNA clones for genes overexpressed in chemically induced rat hepatomas, *Cancer Res.*, 46, 5119, 1986.
325. **Prosch-Hallstrom, I., Blank, A., Eriksson, L. C., and Gustafsson, J. A.**, Expression of the c-*myc*, c-*fos*, and c-ras^{Ha} protooncogene during sex-differentiated rat liver carcinogenesis in the resistant hepatocyte model, *Carcinogenesis*, 10, 1793, 1990.
326. **Walker, G. J., Hayward, H. K., Falvey, S., and Cooksley, W. G. E.**, Loss of somatic heterozygosity in hepatocellular carcinoma, *Cancer Res.*, 51, 4367, 1991.
327. **Arbuthnot, P., Kew, M., and Fitschen, M.**, c-*fos* and c-*myc* oncoprotein expression in human hepatocellular carcinoma, *Anticancer Res.*, 11, 921, 1991.
328. **Means, A. R., Van Berkum, M. P. A., and Bagchi, I.**, Regulatory functions of calmodulin, *Pharmacol. Ther.*, 50, 255, 1991.
329. **Van Eldik, L. J., Zimmer, W. E., Barger, S. W., and Watterson, D. M.**, Perturbation of the calmodulin system in transformed cells, in *Calcium-Binding Proteins in Normal and Transformed Cells*, Prochet, R., Lawson, D.E.M., and Heizmann, C.W., Eds., Plenum Press, New York, 1990, 111.
330. **Palmer, E. J., MacManus, J. P., and Mutus, B.**, Inhibition of glutathione reductase by oncomodulin, *Arch. Biochem. Biophys.*, 277, 149, 1990.
331. **Rasmussen, C. D. and Means, A. R.**, The presence of parvalbumin in a nonmuscle cell line attenuates progression through mitosis, *Mol. Endocrinol.*, 3, 588, 1989.

332. **Hayward, N. K., Walker, G. J., Graham, W. et al.,** Hepatocellular carcinoma mutation, *Nature*, 352, 764, 1991.
333. **Hsu, I. C., Metcalf, R. A., Sun, T., Welsh, J. A., Wang, N. J., and Harris, C. C.,** Mutational hotspot in p53 gene in human hepatocellular carcinomas, *Nature*, 350, 427, 1991.
334. **Bressac, B., Kew, M., Wands, J., and Ozturk, M.,** Selective G to T mutations of p53 gene in hepatocellular carcinoma from southern Africa, *Nature*, 350, 429, 1991.
335. **Birchmeier, W. and Behrens, J.,** Cadherin expression in carcinomas: role in the formation of cell junctions and the prevention of invasiveness, *Biochim. Biophys. Acta*, 1198, 11, 1994.
336. **Takeichi, M.,** Cadherin cell adhesion receptors as a morphogenetic regulator, *Science*, 251, 1451, 1991.
337. **Hedrick, L., Cho, K. R., and Vogelstein, B.,** Cell adhesion molecules as tumor suppressors, *Trends Cell Biol.*, 3, 36, 1993.
338. **Shimoyama, Y. and Hirohashi, S.,** Cadherin intercellular adhesion molecule in hepatocellular carcinomas: loss of E-cadherin expression in an undifferentiated carcinoma, *Cancer Lett.*, 57, 131, 1991.
339. **Schulte-Hermann, R., Ohde, G., Schuppler, J., and Timmermann-Trosiener, I.,** Enhanced proliferation of putative preneoplastic cells in rat liver following treatment with tumor promoters phenobarbital, hexachlorocyclohexane, steroid compounds, and nafenopin, *Cancer Res.*, 41, 2556, 1981.
340. **Schulte-Hermann, R., Schuppler, J., Timmermann-Trosiener, L., Ohde, G., Bursch, W., and Berger, H.,** The role of growth of normal and preneoplastic cell populations for tumor promotion in rat liver, *Environ. Health Perspect.*, 50, 185, 1983.
341. **Tsai, W.-H. and Michalopoulos, G. F.,** Responsiveness of hepatocytes to epidermal growth factor, transforming growth factor-beta and norepinephrine during treatment with xenobiotic hepatic tumor promoters, *Cancer Lett.*, 57, 83, 1991.

3 Cyclic Amp as a Mitogen

I. INTRODUCTION

In the last chapter we learned that removing part of a rat's liver mobilizes a veritable swarm of hormones and factors that surge into the blood and redundantly signal the remaining quiescent hepatocytes to start the G_1 build-up to chromosome replication. Probably the most important of these is hepatocyte growth factor (HGF), which activates its protein tyrosine kinase (YPK) velcroceptors displayed on the surfaces of the waiting hepatocytes. Other mitogenic receptors are also YPK velcroceptors. A receptor-specific team of signaling enzymes and adaptors stick with their SH2 domains to autophosphorylated Y residues of the receptors' cytoplasmic domains. These velcroceptors send multicomponent signals which do such key things as trigger a protein kinase cascade that enhances the translatability of translationally handicapped growth-related messages, stimulates the expression of early cell cycle genes, and stimulates the synthesis of key growth-related proteins.

Because activated velcroceptors' velcroclusters commonly, though not always, include PLCγ-$_1$, which hydrolyzes membrane PtdIns(4,5)P_2 into Ca^{2+}-mobilizing Ins(1,4,5)P_3 and the PKC-stimulating DGs (ΔGs), and because the potent PKCs-stimulating DG analog, TPA, can stimulate the proliferation of various cultured cells, it is commonly believed that a Ca^{2+} surge and a burst of membrane-associated PKCs activity (a Ca^{2+}/PKC signal) are the *sine qua non* of mitogenic signaling. But things are not so simple! In the next two chapters we will learn that Ca^{2+} and PKCs actually *stop* proliferation and trigger the terminal differentiation of normal keratinocytes and colon cells. In this chapter we will meet cells whose cycles are normally started by a simple cyclic AMP/PKA signal and not a multicomponent signal from velcroceptor teams.

II. THE THYROID

A. Normal Thyrocytes

The thyroid consists mainly of thyrocytes arranged in the cubical epithelia of spherical follicles enclosing colloid-filled lumens.[1–2a] Of course there are a few calcitonin-producing cells attached to these follicles, but we will not consider these C (clear) or parafollicular cells here. The thyrocytes (or follicular cells) are normally proliferatively quiescent, but they are busy making TG (thyroglobulin) which they load into the luminal colloid by exocytosis.[1] They also pluck I^- (iodine) from the blood at their basal membranes and ship it to their apical (lumen-facing) membranes where an enzyme, thyroperoxidase, oxidizes it to I°.[1] The I° then binds to Y-residues on colloid TG to form mono- or diiodotyrosine.[1] The mono- and diiodotyrosines are coupled to form T_3 (triiodothyronine) and T_4 (tetraiodothyronine or thyroxine) on the colloid TG.[1] T_3 and T_4 release starts when thyrocytes macropinocytose T_3- and T_4-bearing TG from the colloid.[1] Lysosomes move toward, and fuse with, the macropinocytotic vesicles and the lysosomal enzymes release T_3 and T_4 from TG in the secondary lysosomes. The liberated T_3 and T_4 are then released from the fused vesicles into the blood at the basal cell surfaces. The synthesis of TG, active I^- uptake and peroxidation, the resorption of follicular T_3- and T_4-bearing TG, and the migration of lysosomes from the basal regions of the thyrocytes to the apically endocytosed TG-containing vesicles are coordinately stimulated by the large, dimeric (14-kDα- and β-subunits) TSH, the thyroid-stimulating hormone.[1,2]

TSH is produced by the basophilic thyrotrope cells of the anterior pituitary.[1,2] These cells are stimulated by the hypothalamic TRH (thyrotropin-releasing hormone) to make and release TSH. The TRH-triggered TSH production is mediated by cyclic AMP. TSH production and release is inhibited by the thyroid hormones, which form the negative arm of a feedback loop. The hormones (particularly T_3, which is ten times more potent than T_4) stimulate the expression of a gene coding for a protein that blocks the responses of the TSH gene to TRH stimulation.

Thyrocytes can be stimulated to proliferate by partially removing the gland, or by dietary iodine deficiency.[2] Either of these reduces the amount of circulating T_3, which releases the brakes on TSH production and release by the pituitary thyrotropes. The released TSH homes onto its receptors displayed on the basal surfaces of the thyrocytes.

The glycoprotein TSH receptor is a modular structure that belongs to the "7-TM" superfamily of serpentine G protein-coupled receptors (Figure 1).[3–5] These receptors have an external ligand-binding domain, seven transmembrane helical domains which aggregate to form a "pocket" in the cell membrane, and an amino-terminal signal-transducing domain consisting of intracellular loops I, II, and III and the carboxyl-terminal region, the "C-tail", which may be anchored to the cell membrane by a palmitoylated cysteine residue to form a fourth loop (Figure 2).[3–5]

It seems that the receptor's extracellular domain binds the β-subunit of the TSH dimers. This enables the α-subunit to reach into, and bind to, the transmembrane core,[5] which somehow stimulates the exchange of GTP for GDP bound to the α-subunits of associated heterotrimeric G proteins.[3] This causes the GTP·G-protein to separate into GTP·G_α and $\beta\gamma$ subunits, both of which can

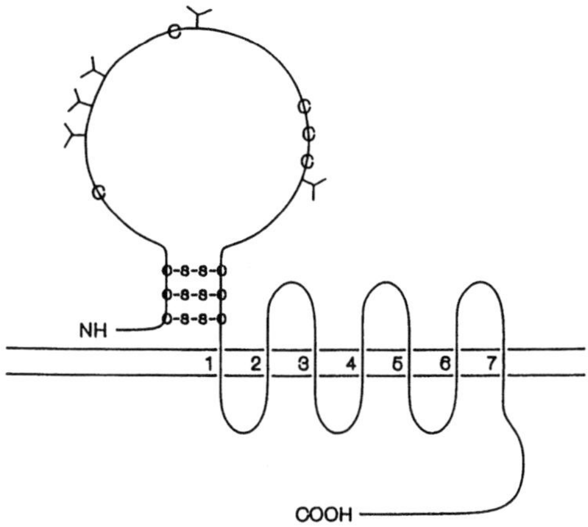

FIGURE 1 The glycosylated TSH receptor.[4] The receptor's TSH-binding external N-terminal domain has 11 C residues, 4 glycosyl groups, and it is put into a stem-loop configuration by 3 cystinyl disulfide bonds. It also has the 7 transmembrane domains that are characteristic of the 7-TM family of receptors.

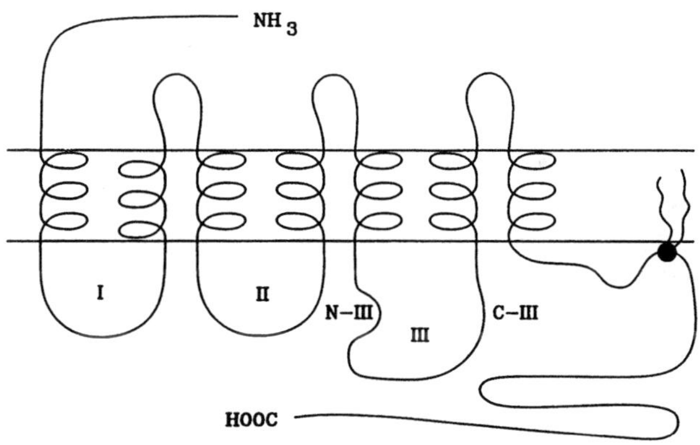

FIGURE 2 The general structure of 7-TM receptors. The cytoplasmic loop III domain can be divided into the N-III and C-III subdomains, which bear the critical motifs for the binding and activation of specific G proteins.[3] There may be a fourth loop produced by the insertion of a palmitoylated C-terminal C-residue (●) into the cell membrane.[3]

affect the activities of target enzymes. The activity of the GTP·G_α subunits is self-limiting: they rather slowly shut themselves off with their rather low GTPase activity which converts them into GDP·G_α complexes.[3] The GDP·G_α subunits reassociate with β dimers to reconstitute inactive GDP·G heterotrimers which can recycle to the receptor for reactivation.

The PTX (pertussis toxin)-insensitive α-subunits of the G_q-class proteins (α_q, α_{11}, α_{14}, $\alpha_{15/16}$) strongly stimulate PLC-βs which trigger a Ca^{2+}/PKC signal by hydrolyzing PtdIns(4,5)P_2 into Ca^{2+}-mobilizing Ins(1,4,5)P_3 and PKCs-stimulating DGs.[3] The PTX-sensitive α-subunits of the G_i class (α_{i1}, α_{i2}, α_{i3}) inhibit adenylyl cyclase and only weakly activate PLCs.[3] The adenylyl cyclase activators are the G_s and olfactory G_{olf} proteins.[3] The βγ heterodimers released by replacing GDP with GTP on G_α subunits are not meant just to bind and inactivate G_α subunits. They can by themselves activate K_{ACh} channels and stimulate adenylyl cyclase II, PLA_2, PLC-β_2, and trigger the translocation of cytosolic β-ARK (the β-adrenergic receptor kinase) to phosphorylate key S and T residues in the intracellular loops (particularly loop III) of β-adrenergic receptors, and thus switch the receptors off.[3]

At physiologically optimal concentrations, the TSH receptor on dog thyrocytes and FRTL-5 rat thyrocytes activates only adenylyl cyclase, but at 1000-fold higher, very supraoptimal concentrations it can also stimulate PtdIns(4,5)P_2 breakdown.[2,5] However, in humans it is a dual-function receptor that can stimulate both adenylyl cyclase and PLC, although it still takes five- to tenfold more TSH to stimulate the phospholipase.[5] Thus, the TSH receptor, like the calcitonin, glucagon, glucagon-like peptide 1, murine LH, low-affinity rat PTH, secretin, and vasoactive intestinal peptide (VIP) receptors, can activate different G proteins.[5–7] The specificity for a particular G protein requires several intracellular domains, but the key ones appear to be N-terminal (N) and C-terminal (C) subregions of loop III (i.e., N-III, which is proximal to transmembrane helix 5, and C-III, which is proximal to transmembrane helix 6) (Figure 2).[3] Indeed, a dual-functional receptor, which can stimulate both adenylyl cyclase and PLC, can be built by putting the N-III region of a β-adrenergic receptor into the loop III of a m1 muscarinic acetylcholine receptor, or by putting the N-III region of an $\alpha_{1\beta}$-adrenergic receptor into a β-adrenergic receptor.[3] This suggests that the dog TSH receptor is different from the human receptor, but this seems not to be the case.[5] In fact, when the dog receptor is expressed in the normally naïve CHO hamster cells, TSH stimulates both adenylyl cyclase and PLC (i.e., the Ca^{2+}PtdIns(4,5)P_2 breakdown mechanism).[5] Thus, it seems likely that the reason for the failure of the TSH receptor to stimulate PLC and PtdIns(4,5)P_2 breakdown in dog cells at physiologically optimal doses is a reflection of the levels or kinds of G_q-class proteins expressed in these cells.[5]

Adding an optimally effective amount of TSH to a primary culture of quiescent dog (or human) thyrocytes triggers a prolonged cyclic AMP surge with no detectable PtdIns(4,5)P_2 breakdown.[5,8] It appears that the cyclic AMP operates mainly by activating PKAI,[2] the PKA associated with proliferation.[8a] The TSH receptor, unlike the 7-TM β-adrenergic receptor, desensitizes slowly and the cellular cyclic AMP level remains high for about 48 h.[2,5] The basis of 7-TM receptor shut-off is the phosphorylation by receptor-targeting PKs (e.g., the β-ARK PK for β-adrenergic receptors) of S or T residues in the receptors' intracellular loops III or C-tails, but the TSH receptor has few such phosphorylatable S or T residues, and thus it stays active for a long time.[2,5]

The burst of PKAI activity somehow starts a chain of events, which, as was the case with the HPX-activated rat hepatocytes in the previous chapter, culminates 16 to 20 h later in the initiation of chromosome replication.[2,8] The cyclic AMP/PKAI signal, like the mitogenic velcroceptor team-signals discussed in the last chapter, stimulates the expression of the immediate-early c-*fos* gene, but unlike the velcroceptor signals, it *inhibits* the expression of another immediate-early gene, c-*jun,* whose product heterdimerizes with the c-*fos* product to form the AP-1 transcription factor that switches on growth-related genes.[2,9] Therefore, there can be no gene-activating surge of c-Fos/ c-Jun AP-1 complexes in cyclic AMP/PKAI-stimulated thyrocytes. Another early gene whose expression is a common prelude to the initiation of the G_1 build-up is c-*myc*. As expected, the TSH-induced cyclic AMP-PKAI signal stimulates c-*myc* expression, but the signal is also self-limiting because it activates a mechanism that shuts c-*myc* expression off after about 1 h.[2,9]

A proliferogenic signal, be it from a velcroceptor team or an adenylyl cyclase-stimulating 7-TM receptor, must derepress the translationally disadvantaged growth-related messages (see the previous chapter); drive growth-triggering transcription factors into the nucleus and onto their target genes; and stimulate the synthesis of key proteins and eventually the production of the first of the several cycle-driving, cyclin-dependent PKs of the cell cycle engine (see the previous chapter).

In the previous chapter, we learned that the complex signal from the cluster of enzymes and adaptors sticking to activated autophosphorylated velcroceptors activates Ras proteins, which trigger a cascade of PKs that includes Raf-1, MAP kinases, and S6PKs, which move the growth-related messages to the head of the translation line and stimulate the synthesis of key cell cycle engine proteins such as cyclins and Cdk2. Obviously, we would expect that the cyclic AMP/PKAI signal from the TSH receptors would do the same thing. But they DO NOT!! The panel of five proteins phosphorylated in thyrocytes stimulated by EGF/TGF-α velcroceptors do not overlap with the panel of eleven proteins phosphorylated in cells stimulated by TSH.[2,10] Thus, for example, EGF or a mitogenic dose of TPA stimulates the phosphorylation of 42- and 44-kDa MAP kinases on S and T residues, which peak after 5 and 15 min, respectively, and then slowly decline.[10] The phosphoenzymes were then translocated to the nucleus.[10] By contrast, TSH or forskolin did not stimulates the phosphorylation or nuclear translocation of the MAP kinases.[10] This means that thyrocytes can arrive at the common cyclin-dependent PK-driven program leading to DNA replication by different routes.[2]

The difference between the initial pathways followed by EGF- and TSH-treated thyrocytes to build the same functioning cell cycle engine is further and dramatically indicated by the fact that TSH stimulates both proliferation and differentiated functions such as TG production while EGF or the PKCs-activating TPA also stimulates proliferation but actually *dedifferentiates* the cells.[2] Quiescent thyrocytes express TSH receptor mRNA, the two

DNAase-sensitive sites in their TG gene promoters are exposed, and the cells transcribe their TG genes and maintain a pool of TG messages.[2] The cyclic AMP/PKAI signal from TSH receptors stimulate TSH receptor gene expression, TG gene expression, TG production, and the production and secretion of T_3 and T_4. By contrast, EGF or TPA *reduces* TSH receptor gene expression, closes the downstream DNAase-sensitive site II of the TG gene promoter, and thus inhibits TG gene expression and TG production, which, in turn, stops T_3 and T_4 synthesis and secretion.[2]

The question then arises as to whether the Ca^{2+}/PKCs signal generated by phosphoinositide breakdown is mitogenic for normal, nonimmortalized thyrocytes or is needed for prereplicative development. The short answer to these questions is — NO! There is no evidence of a Ca^{2+} surge or PtdIns(4,5)P_2 breakdown being triggered in dog thyrocytes by mitogenic doses of EGF or TSH.[2] Another way this was tested was to expose quiescent dog thyrocytes to carbachol or carbamylcholine, a muscarinic agonist which is the most powerful stimulator of PtdIns(4,5)P_2 breakdown and a Ca^{2+}/PKCs signal in these cells.[10] This signal from muscarinic receptors stimulates the expression of the growth-related c-*fos* and c-*myc* genes and the phosphorylation and translocation to the nucleus of the MAP-PKs, just like the complex signal from EGF/TGF-α velcroceptors and the simple powerful supraphysiological surge of PKCs activity stimulated by TPA, but it does *not* stimulate proliferation.[10,11]

Since the PKCs-activating TPA stimulates dog thyrocyte proliferation it would be expected that a Ca^{2+}/PKCs signal from 7-TM muscarinic receptors would also stimulate thyrocyte proliferation. Why doesn't it? Probably because the potent, indestructible TPA rampages throughout the entire cell, powerfully stimulating PKCs, the phosphorylation of nine proteins and mechanisms which could not be affected by the small ephemeral clusters of short-lived DGs around the muscarinic receptors that generated them. One of these nonphysiological mechanisms might be the phosphorylation and activation of eIF-4E which mobilizes translationally repressed growth-related messages and is normally activated not by PKCs, but by a Ras-induced PK such as insulin-stimulated protamine kinase (IPPK).[12–14]

While a Ca^{2+}/PKCs signal is not involved in starting prereplicative development of primary dog thyrocytes, it may drive, or help drive, the G_1 build-up in the cells of some established lines, such as FRTL-5 rat thyrocytes. It has been reported that the cyclic AMP/PKA1 signal from TSH receptors somehow stimulates DGs accumulation, possibly by activating phospholipase D, and thus triggers a burst of PKCs activity without affecting PtdIns(4,5)P_2 breakdown in FRTL-5 cells.[15,16] However, both dog and FRTL-5 thyrocytes, like regenerating rat liver cells and murine fibroblasts, need a lot of insulin or a much smaller amount of IGF-1 and the signals from one of these velcroceptors to finish the G_1 build-up.[2,15] In fact, TSH-stimulated thyrocytes make IGF-1, which can be an autocrine/paracrine promoter of the G_1 build-up.[2,17] IGF-1 and insulin are not by themselves mitogenic, but are prereplicative *progression factors*. The signal from IGF-1 velcroceptors on FRTL-5, and probably dog, thyrocytes further, and differently (by DGs from PtdIns(4,5)P_2 breakdown instead of from a direct or indirect stimulation of phospholipase D by cyclic AMP/PKA1) stimulate PKCs as well as a Ca^{2+} surge from internal stores and through Ca^{2+} channels to supply the Ca^{2+} signal which, from what we learned in the previous chapter, is probably needed to complete the G_1 build-up and trigger replication.

Presumably these signals, and perhaps other accessory signals from adhesion molecules/receptors such as integrins, trigger the tandem expression, assembly, activation, and timely destruction of the several cyclin-dependent PKs of the cell cycle engine such as the Cdk2·cyclin A or Cdk4·cyclin D PKs that hyperphosphorylate and inactivate the Rb protein, which would otherwise prevent the completion of the G_1 build-up, and that also help trigger replication, and the Cdc2·cyclin A or B PKs that trigger mitosis. Also, presumably, periodic Ca^{2+} transients and waves of Ca^{2+}-binding signal-transducing proteins help drive the key stages of the thyrocyte cell cycle as happens in the cells described in Chapter 2.

B. Goiters and Cancers

One way to produce thyroid hyperplasia and hypertrophy (goiter) is to cause excess production of TSH.[2,8] A simple way of doing this is to restrict dietary iodine intake.[1,2] The resulting iodine deficiency reduces the level of circulating T_3 (and T_4) which induces pituitary thyrotrophs to release more TSH, which stimulates normal, IGF-1-dependent thyrocyte proliferation in a futile effort to restore normal T_3 and (T_4) production.[2] Thyroid hyperplasia can also be induced by inserting an A2 adenosine receptor transgene coupled to a TG promoter into the genome of a mouse.[8] The transgenic mouse thyrocytes selectively (because of the thyroid-specific TSH-responsive TG promoter) express the adenylyl cyclase-stimulating A2 receptors, which are continuously activated by the normally large amount of cellular adenosine.[8]

In these cases, the cells are normal and respond normally to an excessive amount of an exogenous proliferogen. Cancer is the result of a parade of mutations and clonal selections that, in a stepwise manner, dysregulate proliferation, impair differentiation, and disrupt intercellular contacts and communication (Figure 3). Carcinogenesis starts with a carcinogen-induced mutation (which constitutively activates a proliferogenic oncogene product, or eliminates or inactivates a tumor suppressor) in a cell which becomes the progenitor of a clone of mutants that are driven by tumor promoters to proliferate excessively with a reduced ability to terminally differentiate in response to signals to do so. The next mutation(s) in a primary mutant cell gives rise to a new clone of mutants which are increasingly mutation-prone and have a further reduced potential to differentiate. Finally, as further mutant clones appear and expand

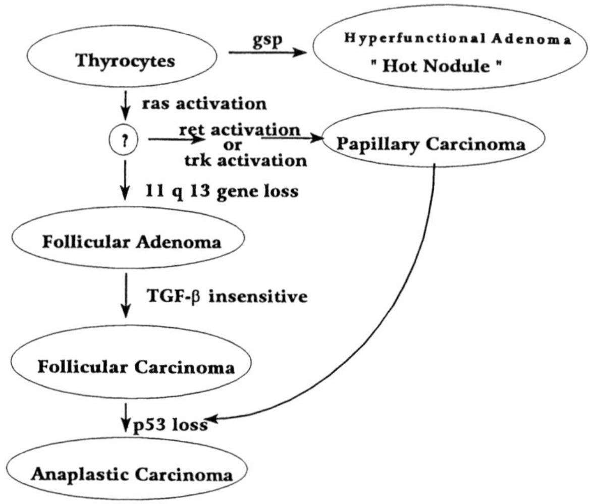

FIGURE 3 The different mutations that have been associated with the various stages of thyroid carcinogenesis[2a,23,28] and are discussed in the text.

a multiple mutant cell may emerge and initiate an expanding clone of malignant cells which do not bind to neighbors, do not communicate normally with neighbors, and are equipped with the angiogenic factor(s), motility factor, and proteases that enable them to invade neighboring tissues and metastasize to other organs. It is important to note that many different genes are specifically involved in initiating carcinogenesis in different tissues while fewer genes are involved in the increasingly common later stages of the process. Thus, gene mutations that start thyroid carcinogenesis are different from those that initiate colon carcinogenesis, but *ras*, *rb*, and p53(TP53) gene mutations are involved in the subsequent stages of carcinogenesis in both tissues.[2a]

One way to deregulate thyrocyte proliferation is to bypass the TSH receptors by constitutively activating adenylyl cyclase. This can be achieved by mutations that eliminate the $G_{s\alpha}$ protein's self-inactivating GTPase activity. The first example of this in mammalian cells were the *gsp* mutations, which were found in a subset of pituitary somatotroph adenomas secreting abnormally large amounts of growth hormone and containing high levels of cellular cyclic AMP.[18] Somatotroph proliferation, like thyrocyte proliferation, is driven by cyclic AMP, and the ability of a constitutively active $G_{s\alpha}$ protein to drive excessive somatotroph proliferation is dramatically demonstrated by the somatotroph hyperplasia and gigantism of mice transgenic for a selectively expressed bacterial cholera toxin gene whose product activates $G_{s\alpha}$ proteins by inhibiting their GTPase activity.[19] It was then found that the McCune-Albright syndrome of multiple endocrine adenomas with its colon and lymphoid (e.g., thymus) hyperplasias is the work of variably (mosaically) expressed $G_{s\alpha}$ mutations (Figure 4).[20,21] One of the common features of the McCune-Albright

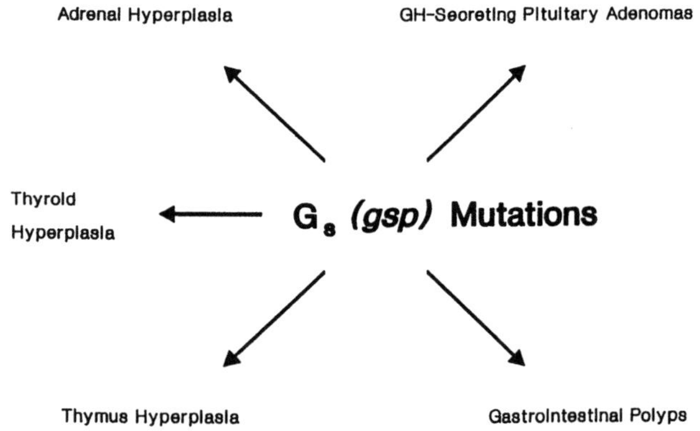

FIGURE 4 *gsp* mutations that eliminate the auto-deactivating GTPase activity of the $G_{s\alpha}$ protein result in the constitutive hyperactivity of adenylyl cyclase. The variable, or mosaic, expression of such a *gsp* mutation in various tissues causes a variety of hyperplasias and hyperfunctioning.

syndrome is hyperthyroidism and goiter (Figure 4).[2,20–22] *Gsp* mutations and their mutant $G_{s\alpha}$ protein products seem specifically to be the initiators of toxic thyroid adenomas ("hot nodules") which are the equivalents of the *gsp*-induced hyperfunctioning pituitary thyrotroph adenomas (Figure 3).[2,8,23]

An early, possibly proliferogenic first mutation in the genesis of 20 to 50% of all types of thyroid tumor, is a mutation of the H-*ras* gene (in carcinogen-induced tumors) or K-*ras* gene (in radiation-induced tumors) which results in the expression of a GTPase-less, constitutively active Ras protein (Figure 3).[2a,23,24] The profound impact of such a mutant Ras protein was demonstrated by the observation of Spina et al.[25] that rat thyrocytes transformed by a viral, GTPase-less H-Ras or K-Ras protein have an elevated basal adenylyl cyclase activity and cyclic AMP content which eliminate the cells' proliferative dependence on TSH. Normal activated Ras proteins are principal mediators of the signal from mitogenic velcroceptors. They trigger a Raf-1-mediated cascade of MAP and other PKs which mobilize repressed messages from genes coding for transcription factors such as the cycle-starting c-Myc and its partner c-Max. Mutant Ras proteins may also promote the production of various autocrine factors, such as IGF-1, which would release the mutant thyrocytes from dependence on exogenous IGF-1 to complete the G_1 build-up.[23]

In some transforming human thyrocytes an inversion occurs in chromosome 10 q which couples part of the *ret* YPK receptor gene to a functionally obscure gene known as H4 (D10S170) (Figure 3).[2a,23] This novel rearranged gene encodes a chimeric protein with the YPK domain of the *Ret* receptor. Since the ligand-binding regulatory domain of the *Ret* velcroceptor is missing, the YPK domain is constitutively active.[2a,23] This chimeric Ret-H4 YPK is associated specifically with papillary thyroid carcinomas (Figure 3).[23] Some papillary thyroid carcinomas are associated with the intrachromosomal rear-angement of another velcroceptor gene, the *trk* (NGF receptor) gene on chromosome 1q23–24.[2a,23] In this case the *trk* gene is linked to the *trp* gene, which codes for a protein with a cytoskeleton function.[2a] The resulting hybrid protein contains the constitutively active YPK domain of the Trk(NGF) velcroceptor and the TPR cytoskeleton-associated protein.[2a] These two chimeric velcroceptor YPKs together account for the constitutively activated YPK activity in 50% of papillary thyroid carcinomas.[2a,23]

Yet another YPK velcroceptor, our old friend the HGF receptor from the last chapter, is involved in some thyroid carcinomas. Thus, in 20% of follicular carcinomas, and in as many as 75% of papillary and poorly differentiated carcinomas, the *met* (HGF receptor) gene is amplified.[2a] As we have seen the signal from the team of signalers stuck to the activated, autophosphorylated HGF velcroceptor is both mitogenic and motogenic, which is why carcinomas hyperexpressing their *met* gene and bristling with HGF receptors are particularly aggressive because of the widespread production of HGF by various nonepithelial cells.[2a]

A common feature of neoplastic cells is a loss of responsiveness to TGFs-β.[26] Normal thyrocytes and thyrocytes from benign adenomas are as sensitive to proliferative inhibition by TFGs-β as other epithelial cells.[26] We learned in the previous chapter that TGFs-β stop the completion of the G_1 build-up in normal epithelial cells by preventing the hyperphosphorylation and inactivation of the Rb suppressor protein by cyclin-dependent PKs. We would expect that there would also be a loss of sensitivity to TGFs-β during the later stages of thyroid carcinogenesis. Indeed some thyroid carcinoma cells are insensitive to TGFs-β.[2a,23,26] But about 50 percent of all types of thyroid carcinoma cells make TGF-β1, while thyroid adenoma cells and normal thyrocytes *do not.*[23,26] In other words, TGF-β1 expression is positively correlated to malignancy! It has been suggested[23] that transforming thyrocytes themselves become insensitive to TGF-β1, and can then take advantage of the factor's ability to be a paracrine stimulator of the proliferation of the nonepithelial, mesenchymally derived supporting stromal cells and a repressor of neighboring normal thyrocytes.

Unlike the cells of colorectal tumors, to be discussed in Chapter 5, the cells of thyroid tumors seem to have an extremely low frequency of allelic loss, i.e., germline inactivation of one allele of a tumor suppressor gene followed some time later in postembryonic life by deletion of the remaining allele. However, in about 14% of human thyroid follicular tumors, there is a deletion of what may be a tumor suppressor gene(s) on human chromosome 11q13 which is close to a gene responsible for type 1 multiple endocrine neoplasia, a disease characterized by anterior pituitary, endocrine pancreatic, and parathyroid tumors (Figure 2).[27] Fagin et al.[28] have suggested that loss of the 11 q 13 gene(s) is a post-*ras* mutation event which triggers the follicular adenoma stage of some thyroid cancers (Figure 3).

As we learned in the previous chapter and will see again even more dramatically in the next two chapters, one of the most common occurrences in the later, premalignant stages of carcinogenesis is the deletion of the p53 gene or a loss-of-function mutation of this gene. We have learned that the p53(TP53) protein is normally expressed in the middle or late stage of the G_1 build-up where it runs a safety-check mechanism that detects mutation-generating chromosome breaks and delays the initiation of replication until the damage is fixed. But if this cannot be done within a certain time, p53 will trigger apoptosis. Thus, p53 reduces (but does not eliminate) the probability of appearance of malignant clones, and the loss or a loss-of-function mutation of the p53 gene facilitates their emergence and expansion. Fagin et al.[28] have suggested that a loss of p53 function is associated with the transition of differentiated follicular carcinoma to anaplastic carcinoma (Figure 3). Indeed, 25% of differentiated follicular and papillary thyroid carcinomas and 86% of anaplastic carcinomas have point mutations, deletions, or insertions in their p53 genes.[2a]

III. THE PAROTID GLAND

Parotid gland acinar cells can be stimulated to massively mobilize their amylase stores and secrete the enzyme by α_1-adrenergic agonists and muscarinic agonists which specifically activate PtdIns(4,5)P_2 breakdown to give a Ca^{2+}/PKCs signal.[29–34] But while this triggers protein synthesis to replenish amylase stores it *never* stimulates proliferation.[29] On the other hand, activating the cells' 7-TM β-adrenergic receptors with isoproterenol, which does *not* stimulate PtdIns(4,5)P_2 breakdown, a Ca^{2+} surge, or the translocation of PKCs to membrane fraction,[31,32,35,36] stimulates both massive amylase secretion and the initiation of DNA replication 20 h later.[29] Therefore, as in thyrocytes, the PtdIns(4,5)P_2 breakdown mechanism with its resulting Ca^{2+}/PKCs signal are not directly proliferogenic. Instead, as in thyrocytes, the proliferogenic signal seems to be the more than sufficient, massive (up to 400-fold), but transient (between 5 and 30 min) cyclic AMP surge (Figure 5).[29,37,38]

The massive first cyclic AMP surge triggers a degranulation-secretion stage during which amylase is released and the first, immediate-early, cell cycle genes (c-*fos*, c-*myc*, and p53) are expressed, mRNA production increases, ODC activity surges upward and putrescine is made, the ribosomal S6 protein is phosphorylated and functionally activated (presumably by $pp90^{rsk}$), and there is a dramatic increase in the synthesis of proteins, among which is CaM.[29,39–41]

This early stage is followed by a second stage with a redistribution of CaM from the particulate to the soluble fraction and the universal, much smaller (twofold) G_1 cyclic AMP surge which peaks at 12 h (Figure 5).[38,40] This cyclic AMP surge activates the perinuclear, gene-activating PKAII which is positioned

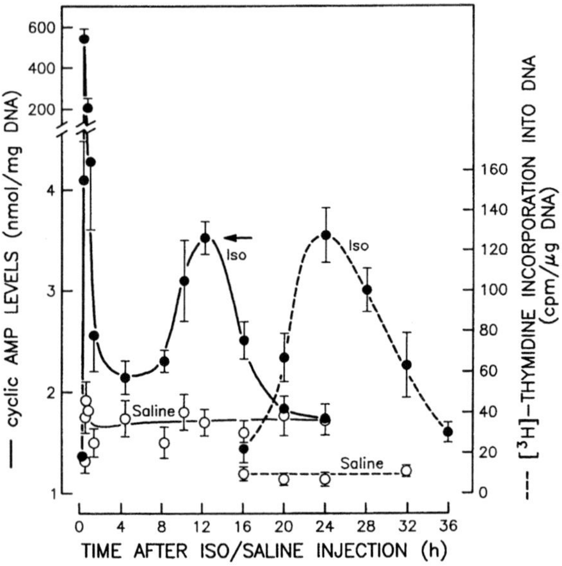

FIGURE 5 Injecting the adenylyl cyclase-stimulating isoproterenol (ISO) first triggered massive cyclic AMP transients and started the prereplicative development of parotid gland acinar cells in Sprague-Dawley rats. As the cells progressed into the G_1 build-up, they generated the smaller, universal G_1 cyclic AMP surge *(arrow)*, which subsided before the onset of DNA replication. Full experimental details are provided by Tsang et al.[38]

to get into the nucleus and phosphorylate key nuclear proteins such as 110- and 130-kDa nucleolar proteins and gene transcription factors that were inacessible to the more peripherally sited PKAI, the main target of the massive isoproterenol-triggered first cyclic AMP surge.[42] Blocking this second cyclic AMP surge by injecting *d,l*-propranolol at 8 h after isoproterenol injection prevented the initiation of DNA replication.[38] This block was lifted completely simply by injecting dibutyryl cyclic AMP along with propranolol at 8 h.[38] Clearly, this second cyclic AMP/PKAII signal triggers something essential. As suggested in the previous chapter, this essential something might be the expression of the first cyclin-dependent PKs of the cell cycle engine, as in *Saccharomyces cerevisiae*.

As in the hepatocytes of the previous chapter, the second stage is followed by the final stage of prereplicative development in which the cyclic AMP and soluble CaM surges subside and DNA replication starts.[38,40]

Again as in hepatocytes, the massive first cyclic AMP surge, the smaller second cyclic AMP surge and the first two stages are largely unaffected by the hypercalcemia and $1,25(OH)_2D_3$ deficiency in the 72-h TPTX rat.[43] But, the surge of soluble CaM does not occur.[40] A final key event(s), maybe the assembly and activation of Cdk2·cyclin A, fails as it does in all other external Ca^{2+}-deprived non-neoplastic cells and DNA replication does not start.[43] However, this critical event does not fail and DNA replication starts on time

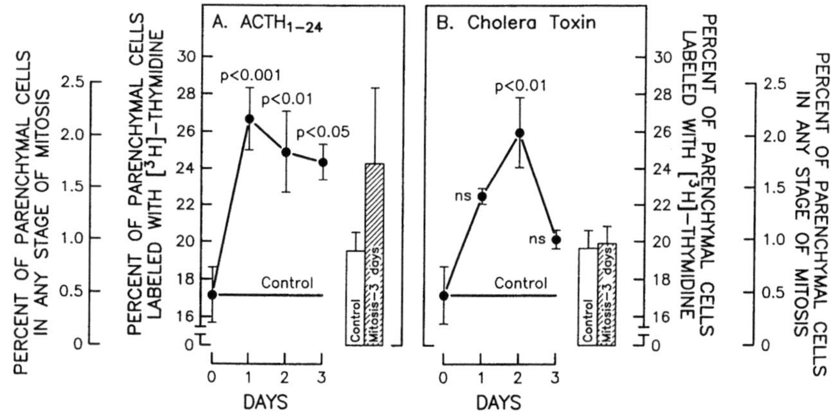

FIGURE 6 The Ca^{2+}/cyclic AMP signal from ACTH receptors activated by ACTH$_{1-24}$ stimulated the cells in primary rabbit adrenocortical cultures to replicate DNA, enter mitosis, and then continue cycling and making DNA. The cyclic AMP-only signal stimulated by the adenylyl cyclase-stimulating cholera toxin stimulated the cells to make DNA, but they could not enter mitosis (bars) and therefore stopped cycling and making DNA. Full statistical and experimental details are provided by Menapace et al.[45]

when PTH and $1,25(OH)_2D_3$ are injected at the same time as isoproterenol.[43] Together, these cause the plasma Ca^{2+} concentration to rise and a $1,25(OH)_2D_3$-stimulated component to accumulate to normal values by 12 h, just as the cells are reaching this critical final stage of the G_1 build-up.

IV. ADRENAL CORTEX

Adrenal cortical parenchymal cells can be stimulated by adrenocorticotropin (ACTH) to secrete steroids and proliferate both in the animal and in primary cultures.[44,45] This hormone acts by binding to two separate 7-TM receptors, one of which is low-affinity, high-capacity, and linked to adenylyl cyclase, and the other of which is high-affinity, low-capacity, and somehow coupled to Ca^{2+} channels.[46,47] ACTH activation of the high-affinity receptor increases Ca^{2+} influx and triggers a transient internal Ca^{2+} surge, seemingly without activating PLCs or PKCs.[46,47] Here again it is a Ca^{2+}/cyclic AMP signal, not a Ca^{2+}/PKCs signal that sparks a secretion-proliferation response in the animal. It seems that it is the cyclic AMP surge triggered by the low-affinity ACTH receptor that starts the events leading to the G_1 build-up and DNA replication while both receptors are required for the events leading to the initiation of mitosis.[45] Thus, treating primary parenchymal cells in primary rabbit adrenocortical cultures with the adenylyl cyclase-stimulating cholera toxin or dibutyryl cyclic AMP stimulates them to initiate DNA replication, but they cannot enter mitosis (Figure 6).[45] Evidently it is the Ca^{2+} signal from the high-affinity receptors that triggers something in early prereplicative development that is needed to complete the G_2 build-up and initiate mitosis after the cell has finished replicating its chromosomes. In the previous chapter we learned that regenerating rat liver cells also must start making something in early prereplicative development that is specifically needed for them to complete the G_2 build-up and enter mitosis. Undoubtedly this something is somehow related to the accumulation and activation of the Cdc2·cyclin B complexes that trigger mitosis.

REFERENCES

1. **Despopoulous, A. and Silbernagl, S.,** *Color Atlas of Physiology*, George Thieme Verlag, Stuttgart, 1984.
2. **Dumont, J.E., Lamy, F., Roger, P., and Maenhaut, C.,** Physiological and pathological regulation of thyroid cell proliferation and differentiation and thyrotropin and other factors, *Physiol. Rev.,* 72, 667, 1992.

2a. **Farid, N.R., Shi, Y., and Zou, M.,** Molecular basis of thyroid cancer, *Endocr. Rev.,* 15, 202, 1994.

3. **Hedin, K.E., Duerson, K., and Clapham, D.E.,** Specificity of receptor-G protein interactions: searching for the structure behind the signal, *Cell Signalling*, 5, 505, 1993.
4. **Nagayama, Y. and Rapoport, G.B.,** The thyrotropin receptor 25 years after its discovery: new insights after its molecular cloning, *Mol. Endocrinol.*, 6, 145, 1992.
5. **Vassart, G. and Dumont, J.E.,** The thyrotropin receptor and the regulation of thyroid function and growth, *Endocr. Rev.,* 13, 596, 1992.
6. **Jouishomme, H., Whitfield, J.F., Chakravarthy, B., Durkin, J.P., Gagnon, L., Isaacs, R.J., MacLean, S., Neugebauer, W., Willick, G., and Rixon, R.H.,** The protein kinase-C activation domain of the parathyroid hormone, *Endocrinology*, 130, 53, 1992.
7. **Segre, G.V. and Goldring, S.R.,** Receptors for secretion, calcitonin, parathyroid hormone (PTH)/PTH-related peptide, vasoactive intestinal peptide, glucagonlike peptide 1, growth hormone-releasing hormone, and glucagon belong to a newly discovered G-protein-linked receptor family, *Trends Endocrinol. Metab.*, 4, 309, 1993.
8. **Ledent, C., Parmentier, M., Maenhaut, C., Taton, M., Pirson, I., Lamy, F., Roger, P., and Dumont, J.E.,** The TSH cyclic AMP cascade in the control of thyroid cell proliferation: the story of a concept, *Thyroidology,* 3, 97, 1991.

8a. **Cho-Chung, Y. and Clair, T.,** The regulatory subunit of cAMP-dependent protein kinase as a target for chemotherapy of cancer and other cellular dysfunctional-related diseases, *Pharm. Ther.,* 60, 265, 1994.

9. **Reuse, S., Maenhaut, C., and Dumont, J.E.,** Regulation of protooncogenes c-*fos* and c-*myc* expressions by protein tyrosine kinase, protein kinase C, and cyclic AMP mitogenic pathways in dog primary thyrocytes: a positive and negative control by cyclic AMP on c-*myc* expression, *Exp. Cell Res.*, 189, 33, 1990.
10. **Lamy, F., Wilkin, F., Baptist, M., Posada, J., Roger, P.P., and Dumont, J.E.,** Phosphorylation of mitogen-activated protein kinases is involved in the epidermal growth factor and phorbol ester, but not in the thyrotropin/cAMP, thyroid mitogenic pathway, *J. Biol. Chem.*, 268, 8398, 1993.

11. **Raspé, E., Reuse, S., Roger, P.P., and Dumont, J.E.,** Lack of correlation between the activation of the Ca^{2+}-phosphatidylinositol cascade and the regulation of DNA synthesis in the dog thyrocyte, *Exp. Cell Res.*, 198, 17, 1992.
12. **Frederickson, R.M. and Sonenberg, N.,** Signal transduction and regulation of translation initiation, *Semin. Cell Biol.*, 3, 107, 1992.
13. **Rhoads, R.E.,** Regulation of eukaryotic protein synthesis by initiation factors, *J. Biol. Chem.*, 268, 3017, 1993.
14. **Rhoads, R.E., Joshi-Barve, S., and Rinker-Schaeffer, C.,** Mechanism and regulation of protein synthesis initiation factor 4E: effects on mRNA discrimination, cellular growth rate, and oncogenesis, *Prog. Nucleic Acids Res.*, 46, 183, 1993.
15. **Fujimoto, J. and Brenner-Gati, L.,** Protein kinase-C activation during thyrotropin-stimulated proliferation of rat FRTL-5 thyroid cells, *Endocrinology*, 130, 1587, 1992.
16. **Takada, K., Amino, N., Tada, H., and Miyai, K.,** Relationship between proliferation and cell cycle-dependent Ca^{2+}-influx induced by a combination of thyrotropin and insulin-like growth factor-I in rat thyroid cells, *J. Clin. Invest.*, 86, 1548, 1990.
17. **Gartner, R.,** Thyroid growth in vitro, *Exp. Clin. Endocrinol.*, 100, 32, 1992.
18. **Spada, A., Vallar, L., and Faglia, G.,** G protein oncogenes in pituitary tumors, *Trends Endocrinol. Metab.*, 3, 355, 1992.
19. **Burton, F.J., Hasel, K.W., Bloom, F.E., and Sutcliffe, J.G.,** Pituitary hyperplasia and gigantism in mice caused by a cholera toxin transgene, *Nature*, 350, 74, 1991.
20. **DiGeorge, A.M.,** Albright syndrome: is it coming of age? *J. Pediatr.*, 87, 1018, 1975.
21. **Schwindinger, W.F., Francomano, C.A., and Levine, M.A.,** Identification of a mutation in the gene encoding the α subunit of the stimulating G protein of adenylyl cyclase in McCune-Albright syndrome, *Proc. Natl. Acad. Sci. U.S.A.*, 89, 5152, 1992.
22. **Feuillan, P.P., Shawker, T., Rose, S.R., Jones, J., Jeevanram, R.K., and Nisula, B.C.,** Thyroid abnormalities in the McCune-Albright syndrome: ultrasonography and hormonal studies, *J. Clin. Endocrinol. Metab.*, 71, 1596, 1990.
23. **Wynford-Thomas, D.,** Molecular basis of epithelial tumorigenesis: the thyroid model, *Crit. Rev. Oncogenesis*, 4, 1, 1993.
24. **Lemoine, N.R., Mayall, E.S., Williams, E.D., Thurston, V., and Wynford-Thomas, D.,** Agent-specific *ras* oncogene activation in thyroid tumors, *Oncogene*, 3, 541, 1988.
25. **Spina, A., DiDonato, A., Colella, G., Illiano, G., Berlingieri, M.T., Fusco, A., and Grieco, M.,** Increased adenylate cyclase activity in rat thyroid epithelial cells expressing viral RAS gene, *Biochem. Biophys. Res. Commun.*, 142, 527, 1987.
26. **Fynan, T.M. and Reiss, M.,** Resistance to inhibition of cell growth by transforming growth factor-β and its role in oncogenesis, *Crit. Rev. Oncogenesis*, 4, 439, 1993.
27. **Matsuo, K., Tang, S.-H., and Fagin, J.A.,** Allelotype of human thyroid tumors: loss of chromosome 11q 13 sequences in follicular neoplasms, *Mol. Endocrinol.*, 5, 1873, 1991.

28. **Fagin, J.A., Matsuo, K., Karmakar, A., Chen, D.L., Tang, S.-H., and Koeffer, H.P.,** High prevalence of human thyroid carcinomas, *J. Clin. Invest.*, 91, 179, 1993.
29. **Whitfield, J.F.,** Adrenergic agents, calcium ions, and cyclic nucleotides in the control of cell proliferation, *Handb. Exp. Pharmacol.*, 54, 267, 1980.
30. **Baum, B.J. and Ambudkar, I.S.,** Regulation of calcium handling by rat parotid acinar cells, *Mol. Cell. Biochem.*, 82, 67, 1988.
31. **Hincke, M.T.,** Characterization of rat parotid protein kinase C, *Int. J. Biochem.*, 20, 303, 1988.
32. **Michell, R.H.,** Inositol phospholipids and cell surface receptor function, *Biochim. Biophys. Acta*, 415, 81, 1975.
33. **Söling, H.-D., Padel, U., Jahn, R., Thiel, G., Kricke, P., and Fest, W.,** Regulation of protein kinases in exocrine secretory cells during agonist-induced exocytosis, *Adv. Enzyme Regul.*, 23, 141, 1985.
34. **Tennes, K.A. and Putney, J.W.,** Phosphoinositides and exocrine gland receptors, in *Phosphoinositides and Receptor Mechanisms*, Putney, J.W., Ed., Alan R. Liss, New York, 1986, 363.
35. **Michell, R.H.,** Inositol lipids and their role in receptor function: history and general principles, in *Phosphoinositides and Receptor Mechanisms*, Putney, J.W., Ed., Alan R. Liss, New York, 1986, 1.
36. **Söling, H.-D., Fest, W., Machoczek, K., Schmidt, T., Esselmann, H., and Fischer, N.,** Mechanisms of short-term (second range) regulation of the activities of enzymes of lipid and phospholipid metabolism in secretory cells, *Adv. Enzyme Regul.*, 29, 35, 1989.
37. **Guidotti, A., Weiss, B., and Costa, E.,** Adenosine 3′,5′-monophosphate concentrations and isoproterenol-induced synthesis of deoxyribonucleic acid in mouse parotid gland, *Mol. Pharmacol.*, 8, 521, 1972.
38. **Tsang, B.K., Rixon, R.H., and Whitfield, J.F.,** A possible role for cyclic AMP in the initiation of DNA replication by isoproterenol-activated parotid gland cells, *J. Cell. Physiol.,* 102, 19, 1980.
39. **Boynton, A.L., Whitfield, J.F., and Walker, P.R.,** The possible roles of polyamines in prereplicative development and DNA synthesis: a critical assessment of the evidence, in *Polyamines in Biomedical Research,* Gaugas, J.M., Ed., John Wiley & Sons, New York, 1980, 63.
40. **Campos-Gonzales, R., Whitfield, J.F., Boynton, A.L., MacManus, J.P., and Rixon, R.H.,** Prereplicative changes in the soluble calmodulin of isoproterenol-activated rat parotid cells, *J. Cell. Physiol.,* 118, 257, 1984.
41. **Kousvelari, E.E., Louis, J., Curran, T., and Baum, B.J.,** Induction of c-*fos*, c-*myc* and p53 by β-adrenergic receptor (β-AR) stimulation of rat parotid acinar cells (RPAC), *Fed. Proc.*, 46, 2068, 1987.
42. **Hoffmann, J. and Schwoch, G.,** Co-ordinated changes in the cyclic AMP signalling system and the phosphorylation of two nuclear proteins of M_r 130,000 and 110,000 during proliferative stimulation of the rat parotid gland by isoprenaline, *Biochem. J.*, 263, 785, 1989.
43. **Tsang, B.K., Whitfield, J.F., and Rixon, R.H.,** The reversible inability of isoproterenol-activated parotid gland cells to initiate DNA synthesis in the hypocalcemic thyroparathyroidectomized rat, *J. Cell. Physiol.*, 107, 41, 1981.

44. **Boynton, A.L. and Whitfield, J.F.,** The role of cyclic AMP in cell proliferation: a critical assessment of the evidence, *Adv. Cyclic Nucleotide Res.*, 15, 193, 1983.
45. **Menapace, L., Armato, U., and Whitfield, J.F.,** The effects of corticotrophin ($ACTH_{1-24}$) cyclic AMP and TPA (12–*O*-tetradecanoyl phorbol-13-acetate) on DNA replication and proliferation of primary rabbit adrenocortical cells in a synthetic medium, *Biochem. Biophys. Res. Commun.*, 148, 1295, 1987.
46. **Hornsby, P.J.,** The mechanism of action of ACTH in the adrenal cortex, in *Hormones and their Actions*, Part III, Cooke, B.A., King, R.J.B., and van der Molden, H.J., Eds., Elsevier, Amsterdam, 1988, 193.
47. **Rasmussen, H., Apfeldorf, W., Barret, P., Jakuwa, N., Zawalich, W., Kreutter, D., Park, S., and Takuwa, Y.,** Inositol lipids: integration of cellular signalling systems, in *Phosphoinositides and Receptor Mechanisms*, Putney, J.W., Ed., Alan R. Liss, New York, 1986, 109.

4 Calcium as Differentiator and Killer — Keratinocytes

I. NORMAL KERATINOCYTES

The skin is maintained by the proliferation of the cells in its basal layer which is balanced by differentiation and by a special kind of programmed suicide (apoptosis) of the precursor cells' progeny after they lose contact with the basal lamina and accumulate key Ca^{2+}-activated enzymes and precursors.[1–5] The epidermal cells are grouped into epidermal proliferative units or EPUs which are stacks or columns of differentiating cells topped by the terminal cornified squames (Figure 1).[3–4] Each column is stacked on top of 10 to 11 cycling basal cells, 6 or 7 of which surround a central group of 3 to 5 cells, one of which is a dendritic macrophage-like (it can present processed antigen to lymphocytes) Langerhans cell, and one of which is the niche-bound mother keratinocyte, a slowly or intermittently cycling, self-renewing stem cell. Stem cells have a lot of Ca^{2+}-binding/activated $\alpha_5\beta_1$ and hemi-desmosomal, keratin-coupled $\alpha_6\beta_4$ integrins on their surfaces, which they use to bind themselves strongly to basal lamina components such as collagens I and IV, fibronectin, and laminin.[6,7] The outer six or seven cells at the base of the EPU are transit amplifying precursor cells which, because they have fewer integrins on their surfaces, adhere less strongly to the basal lamina, are not niche-bound, and cycle faster than the parental stem cell.[8] While the self-renewing stem cell is a high-efficiency colony former in culture, the less adherent transit amplifying cells have far less proliferative potential. After only a few (e.g., five) cycles, all of the progeny of a cultured transit amplifying cell stop expressing key cell cycle genes such as c-*myc*, dismantle their cell cycle engines, stop cycling, and differentiate.[8]

The maintenance of functioning cell cycle engines and proliferation of the basal EPU cells require signals from the basal lamina-activated integrin receptors. The basal stem cells are also protected from the potential apoptogenic action of their expressed c-Myc protein[8a] and signal-induced reactive oxygen intermediates by producing the anti-apoptosis Bcl-2 protein.[8b] When activated by binding to the substrate, integrins activate a Src-family protein tyrosine kinase, YPK (c-Fyn, c-Lyn, c-Yes) which phosphorylates, and thus activates,

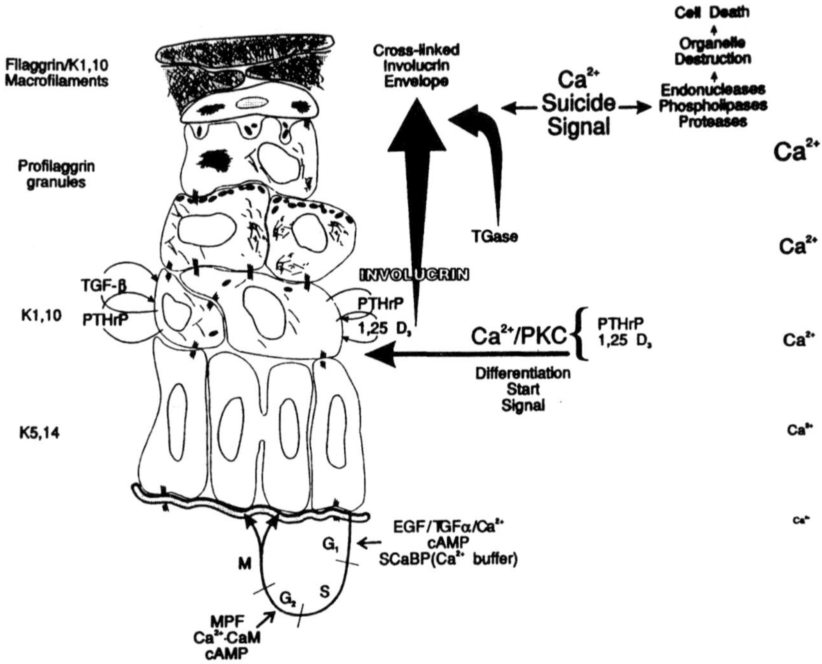

FIGURE 1 The main features of the keratinocyte differentiation program that unfolds within an upward Ca^{2+} gradient and starts with the cycling progeny of a self-renewing stem cell in the presence of a very low Ca^{2+} concentration at the base of the EPU (epidermal proliferative unit). The proliferation of the stem and precursor basal cells is driven by adhesion-dependent signals from lamina-binding receptors and signals from receptors such as autocrine/paracrine KGF (keratinocyte growth factor), TGF-α, and by cyclin-dependent PKs (one of which, Cdc2-cyclin B, is specifically recognized in the scheme by MPF). The proliferating stem cell and its precursor cell progeny protect themselves from apoptogenic, signal-generated reactive oxygen intermediates by making the Bcl-2 protein. The most mature basal precursor cells start detaching from the basal lamina as proliferation-promoting integrin receptors lose their affinity for their basal lamina fibronectin ligands. This stops the adhesion-generated signals that stimulate the production of the Cdk2-cyclin A PKs needed to finish the G_1 build-up and initiate DNA replication. Thus, the cell stops cycling as it lifts off the basal lamina and starts climbing up the Ca^{2+} gradient. Then the cell also stops making the protective anti-apoptosis Bcl-2 protein and the protective Ca^{2+}-buffering SCaBP protein and starts making autocrine/paracrine factors such as PTHrP and its fragments, TGF-βs, and $1,25(OH)_2D_3$ that drive the expression and build-up of differentiation-specific keratins, cell-adhesion molecules, TGase, and the several cornified envelope components such as involucrin. When the rising, maturing cell reaches a critical point in the Ca^{2+} gradient, its surface Ca^{2+} receptor/sensors are saturated and fire to trigger an apoptogenic internal Ca^{2+} surge that guts and kills the cell and converts it into a tough cornified shell reinforced by keratin struts. The key to the several abbreviations are to be found in the Abbreviations section at the end of this book. From Whitfield.[106a]

the pivotal 125-kDa FAK YPK. FAK phosphorylates, and thus velcroizes, the 80-kDa focal adhesion complex component, paxilin.[8c] c-Crk, a universal adaptor,[8d] then binds with its SH2 domains to the phosphorylated Y residues of the velcroized paxilin and collects a second team of signalers with its SH3 domains. Thus, the integrin-activated focal adhesion complex becomes a functional modular velcroceptor (like those of the lymphocytes of Chapter 6), which organizes a team of signalers that trigger Ca^{2+} surges, bursts of protein kinase C (PKCs) activity, and the activation of "matrix-responsive" genes, such as the gene coding for the cell cycle engine's[6,7,8c,9–9b,10] cyclin A. Proliferation of basal cells is also driven and controlled by paracrine and autocrine factors, direct contact of the stem cell and its progeny with each other and the Langerhans cell, and by signals from the underlying dermal cells without which the basal keratinocytes rapidly lose their self-renewing capability.[3] It is also likely that the proliferation of the basal cells of an EPU is regulated by the passage of growth factors and second messengers through gap junctions.

We do not know what signals drive the keratinocyte cycle, but they are likely to be the same ones that drive the G_1 build-up, the initiation of DNA replication and mitosis in regenerating rat liver cells primed by the cascade of competence-establishing events. However, we do know that cultured skin keratinocytes, unlike hepatocytes and fibroblasts, need only a very small amount (0.0l to 0.1 m*M*) of Ca^{2+} and cyclic AMP surges to proliferate indefinitely and maintain their morphological identity.[1] Signals triggered by the binding of cytoskeleton-coupled adhesion receptors such as $\alpha_6\beta_4$ integrins to their counter-receptors in the basal lamina combine with similar or different signals and signal patterns from activated 7-TM receptors for β-adrenergic agonists, and the velcroceptors for epidermal growth factor (EGF), extracellular matrix (ECM)-bound paracrine basic fibroblast growth factor (bFGF), ECM-bound paracrine FGF-related KGF (keratinocyte growth factor), autocrine insulin-like growth factor-1 (IGF-1), and probably most important of all, autocrine transforming growth factor (TGF)-α, drive keratinocyte proliferation by stimulating the sequential appearance and destruction of the different stage-specific cyclin-dependent PKs of the cell cycle engine,[11] and suppressing the activation of the differentiation gene cassettes. It is noteworthy that the very important TGF-α is also active in its large transmembrane precursor form,[12,13] and thus may stimulate proliferation by binding to EGF/TGF-α receptors on adjacent cells, a "juxtacrine" binding to EGF/TGF-α velcroceptor receptors on the same cell surface, or by paracrine secretion and intercellular diffusion or extracellular matrix-mediated transfer of the small liberated TGF-α peptide. The continuous cycling of keratinocyte precursors is maintained by a very low, though essential, external Ca^{2+} concentration, but proliferation stops and terminal differentiation starts if the external Ca^{2+} concentration should exceed a critical threshold value (Figure 2). Keratinocytes cycling on the basal lamina make a skin-specific Ca^{2+}-binding buffer protein (SCaBP) that would reduce

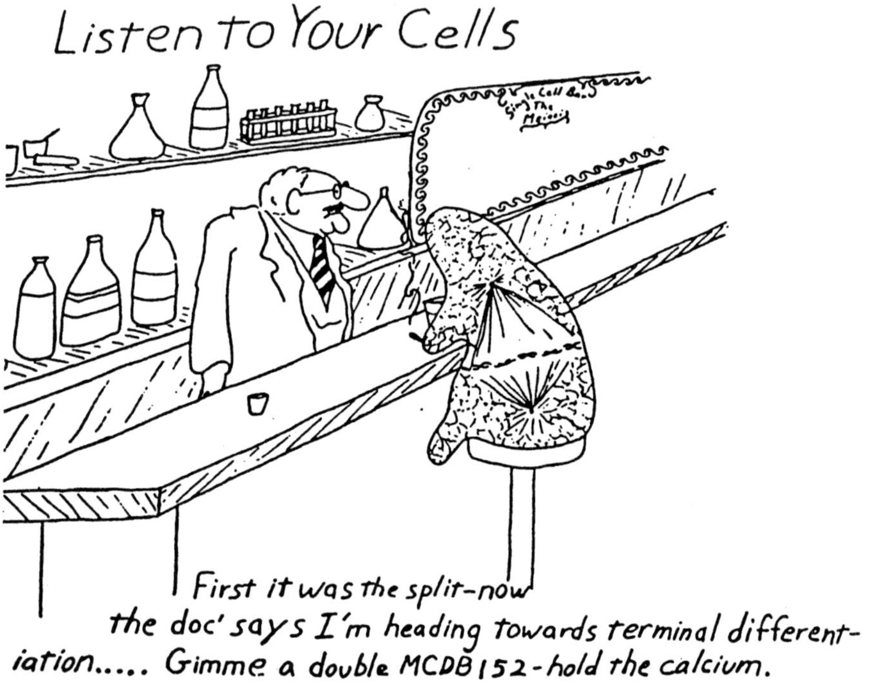

FIGURE 2 A cultured keratinocyte's sad lament to the bartender as she tries to escape the deadly apoptogenic Ca^{2+} signal by getting drunk on proliferogenic, low-Ca^{2+} medium. (From *TCA Report,* courtesy of Richard G. Ham and Gretchen H. Stein, drawn by David C. Hamlin, University of Colorado at Boulder.)

the size of any unscheduled Ca^{2+} surge, which, without this protective protein, would prematurely stop their proliferation.[1]

How does raising the external Ca^{2+} concentration trigger the differentiation of cultured keratinocytes? The key to answering this question clearly lies in the keratinocyte's Ca^{2+}-signaling machinery, parts of which, such as the Ins(1,4,5) P_3-receptor-bearing Ca^{2+} storage vesicles and the Ca^{2+}/Mg^{2+}-ATPase pumps, are concentrated around membrane pockets called "caveolae".[14,15] Stimulating this machinery by raising the Ca^{2+} concentration (e.g., from 0.05 to 1.8 m*M*) in the necessarily serum-supplemented medium of primary human or mouse epidermal keratinocytes stops the cells from cycling and stimulates them to stratify and ultimately become corneocytes. This "Ca^{2+} switch" seems to be the signal from Ca^{2+}-*receptors* on the cycling keratinocytes' surfaces which, when primed by a serum factor and activated by a rise in the external Ca^{2+} concentration, generate the multitrack differentiation signal.[16–17a] The receptors on some keratinocytes can be activated by Sr^{2+} and micromolar concentrations of La^{3+} (higher concentrations of La^{3+} block the differentiation response).[17,18] But, we cannot get a response, in this case a burst of membrane-associated PKCs activity, from BALB/MK mouse keratinocytes by adding 1.8 m*M* Sr^{2+} to medium containing 0.05 m*M* Ca^{2+} with or without serum. The keratinocyte's polyvalent cation receptor might be at least functionally the same as the polyvalent cation-responsive (e.g., Ba^{2+}, Ca^{2+}, Gd^{3+}, La^{3+}, Mg^{2+}, Sr^{2+}, neomycin, polyarginine) receptors that PTH-producing parathyroid gland chief cells use to measure the blood Ca^{2+} concentration and accordingly regulate PTH secretion.[19] The parathyroid cell's 120-kDa, low-affinity, 7-TM receptor is set to respond and stop PTH secretion only when the external Ca^{2+} concentration reaches 1.0 to 1.25 m*M*.[19] Although it can respond to several cations, Ca^{2+} is the main polyvalent cation it sees in the extracellular environment in an animal. Brown et al.[19] have found that this receptor is a glycoprotein, which resembles neuronal metabotropic receptors and glycoprotein hormone receptors like the TSH receptor we met in the last chapter. It is coupled to several signal components: a pertussis toxin-sensitive, adenylyl cyclase-inhibiting G_i protein which activates a PLC, and a Ca^{2+} channel. The signal from this receptor is a prompt Ca^{2+} surge from internal stores, which drops to an extended suprabasal plateau as external Ca^{2+} flows in, and a burst of PLC-induced phosphoinositide breakdown.[19] As we shall see, this is exactly like the Ca^{2+}-triggered signal in keratinocytes. But while the keratinocyte Ca^{2+} receptor may be functionally the same as the parathyroid cell receptor it is not the same receptor because we have found BALB/MK murine keratinocytes to be insensitive to Gd^{3+} and neomycin, responsiveness to which are hallmarks of the parathyroid Ca^{2+} receptor. Perhaps it is a Ca^{2+}-binding adhesion signaler such as an integrin.[20]

When the external Ca^{2+} concentration exceeds the 0.5- to 0.8-m*M* threshold, the keratinocyte's Ca^{2+} receptors trigger Ca^{2+} influx through voltage-independent, nonspecific 14-pS channels.[21] The receptor stimulates an isoform of PLC. This PLC isoform is not PLC-γ_1. PLC-γ_1 is not phosphorylated on Y residues when Ca^{2+} switches off keratinocyte proliferation as it is when EGF stimulates keratinocyte proliferation.[22] This suggests that the keratinocyte's Ca^{2+} receptor is also not a YPK velcroceptor like the EGF/TGF-α receptor. The apparently slowly inactivated receptor causes a remarkably sustained (longer than 24 h) hydrolysis of phosphoinositides to produce inositol phosphates and diacylglycerols (DGs). The inositol phosphates release Ca^{2+} from inositol phosphate-responsive caveolar-sited stores, and within 30 s it peaks at a level four to five times the basal value and then drops to a still-suprabasal (twice the basal level) plateau for more than 24 h.[16,23–26] The DGs would be expected to stimulate the α, δ, ϵ, η (but not the ζ) PKC isoforms of the keratinocytes.[27] Since PKCs activators such as TPA can mimic the Ca^{2+} switch, PKCs activation by DGs should be a key part of the signal from the Ca^{2+} receptors. Indeed, PKC activators stimulate, and a PKC blocker, such as RO31–7549 blocks, keratinocyte differentiation.[28–30] But Moscat et al.[31] claimed that PKCs do not in fact mediate the response of BALB/MK mouse keratinocytes to Ca^{2+} because they could find no phosphorylation of the 80-kDa MARCKS protein, the specific intracellular PKCs substrate that was described in Chapter 2, and there was no inactivation of the EGF receptor, another specific PKCs substrate. Indeed, Ca^{2+} seemed to trigger an unconventional, and thus deceptive, pattern of phosphoinositide breakdown, because the initial product of the Ca^{2+}-stimulated phosphoinositide breakdown was Ins(1,3,5)P_3 instead of the usual Ins(1,4,5)P_3 and a significant fraction (30%) of the generated DGs were alkyl ether glycerides rather than alkyl ester glycerides.[31] However, Moscat et al.[31] were wrong about there being no burst of membrane-associated PKC activity in Ca^{2+}-stimulated BALB/MK cells. We have found,[31a] with our highly specific assay for *in situ* membrane-associated PKCs activity,[32] that raising the external Ca^{2+} concentration from a proliferation-supporting 0.05 m*M* to a proliferation-suppressing, differentiation (cornified envelope)-triggering, 1.8 m*M* promptly and powerfully stimulates membrane-associated PKCs in the same BALB/MK keratinocytes. But the highly selective PKCs substrate used in this assay is the FKKSFKL peptide, which corresponds to part of the MARCKS protein's phosphorylation site-domain.[32] In other words, Ca^{2+} strongly stimulates an enzyme in BALB/MK keratinocyte membranes that certainly should have phosphorylated the cell's intact MARCKS protein. So why didn't it in the experiments of Moscat et al.? The answer is simple. The activated Ca^{2+} receptor triggers a $[Ca^{2+}]_i$ surge which generates Ca^{2+}•CaM complexes, and it is these complexes that prevent activated PKCs from phosphorylating the MARCKS protein,[31a] probably by binding to the protein's phosphorylation site-domain.[33]

The differentiation-triggering Ca^{2+} receptor is not a YPK velcroceptor, like the EGF/TGF-α receptor. As noted above, unlike the EGF/TGF-α receptor, the Ca^{2+} receptor does not phosphorylate PLC-γ_1 or the 85-kDa subunit of PI-3K.[22] Nevertheless, a burst of YPK activity is a very early and specific part of the differentiation program triggered by Ca^{2+} or the PKCs-activating TPA, and YPK inhibitors prevent differentiation.[34] The receptor operates indirectly by stimulating nonreceptor YPKs. It stimulates a protein tyrosine phosphatase that activates both the c-Src YPK ($pp60^{c\text{-}src}$) and the closely related c-Yes kinase by dephosphorylating a C-terminal tyrosine residue (Y^{527} in the case of the c-Src YPK) which shifts the enzymes from the curled-up inactive configuration to the uncurled active configuration, but the c-Yes YPK promptly associates with proteins that suppress its new-found activity.[35,36] By contrast, c-Src associates with a set of nonsuppressing proteins and some of the kinase molecules move from the cell membrane and surface adhesion sites to the nucleus.[35,36] At the same time, Y residues of a cytoplasmic and membrane-associated 62-kDa protein are phosphorylated and plug into the SH2 pockets of adaptor proteins which collect on the cell membrane to activate Ras proteins, which stimulate Raf-1 STPKs and thus set off an explosion of STPK and STYPK activities.[35,36a]

The cycling basal keratinocytes have a very small complement of PKCs and very low PKCs activity. But the signal from the Ca^{2+} receptor stimulates the translation of mRNAs for all of the cell's several PKC isoforms which, because of the sustained generation of DGs, results in escalating PKCs activity.[27,37,38] The mounting PKCs activity stimulates the expression of the involucrin gene which has a TPA (i.e., PKCs)-responsive element in its 5′-regulatory region.[39] The signal also stimulates the expression of the gene coding for loricrin, another major component of the cornified envelope, the gene coding for filaggrin, which bundles keratin filaments, and the gene coding for the Ca^{2+}-stimulatable transglutaminase, TGase 1(K).[39–41]

The Ca^{2+} signal does four things to cultured keratinocytes: stops cycling, stimulates cell-cell adhesion, stimulates the formation of a tough cornified envelope, and stimulates the bundling of keratin filaments into tonofilaments and ultimately a lattice of keratin cables. In cycling keratinocytes, with their active cell cycle genes, the stage-specific cyclin-dependent PKs of the cell cycle engine, and very low PKCs activity, urokinase-plasminogen activator, E-cadherin, and 93-kDa β1 integrins are located in the cytoplasm and often cluster around the nucleus.[42,43] The signal from the Ca^{2+} receptor triggers a dramatic stampede of signaling adhesion molecules to the cell surface. The 93-kDa β1 integrins are converted into 113-kDa β1 integrins which are displayed on the cell surface.[42] Urokinase-plasminogen activator and E-cadherin move together to the cell surface, specifically cell-cell interfaces.[43] Fodrin is phosphorylated and also moves to the plasma membrane.[43] The Ca^{2+} receptor signal, particularly its PKCs component, triggers, within 1 to 2 h, the assembly of desmoplakin, vinculin, and other components into the desmosomes that "spot-weld" the cells together, and whose appearance marks the start of stratification.[42,44–47] It seems that a sustained Ca^{2+} signal plus the establishment of cell-cell contacts are the prerequisites for the optimal coordinate increase of involucrin and the TGase 1(K) activity required to start stratification and ultimately cornified envelope formation.[48]

A Ca^{2+} surge and a burst of membrane-associated PKCs activity trigger the keratinocyte's terminal differentiation program, but it is also known that stimulating membrane-associated PKCs activity with TPA stimulates the proliferation of basal keratinocytes.[3] How can such seemingly opposite findings be reconciled? It is really quite simple! The TPA-induced burst of membrane-associated PKCs activity stimulates the peripheral transit cells in an EPU to move up into the suprabasal differentiation zone. The EPU's less mature central transit cells sense this movement and increase their proliferative activity to fill the vacated niche.

We don't know what triggers stratification and cornification in the real world of the skins of beasts and humans. However, Ca^{2+} is almost certainly a major player, because the body maintains a Ca^{2+} gradient in the epidermis with the highest level being at the top of the differentiation zone (Figure 3).[49,52] This gradient is an elegant device to ensure that each stage of differentiation is triggered at the correct level in the epidermis. There is an optimal Ca^{2+} concentration for each part of the differentiation program.[24,53,54] As the cell moves up the gradient one set of genes is switched off and the next set is switched on. Thus, when the external Ca^{2+} concentration is 0.05 m*M*, cultured keratinocytes, like basal cells, express cell cycle genes and the genes coding for keratins 5 and 14 (Figure 1).[53,54] When the external Ca^{2+} concentration is raised

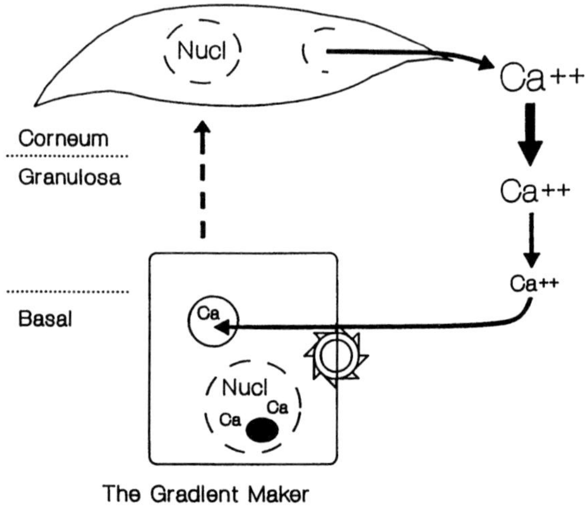

FIGURE 3 A possible epidermal Ca^{2+} gradient maker. Basal cells accumulate and sequester Ca^{2+} to produce the very low Ca^{2+} concentration in the basal layer. This accumulated Ca^{2+} is ultimately released in response to signals from surface Ca^{2+} sensors on the upper granulosa cells to trigger the cornifying, apoptoid process, and it is ultimately pumped out of the developing corneocytes to prevent a serious loss of body Ca^{2+} and maintain the top of the gradient for activating the Ca^{2+} sensors on upcoming granulosa cells.

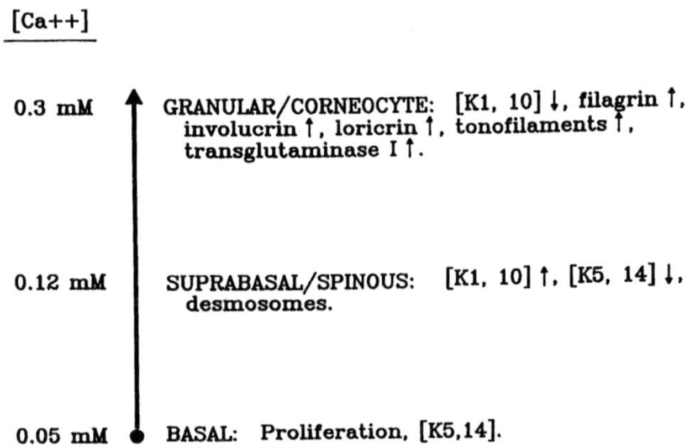

FIGURE 4 Increasing the Ca^{2+} concentration in the medium triggers increasingly advanced stages, and suppresses the preceding stages, of the differentiation program of cultured keratinocytes.

to 0.12 m*M*, they switch off the cell cycle genes and the genes coding for keratins 5 and 14 and switch on the genes coding for keratins 1 and 10 and other genes responsible for the spinous phenotype (Figure 4).[53,54] Raising the external Ca^{2+} concentration to 0.3 m*M* switches off the genes coding for keratins 1 and 10 and stimulates the expression of the genes of the granular/corneocyte program that code for the enzymes and structural components needed for corneocyte formation (Figure 4). Raising the external Ca^{2+} concentration from 0.05 to 1.0 m*M* causes the cells to directly and prematurely trigger the lethal terminal events (differentiation-apoptosis or "diffpoptosis") and convert themselves into abnormally cornified corpses.[53,54] The need for this gradient explains why keratinocytes stratify and differentiate better in floating cultures (liquid medium/air interface) than in submerged cultures.[54] The floating cultures can establish the physiological or near-physiological Ca^{2+} gradient while the Ca^{2+} concentration throughout submerged cultures would tend to equalize with the concentration in the medium.

The importance of Ca^{2+}, CaM, and the epidermal Ca^{2+} gradient for the control of keratinocyte proliferation and terminal differentiation is demonstrated by one of Nature's nastier experiments with Ca^{2+} controls — psoriasis. Normally, basal keratinocytes are exposed to low external Ca^{2+} concentrations and they have a high level of nuclear Ca^{2+} which is precipitable by the ion-trapping oxalate-pyroantimonate probe, but the proliferatively suppressed suprabasal keratinocytes *do not.*[50a] Basal keratinocytes in psoriatic lesions are also exposed to very low levels of Ca^{2+}, but they have twice as much CaM and cycle as much as eight times faster than normal keratinocytes, and they have large amounts of nuclear Ca^{2+} *at all levels* (except, of course, the stratum corneum).[50a] What might these striking differences mean? As we learned in Chapter 2, the final events in the G_1 build-up in other cells include the transport of Ca^{2+} into the nucleus and the accumulation and redistribution of CaM inside the nucleus, all for the assembly and operation of the DNA replication machinery. The higher CaM content is probably at least partly responsible for the rapid cycling of the psoriatic cells, because Rasmussen and Means[54a] have shown that hyperexpression of CaM speeds up the cycle of other cells such as C127 mouse fibroblasts. It also follows that the high nuclear content of precipitable Ca^{2+} in suprabasal regions indicates extension of the proliferative pool into the suprabasal levels in psoriatic lesions. This also indicates the failure of a differentiation signal, or the failure of the cells to respond to the signal.

While graded rises in the extracellular Ca^{2+} concentration can stop proliferation and drive the stepwise expression of the sequential stages of differentiation of *cultured* keratinocytes, it is likely that external Ca^{2+} and signals from Ca^{2+} receptors come into play only some considerable time after the differentiation program has been started, when the cell detaches from the basal lamina in the low-Ca^{2+} basal layer. As stated above, the most immature precursor keratinocytes in the EPU are located around the stem cell and the Langerhans cell.[3] The cycling of these amplifying precursor cells is driven by signals from ECM-binding/activated receptors such as the Ca^{2+}-binding/activated α_5/β_1 and $\alpha_6\beta_4$ integrins as well as bFGF and KGF that have been loaded into the ECM by dermal fibroblasts.[55–58] Normal cells must be anchored to a solid surface and have an external Ca^{2+}-loaded component on their surfaces if they are to complete the G_1 build-up and start replicating DNA.[55,59] This key component could be an integrin. Thus, if the basal keratinocytes should lose contact with the basal lamina, they will automatically stop cycling. Indeed, putting keratinocytes into suspension culture stops them from cycling and they synchronously form cornified envelopes.[60] However, cornified envelope formation can be stopped by adding the integrin-binding/activating fibronectin.[61] Moreover, partially coating the surface of the culture dish with a nonadhesive compound, "Poly-Hema", allows keratinocytes to attach to, but not spread out on, the surface. These cultured cells cannot cycle, but instead start making involucrin.[61] It is likely that the first step in the differentiation program is taken when a basal cell loses its affinity for ECM molecules and then stops displaying $\alpha_2\beta_1$, $\alpha_3\beta_1$, $\alpha_5\beta_1$, and $\alpha_6\beta_4$ receptors on its surface.[62,63] In human keratinocytes the $\alpha_5\beta_1$ integrin, along with the $\alpha_6\beta_4$ integrin, is a key player in maintaining attachment to the basal lamina and proliferative activity. And $\alpha_5\beta_1$ loses its ability to bind to the basal lamina's fibronectin and thus stops signaling several hours before it disappears from the cell surface.[62] When the contact with the basal lamina lessens and the Ca^{2+} and other signals from the integrin receptors[9,9a,64] stop, the cell can no longer express the cyclin A and assemble the active Cdk2·cyclin A PKs needed to initiate DNA replication,[65,66] and it now stops cycling and starts differentiating.

By the time the cell has been pushed up from the basal lamina, it has initiated the spinous program. Although it has stopped cycling, the cell is still growing and reconfiguring itself.[67] It welds itself to its neighbors with desmosomes. Cell-adhesion molecules move from the cytosol to the cell surface and new signals and cytoskeletal changes are triggered by the homophilic binding of the external domains of the E- and P-cadherins that have sprouted on the cell surface and the binding of stage-specific integrins to their counter-receptors on neighboring cells.[43,55–57,67] The cells now stop, or reduce, the production of the light, proliferation-compatible K5 and K14 keratins and start making the heavier, proliferation-incompatible type II K1 keratins which trigger the production of K10 keratin when it reaches a critical level. In fact, in human keratinocytes, the keratin profile changes from K5, 14 to K1, 5, 10, 11, 14 during spinous development.[68]

Whatever the signal that starts the spinous program might be, it cannot be a Ca^{2+} signal because of the very low Ca^{2+} level in both the basal and immediately suprabasal layers. It also should not include a *large* burst of all of the keratinocyte's five main PKC activities, because strongly stimulating α, δ, ϵ, and η PKCs with TPA (PKC-ζ does not have a TPA-binding domain) rapidly and selectively *inhibits* rather than stimulates the transcription of the K1 gene and destabilizes K1 mRNA.[69] Another reason for believing that the "spinous" signal does not include PKCs is the very low PKC content of cells freshly emerging from the very low-Ca^{2+} basal lamina. However, this signal probably selectively stimulates the expression and accumulation of the PKCs needed later on to stimulate the "granular" program.[69]

At this point we should stop and briefly look at the influence of PKCs on keratinocyte proliferation and differentiation. PKCs are key players in the Ca^{2+}

triggering of the differentiation (diffpoptosis) program, as opposed to sustaining cycling, in untransformed keratinocytes. Stimulating PKC activity with TPA inhibits the expression of the c-*myc* cell cycle gene and at the same time strongly stimulates involucrin gene expression and involucrin synthesis.[70] Reducing PKCs activity by 50% in human foreskin keratinocytes with PKCs inhibitors (10^{-9} *M* calphostin or 10^{-13} to 10^{-10} *M* staurosporine) or downregulating PKCs by prolonged preexposure to TPA enhances the proliferative response to EGF.[71] This enhancement may be due to the activation of EGF/TGF-α receptors which are switched off by PKCs.[72–74] TPA-induced PKCs activity has two other negative and positive actions. It stimulates the production of TGF-β mRNA and TGF-β protein,[72,75,76] which inhibits the proliferation of untransformed human and murine keratinocytes.[72,77–81] TGF-βs stop precursor cells cycling without stopping their growth by preventing the inactivation (by hyperphosphorylation) of Rb protein and thus the completion of the late stages of the G_1 build-up. They may also indirectly stimulate differentiation by enhancing Ca^{2+} signal surges by turning off the caveolar Ca^{2+}/Mg^{2+} ATPase efflux pumps.[14,18] TPA also stimulates the production of TGF-α,[72,82] whose initial proliferogenic thrust would later be aborted by the accumulation of TGF-β.

Budtz[5] and Whitfield[82a] have proposed that the terminal differentiation of keratinocytes is really just a special kind of apoptosis, diffpoptosis, a marriage of differentiation and apoptosis. Indeed, in preparation for this, suprabasal cells stop making the anti-apoptosis Bcl-2 protein,[8b] which protects the cycling basal cells from killing themselves with the reactive oxygen intermediate by products of their proliferogenic signaling. Thus, a key to understanding the triggering of the later part of the differentiation program may be the expression in the spinous cells of the genes coding for peculiar 10-kDa Spr (*S*mall, *P*roline-*R*ich) proteins.[83–86] The *spr* genes belong to the group of genes that are expressed in response to the withdrawal of growth factors, chromosome breakage, and genotoxic stress.[86] We have already met one of the genes in Chapter 2, GADD45, whose product mediates the proliferation-suppressing, chromosome safety-checking, apoptogenic actions of the p53 protein. Expression of *spr* genes might well be triggered by the withdrawal of the growth factor-like signals from integrins as the cell detaches itself from the basal lamina. But how does this happen? One of the *spr* genes, *spr2*–1, has a -TGAGTCA- sequence in its upstream regulatory region which binds the TPA/PKCs-activated AP-1 (c-Fos/c-Jun) transcription-enhancing heterodimer.[83] Consequently, a Ca^{2+}/PKCs (possibly the keratinocytes's nuclear PKC-η (L)[87]) signal can stimulate the *spr2*–1 promoter in cultured keratinocytes.[83] But is this the signal in the skin? The expression of the *spr* genes seems to accompany and may cause, or be caused by, the appearance of a PKC-stimulating autocrine/paracrine factor that might be the initiator of the granular program.

This autocrine/paracrine signaler may well be PTHrP or one of its several intracellularly processed bioactive fragments. In keratinocytes stimulated to differentiate in culture by raising the external Ca^{2+} concentration high enough

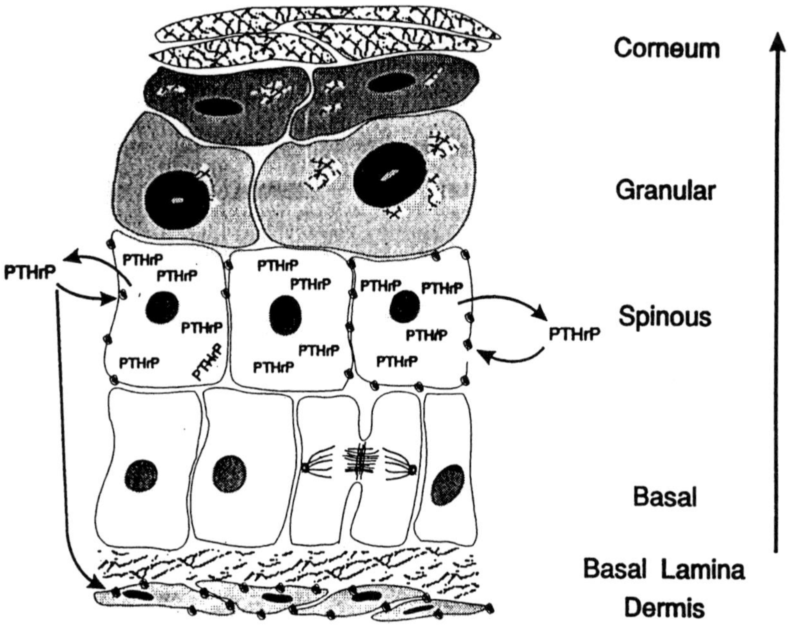

FIGURE 5 Keratinocytes seem to transiently produce PTHrP in the spinous stage of their journey from basal cell to corneocyte. Cellular proteases or convertases break the PTHrP holoproteins into N-terminal fragments that bind to PTH/PRHrP receptors and C-terminal fragments with the 107–111 TRSAW motif which bind to different receptors. Since keratinocytes express both PTH/PTHrP and TRSAW-specific PTHrP receptors, these fragments probably are paracrine modulators of basal cell cycling and autocrine/paracrine modulators or initiators of the spinous-level differentiation program. The PTHrP holoprotein and N-terminal fragments may also bind to the PTH/PTHrP receptors on dermal fibroblasts and stimulate these cells to produce paracrine cytokines that affect basal cell cycling. From Whitfield.[106a]

to massively stimulate the polyvalent cation receptors, involucrin synthesis starts immediately above the basal layer, but in the real world of the skin it usually starts in the upper spinous layer well after the cell has stopped cycling.[2,88] Danks et al.,[89] Hayman et al.,[90] and Martin et al.,[91] using affinity-purified antibody to PTHrP, found that only spinous cells contained immunoreactive protein (Figure 5). (However, Atillasoy et al.[92] claim to have found immunoreactive PTHrP in all layers of the epidermis.) And both PTHrP and PTH have been found to trigger internal Ca^{2+} surges, stop proliferation, and stimulate involucrin production and cornified envelope formation in primary human keratinocyte cultures and BALB/MK mouse keratinocytes.[93–95] Another indication of PTHrP's cycle-restraining activity is the ability of antisense RNA for PTHrP to stop PTHrP production and secretion and increase the proliferation of transformed (by human papilloma type 16) human HPK1A keratinocytes.[96] We have found that exposing mouse BALB/MK2 keratinocytes to picomolar concentrations of PTHrP-(1–40) in low (0.05 m*M*)-Ca^{2+} medium

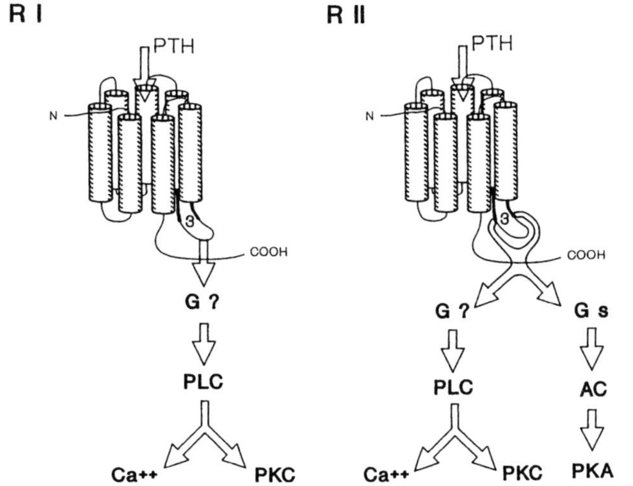

FIGURE 6 The two types of 7-TM or serpentine PTH/PTHrP receptor. RI is a monofunctional variant whose third cytoplasmic loop is structured only to activate a G protein that stimulates a phospholipase (PLC) isozyme. This receptor generates a Ca^{2+}/PKCs signal. RII is the classical dual-function PTH/PTHrP receptor whose third cytoplasmic domain is structured to activate the same G protein that RI uses to stimulate the PLC isozyme as well as to activate the G_s protein that stimulates adenylyl cyclase. This receptor generates a Ca^{2+}/PKC/PKA signal. It seems that human and mouse keratinocytes express only the monofunctional, PKCs-stimulating receptor.[95,97]

nearly doubled their membrane-associated PKC.[97] The results of experiments with ROS 17/2 rat osteosarcoma cells have indicated that there are two different receptors which are activated by N-terminal fragments of both PTH and PTHrP, and are differently coupled to adenylyl cyclase- and PKC-activating mechanisms.[97–100] Because they are activated by both PTH and PTHrP, these receptors are called PTH/PTHrP receptors. In the bone cells, the RI high-affinity PTH/PTHrP receptor (activated by low picomolar concentrations of PTH or PTHrP) is monofunctional: it activates only one kind of G-protein which stimulates a PLC isoform (not PLC-γ_1) that hydrolyzes PtdIns(4,5)P_2 into PKCs-stimulating DGs and Ins(1,4,5)P_3 which mobilizes Ca^{2+} from internal stores (Figure 6). The RII low-affinity PTH/PTHrP receptor (activated by much higher nanomolar concentrations of PTH and PTHrP) is bifunctional: it activates two G proteins, one which, Gs, stimulates adenylyl cyclase and the other of which stimulates PLC and ultimately the PKCs (Figure 6). Recent results are consistent with BALB/MK keratinocytes having only one kind of PTH/PTHrP receptor which stimulates membrane-PKCs activity but not adenylyl cyclase.[95,97] But there is more! We have found that BALB/MK keratinocytes also express different receptors specific for PTHrP fragments having the C-terminal PTHrP-(107–111) fragment's TRSAW motif. Indeed, PTHrP-(107–111) can maximally affect BALB/MK keratinocytes at nanomolar concentrations and freshly isolated human keratinocytes at subpicomolar concentrations. Thus, the PTHrP produced by spinous keratinocytes is a potent autocrine/paracrine factor, which is processed by endocellular proteases into different bioactive fragments,[100a] which, when secreted, activate PTH/PTHrP receptors and receptors for C-terminal PTHrP (TRSAW) fragments on the producer keratinocytes themselves and neighboring keratinocytes or other cells expressing one or more of these types of receptor.

The results of experiments by Regalia et al.[101] on primary cultures of EGF-stimulated normal adult human keratinocytes on polyethylene discs floating on top of serum-free MCDB 153 synthetic medium have demonstrated the special power of PTHrP breakdown products as modulators of keratinocyte proliferation. At concentrations between 10^{-12} and 10^{-8} *M*, PTHrP-(1–40) reduced the initiation of DNA replication by 71% when added to the cultures during the second half of the EGF-induced prereplicative period. But the keratinocytes are even more sensitive to PTHrP-(107–111): a concentration as low as 10^{-14} *M* reduced proliferation by about 40%. By contrast, PTH-(1–34) had no effect between 10^{-12} and 10^{-8} *M*, but started to become effective only around 10^{-7} *M*, which inhibited DNA synthesis by 39%. By contrast, 10^{-10} *M* PTH-(1–34), 10^{-10} *M* PTHrP-(1–40), and 10^{-14} *M* PTHrP-(107–111), optimally stimulated DNA synthesis two- to threefold by cells in quiescent keratinocyte cultures 48 h after removing EGF. Therefore, in the skin, processed, secreted PTHrP fragments, especially those with the (107–111) TRSAW motif, could suppress or stimulate basal cell proliferation depending on whether the cells are actively cycling under the influence of EGF/TGF-α receptor signals or have stopped cycling but are still in the proliferative mode.

It is currently impossible to know what triggers PTHrP expression. It could be TGF-β2 which (unlike TGF-β1) is expressed suprabasally. Indeed, it has been shown that TGF-β strongly stimulates PTHrP production by normal human foreskin keratinocytes.[102] Thus, perhaps the silencing of the attachment-dependent proliferation signal from integrin receptors triggers the expression of TGF-β2 which, in turn, triggers PTHrP production. However, PTHrP can also stimulate TGF-β production by stimulating membrane-associated PKC activity. Indeed, the two could be linked into an autorcrine/paracrine positive feedback loop (Figures 1 and 5). TGF-β2 stimulates the expression of differentiation-specific surface components[103] and prevents unscheduled suprabasal proliferation by suppressing $\alpha_6\beta_4$ integrin expression[61] until the cells reach the top of the Ca^{2+} gradient where the Ca^{2+} receptors can swing into action and trigger diffpoptosis, the final apoptosis-like stages of differentiation. The autocrine loop does two things: it shuts off at least one key cell cycle gene, c-*myc*,[104,105] and it blocks the middle to late stages of the G_1 build-up to DNA replication. It does these things by preventing the inactivation of the Rb protein. As discussed in Chapter 2, the TGF-β/TGF-β receptor loop somehow prevents hyperphosphorylation, and hence the inactivation of the Rb suppressor protein by preventing the activation of its phosphorylator, a Cdk2·cyclin

STPK. The loop signals prevent Cdk2 from being primed by phosphorylation and thus from being activated by combining with a G_1 cyclin, cyclin D_1 or E. This failure to make active Cdk2·cyclin D_1 or E PK complexes prevents the phosphorylation and inactivation of the Rb protein that is needed to finish the G_1 build-up and start DNA replication. However, the TGF-β/TGF-β receptor autocrine loop may further stimulate the transcription of the PTHrP gene and the production of PTHrP.[91,106] While the autocrine loops stop the suprabasal cells cycling, they do not stop growth, which is necessary for differentiation: the proliferatively suppressed differentiating cells may grow to as much as 10 times their basal size.[2,106a]

The surge of membrane-associated PKC activity stimulated by the PTH/PTHrP autocrine loops might stimulate the keratinocyte's 1α-hydroxylase, which catalyzes the formation of $1\alpha,25(OH)_2D_3$.[107,108] This, in turn, could further enhance keratinocyte differentiation by modulating specific genes and by contributing to the increase of cytosolic Ca^{2+} concentration and membrane-associated PKCs activity by activating a specific receptor on the cell surface rather than the conventional cytoplasmic/nuclear receptor.[109–112]

PTHrP, like other factors secreted by the spinous cells, may also establish a paracrine loop by binding the adenylyl cyclase-linked PTH/PTHrP receptors on the underlying dermal fibroblasts[1] (Figure 5). The processed PTHrP fragments seeping down from the spinous layer may stimulate the dermal fibroblasts to make and secrete cytokines and other factors which directly regulate the proliferation of the basal keratinocytes and thus ultimately the generation of PTHrP-producing spinous cells. Raising the external Ca^{2+} concentration from 0.05 to 2.0 m*M* doubles the gap-junctional communication between the cells of primary mouse skin keratinocytes within 12 h.[113] In skin, the rising, growing spinous keratinocytes of the EPU continue talking to, and thus modulating the proliferation of, the basal cells through their Ca^{2+}-modulated gap junctions.[106a] As expected from the response of cultured keratinocytes to a differentiation-triggering Ca^{2+} signal, they increase their gap-junctional communication with overlying spinous and granulosa cells.[1]

When the cell has reached the end of spinous development it has a new complement of heavy instead of light keratins; autocrine/paracrine signalers, PTHrP and its processed fragments; two kinds of receptor, each of which can respond to processed PTHrP N-terminal fragments or C-terminal fragments with the TRSAW motif; and with cation receptors whose cation-binding sites are being occupied stepwise as the cell rises in the external Ca^{2+} gradient (Figures 1 and 3). At some point, PTHrP fragments reach a critical level and trigger the "granular" program. The PTHrP-induced burst of PKCs activities might be the stimulator of Spr protein production which, in turn, might stimulate the expression of some of the genes of the granular program. The PTHrP-induced burst of PKC activities probably now stimulates the production of 1α-hydroxylase and thus the production of the secondary autocrine/paracrine signaler $1\alpha,25(OH)_2D_3$.[107,108] The now large burst of specific PKC activities *stops* the production of K1 and K10 keratin mRNAs, but, together with a Ca^{2+} surge from the $1\alpha,25(OH)_2D_3$ and PTHrP receptors,[95] stimulates the expression of the involucrin, loricrin, profilaggrin, and TGase I(K) genes, whose products will be the star players in transforming the granular cell into a dead keratinous squame encased in a tough, horny shell.

The granular cell then stops making PTHrP and gap junctions and loads up with cytoplasmic keratohyalin granules containing the 350-kDa profilaggrin, with its Ca^{2+}-binding, S-100-like N-terminal domain, which is used to pack the protein into the granules.[114] Cornifin, loricrin, and involucrin are collecting on the inner surface of the plasma membrane,[115,116] and membrane-associated, Ca^{2+}-triggered TGase I(K), and the loricrin-selective, Ca^{2+}-triggered transglutaminase E are poised to act.[116] A scenario for what might now happen goes like this. The cell has reached the highest point in the external Ca^{2+} gradient where the external Ca^{2+} concentration is 20 times higher than it was in the basal layer (Figures 1 and 3). Now, the progressively occupied cation-binding sites of the *still silent* receptors become saturated. By analogy with the cooperative Ca^{2+} binding/activation of the (differently set) Ca^{2+} receptors on active osteoclasts and parathyroid gland chief cells,[19] when this happens (possibly when the external Ca^{2+} concentration is between 0.5 and 0.7 m*M*) the receptors switch suddenly into the active mode and fire. The activated receptors generate a massive, killer Ca^{2+} surge (Figure 1), which is the physiological equivalent of the signal given prematurely to cultured keratinocytes by raising the external Ca^{2+} concentration from 0.05 to 1 m*M*. The in-rushing Ca^{2+} ions cut off residual intercellular gap junctions.[106a] The granulosa cell then seems to commit a special kind of suicide, diffpoptosis, based on the Ca^{2+}-mediated apoptosis program. The purpose of apoptosis is to get rid of cells at appropriate times and places without killing their neighbors by spilling out proteases and other destructive internal components. Therefore, apoptosis must always include the destruction of the plasma membrane and its replacement by a horny shell made by Ca^{2+}-activated TGases. This part of the apoptosis program has been cleverly exploited to make the tough corneocytes of skin and scales. The killer Ca^{2+} signal activates the keratinocyte's special plasma membrane-bound TGase 1(K) transglutaminase, possibly in part by stimulating a calcium-dependent neutral protease(s) — CANP(s) or calpain(s) — which is known to activate the transglutaminase by limited proteolysis[117] and in part by directly stimulating this Ca^{2+}-dependent enzyme. The activated transglutaminase cross-links the accumulated involucrin and loricrin molecules into a tough, cornified submembranous envelope.[115] The Ca^{2+} signal also activates the breakdown of the 350-kDa profilaggrins of the keratohyalin granules into the 37-kDa filaggrins that bundle the accumulated heavy keratins into macrofilaments or cables.[106a] So far the cells have been "glued" together by cholesterol sulfate produced by the high cholesterol sulfotransferase activity in the spinous cells, but now the "glue" is destroyed by cholesterol sulfatase and desquamation begins.[118] The Ca^{2+} surge generates Ca^{2+}•CaM complexes which activate the Ca^{2+}-ATPase pumps in the cell's membrane. The Ca^{2+}-activated CANP(s) may also bind via its calmodulin-like domains to the calmodulin-binding regulatory sites of the Ca^{2+} pumps and then maximally stimulate these pumps by limited proteolysis.[119] This double stimulation ensures that once it has done its job most of the exchangeable Ca^{2+} is pumped back into the intercellular space to prevent a continous and dangerous leakage of Ca^{2+} from the body through the skin. This might be the origin of the transepidermal Ca^{2+} gradient as the ejected Ca^{2+} filters downward (Figure 3). The activated Ca^{2+}/Mg^{2+}-dependent endonucleases, phosphatases, phospholipases, and proteases now destroy all of the cell organelles, leaving the final functional form — a dead, cornified envelope purged of its Ca^{2+} and held together by a latticework of keratin cables.

The hypothesis that squame formation is a special case of apoptosis is strongly supported by its similarities to normal senescing hepatocytes and the excess hepatocytes in an involuting (down-sizing), lead nitrate-induced hyperplastic liver which, as we have seen in Chapter 2, are killed by apoptosis. In these hepatocytes there is a burst of Ca^{2+}-dependent transglutaminase activity that encases the cells in a cross-linked protein shell very much like the keratinous squame's cornified envelope. The other of the twin pillars of apoptosis is the cleavage of chromatin into a characteristic "ladder", as opposed to a continous smear, of oligo- and mononucleosomes probably by a Ca^{2+}/Mg^{2+}-dependent endonuclease. And the cells of the stratum corneum do indeed have the cleaved chromatin characteristic of apoptosis, as has been shown by Gavrieli et al.[120] using a procedure in which the free 3′-OH ends of the cleaved chromatin were labeled with biotinylated polydeoxyribonucleotide transferase and stained with biotin-binding avidin-conjugated peroxidase.

II. SUMMARY

At the base of each epidermal EPU is a slowly or intermittently cycling, niche-bound, self-renewing stem cell which sticks tightly to the basal lamina and to the EPU's macrophase-like Langerhans cell. Signals from the stem cell's basal lamina-activated integrin receptors and from its receptors for various autocrine/juxtacrine/paracrine factors produced by itself, the Langerhans cells, and the underlying dermal fibroblasts keep the cell cycle genes, the cyclin-dependent PKs of the cell cycle engine, and the genes coding for keratins 5 and 14 active. The stem cell generates amplifying transit cells, each of which has a much more limited proliferative future than the indefinitely self-renewing stem cell, but cycles more rapidly than the stem cell in this very low-Ca^{2+} basal zone. As these amplifier precursor cells proliferate and move away from the central niche of the EPU, the number and affinity of their integrin receptors drop and they stick increasingly less firmly to the basal lamina. Eventually, their stickiness falls below a critical level and they cannot advance beyond the late, anchorage-dependent, Ca^{2+}-dependent "START" stage of the G_1 build-up. So they switch off their cell-cycle genes, stop making the anti-apoptosis Bcl-2 protein, and stop cycling. But they keep growing and are pushed up into the suprabasal zone by less mature cells. The now-suprabasal cells switch off the genes coding for keratins 5 and 14 and shift into the spinous mode with its expression of keratins 1 and 10, redistribution of cell-cell adhesion-promoting cadherins and differentiation-specific integrins, formation of desmosomes and assembly of tonofilaments from accumulating keratins 1 and 10, and an increasing load of PKC isoforms.

The spinous cell starts making PTHrP fragments and their different receptors which, in the presence of a higher external Ca^{2+} level, trigger a Ca^{2+} surge and burst of PKCs activity. This signal stimulates the production of TGF-β_2 and $1\alpha,(OH)_2D_3$ which switch off gap "junctions" the keratin 1 and 10 genes, and PTHrP production and stimulate the genes of the granular program coding for things such as involucrin, loricrin, profilaggrin, and TGase 1, that the cell will need to make a cornified envelope.

The granular cell eventually reaches the top of the transepidermal Ca^{2+} gradient. Its Ca^{2+} receptors have been progressively occupied during its upward journey from the very low-Ca^{2+} basal region, and are fully occupied and finally switch into the active configuration and fire. The activated Ca^{2+} receptors trigger a large Ca^{2+} surge that activates Ca^{2+}-dependent enzymes such as Ca^{2+}-ATPase pumps, nucleases, proteases, lipases, and TGase 1 which build the cornified envelopes, and gut and kill the cell, and finally pump the accumulated Ca^{2+} out of the cells to maintain the skin Ca^{2+} gradient and prevent a steady massive leakage of Ca^{2+} from the body.

III. CANCEROUS KERATINOCYTES

When normal keratinocytes lift off the basal lamina because of declining affinity and numbers of certain matrix-activated integrin receptors, they switch off cell cycle genes such as c-*myc* and the genes for the cyclin-dependent protein kinases and their cyclins, but they continue growing and do not switch off their safety-checking apoptogenic genes. The progeny of carcinogen-initiated stem cells are different. They overproduce the proliferogenic autocrine/paracrine TGF-α and overexpress EGF/TGF-α receptors because of an amplified receptor gene or an inability to downregulate the receptors.[121–123] They also overproduce the $\alpha_6\beta_4$ integrin receptor even after lift-off, which maintains the late G_1 signaling that normally requires matrix adherence and is needed for the G_1/S transition.[59,63,124] Cultured initiated cells are less responsive to Ca^{2+}, and when confronted by a high Ca^{2+} concentration they reduce, but do not stop, proliferating and express only uncoordinated parts of the differentiation program such as TGase 1(K) and involucrin, but not PVA (pemphigus vulgaris antigen), the desmosomal cadherin that is normally expressed only in differentiating suprabasal cells.[72,125–129] They also become insensitive to the cell cycle-stopping action of the Ca^{2+}- and PKCs-induced TGF-β because: they cannot activate the inactive secreted form of the factor, or they can no longer display TGF-β receptors, or they cannot respond to the TGF-β receptor signal, maybe because they no longer make active Rb protein, which seems to mediate suppression of the c-*myc* cell cycle gene.[72,104,105]

Once the carcinogenic process has been initiated in a niche-anchored basal stem cell, the descendants of that cell may be selectively expanded into a visible papilloma by exposure to a PKCs-activating tumor promoter such as TPA.[72,127] One way by which TPA does this is to override the suppression of the hyperproliferation of initiated keratinocytes by surrounding normal cells.[130] Another way originates from the radically different response of initiated keratinocytes to TPA. TPA stimulates keratinocytes to make both TGF-α and TGF-β.[72,74,75,81] TGF-α would, of course, keep the cells cycling, but TGF-β induces the uninitiated cells to stop cycling and differentiate. In other words, there is a transient burst of proliferation and then a shutdown in TPA-treated normal keratinocyte populations. By contrast, initiated keratinocytes are strongly and persistently stimulated to proliferate by the promoter because they too are induced to make more TGF-α and TGF-β, but they cannot respond to the dampening action of TGF-β. Support for such a scheme is the observation of Furstenberger et al.[78] that simultaneous intradermal injection of TGF-α and TGF-β into NMRI mice mimicked the first of a two-stage tumor promotion protocol. Thus, TPA-promoted, initiated cells, having lost both their need for anchorage to the basal lamina and their responsiveness to Ca^{2+}, proliferate in the suprabasal layers while their uninitiated neighbors, restrained by their reponsiveness to TGF-β, stop proliferating, differentiate, and die.[127]

Neoplastic keratinocytes overproduce and secrete the PKCs-stimulating, hence TPA-like, PTHrP.[90,131] When this protein spills into the blood it mobilizes bone Ca^{2+} to produce the humoral hypercalcemia of malignancy.[1,131] PTHrP, like TPA, might therefore be an autocrine proliferation stimulator for initiated, but not normal cells. But this may not be the case, because PTHrP production is still coupled to differentiation in squamous carcinomas.[90] Thus, PTHrP is not expressed by basal cell carcinoma cells, which do not differentiate, but is expressed in squamous carcinomas, which contain a substantial fraction of differentiated cells.[90] However, the hypersecreted PTHrP, especially its secreted C-terminal breakdown products with TRSAW motif, might enhance tumor growth by inhibiting the proliferation and stimulating the differentiation of neighboring normal cells without affecting the proliferation of more advanced tumor cell clones.

As the tumor promoters selectively stimulate the expansion of the transforming cell populations, they also trigger a shower of active oxygen radicals via the PKCs-stimulated cell membrane-associated NAD(P)H oxidase, which generates clastogenic reactive oxygen intermediates and Ca^{2+}-dependent endonuclease activity that break chromosomes and cause unequal chromosome exchange with resulting gene deletions and rearrangements. At first the apoptogenic p53 safety-check mechanism responds to severe chromosome damage and kills its cell in order to stop it from generating variant, more advanced, and ultimately malignant clones. But eventually, some cells may suffer a loss of function mutation and/or deletion of their p53 gene[132] and with it their safety-check mechanism. It would be interesting to know whether these initiated cells also retain the basal stem cells' ability to make the anti-apoptosis Bcl-2 protein, which would protect them from apoptosis induced by reactive oxygen intermediate byproducts of signals from autocrine mitogens and tumor promoters.[133] This clears the way for the survival and proliferation of mutant cells. This removes an important barrier against the appearance of a malignant clone, the cells of which cannot respond to Ca^{2+} signals to differentiate and die (diffpoptose), which produce autocrine mitogens and angiogenic factors that enable them to proliferate away from the basal lamina, and make the hydrolases, autocrine motility factor angiogenesis factor(s), and the set of surface adhesion molecules that enable them to punch through the basal lamina and preferentially metastasize to certain distant organs.

REFERENCES

1. **Whitfield, J.F.,** *Calcium, Cell Cycles, and Cancer*, 1st ed., CRC Press, Boca Raton, FL, 1990.
2. **Watt, F.M., Jordan, P.W., and O'Neill, C.H.,** Cell shape controls terminal differentiation of human epidermal keratinocytes, *Proc. Natl. Acad. Sci. U.S.A.*, 85, 5576, 1988.
3. **Parkinson, E.K.,** Epidermal keratinocyte stem cells: their maintenance and regulation, *Semin. Cell Biol.*, 3, 435, 1992.
4. **Potten, C.S.,** Cell replacement in epidermis (keratinopoiesis) via discrete units of proliferation, *Int. Rev. Cytol.*, 69, 271, 1981.
5. **Budtz, P.E.,** Epidermal homeostasis: a new model that includes apoptosis, in *Apoptosis II: The Molecular Basis of Apoptosis in Disease*, Tomei, L.D. and Cope, F.O., Eds., Cold Spring Harbor Laboratory Press, Cold Spring Harbor, NY, 1994, 165.
6. **Deluca, M., Tamura, R.N., Kajiji, S., Bondanza, S., Rossino, P., Cancedda, R., Marchisio, P.C., and Quaranta, V.,** Polarized integrin mediates human keratinocyte adhesion to basal lamina, *Proc. Natl. Acad. Sci. U.S.A.*, 87, 6888, 1990.
7. **Ruoslahti, E.,** Integrins, *J. Clin. Invest.*, 87, 1, 1991.
8. **Jones, P.H. and Watt, F.M.,** Separation of human epidermal stem cells from transit amplifying cells on the basis of differences in integrin function and expression, *Cell*, 73, 713, 1993.

8a. **Bissonnette, R.P., Shi, Y., Mahaboudi, A., Glynn, J.M., and Green, D.R.,** c-myc and apoptosis, in *Apoptosis II, The Molecular Basis of Apoptosis in Disease*, Tomei, L.D. and Cope, F.O., Eds., Cold Spring Harbor Laboratory Press, Cold Spring Harbor, NY, 1994, 327.

8b. **Hockenbery, D.M., Zutter, M., Hickey, W., Nahm, M., and Korsmeyer, S.J.,** BCL2 protein is topographically restricted in tissues characterized by apoptotic cell death, *Proc. Natl. Acad. Sci. U.S.A.,* 88, 6961, 1991.

8c. **Giancotti, F.G. and Mainiero, F.,** Integrin-mediated adhesion and signaling in tumorigenesis, *Biochim. Biophys. Acta,* 1198, 47, 1994.

8d. **Cooper, G.M.,** *Oncogenes,* Jones and Bartlett Publishers, Boston, 1990.

9. **Damsky, C. and Webb, Z.,** Signal transduction by integrin receptors for extracellular matrix: cooperative processing of extracellular information, *Curr. Opinion Cell Biol.,* 4, 772, 1992.

9a. **Juliano, R.,** Signal transduction by integrins and its role in the regulation of tumor growth, *Cancer Metastasis Rev.,* 13, 25, 1994.

9b. **Lo, S.H. and Chen, L.B.,** Focal adhesion as a signal transduction organelle, *Cancer Metastasis Rev.,* 13, 9, 1994.

10. **Jones, P.L., Schmidhauser, C., and Bissell, M.J.,** Regulation of gene expression and cell function by extracellular matrix, *Crit. Rev. Eukaryotic Gene Expression,* 3, 137, 1993.

11. **Murray, A. and Hunt, T.,** *The Cell Cycle,* W. H. Freeman, New York, 1993.

12. **Brachmann, R., Lindquist, P.B., Wagashima, M., Khor, W., Lipari, T., Napier, M., and Derynck, R.,** Transmembrane TGF-α precursors activate EGF/TGF-α receptors, *Cell,* 56, 691, 1989.

13. **Wong, S.T., Winchell, L.F., McCune, B.K., Earp, H.S., Teixido, J., Massague, J., Herman, B., and Lee, D.C.,** The TGF-α precursor expressed on the cell surface binds to the EGF receptor on adjacent cells, leading to signal transduction, *Cell,* 56, 495, 1989.

14. **Fujimoto, T.,** Calcium pump of the plasma membrane is localized in caveolae, *J. Cell Biol.,* 120, 1147, 1993.

15. **Fujimoto, T., Nakade, S., Miyawaki, A., Mikoshiba, K., and Ogawa, K.,** Localization of inositol-1,4,5-tris phosphate receptor-like protein in plasmalemmal caveolae, *J. Cell Biol.,* 119, 1507, 1992.

16. **Kruszewski, F.H., Hennings, H., Yuspa, S.H., and Tucker, R.W.,** Regulation of intracellular free calcium in normal murine keratinocytes, *Am. J. Physiol.,* 261, 6767, 1991.

17. **Li, L., Kruszewski, F.H., Punnonen, K., Tucker, R.W., Yuspa, S.H., and Hennings, H.,** Strontium induces murine keratinocyte differentiation in vitro in the presence of serum and calcium, *J. Cell. Physiol.,* 154, 643, 1993.

17a. **Brown, E.M.,** Extracellular calcium sensing, regulation of parathyroid cell function, and role of calcium and other ions as extracellular (first) messengers, *Physiol. Rev.,* 71, 371, 1991.

18. **Reiss, M., Lipsey, L.R., and Zhou, Z.-L.,** Extracellular calcium-dependent regulation of transmembrane calcium fluxes in murine keratinocytes, *J. Cell. Physiol.,* 147, 281, 1991.

19. **Brown, E.M., Gamba, G., Riccardi, D., Lombardi, M., Butters, R., Kifor, O., Sun, A., Hedlger, M.A., Lytton, J., and Hebert, S.C.,** Cloning and characterization of an extracellular Ca^{2+}-sensing receptor from bovine parathyroid, *Nature,* 366, 575, 1993.

20. **Schwartz, M.A.,** Signalling by integrins: implications for tumorigenesis, *Cancer Res.,* 53, 1503, 1993.

21. **Maura, T.M., Isseroff, R.R., Lasarow, R., and Pappone, P.A.,** Ion channels are linked to differentiation in keratinocytes, *J. Membr. Biol.,* 132, 201, 1993.

22. **Filvaroff, E., Calautti, E., McCormick, F., and Dotto, P.,** Specific changes of RAS GTPase-activating protein GAP and a GAP-associated p62 protein during calcium-induced keratinocyte differentiation, *Mol. Cell. Biol.,* 12, 5319, 1992.
23. **Kruszewski, F.H., Hennings, H., Tucker, R.W., and Yuspa, S.H.,** Differences in the regulation of intracellular calcium in normal and neoplastic keratinocytes are not caused by ras gene mutations, *Cancer Res.,* 51, 4206, 1991.
24. **Lee, E. and Yuspa, S.H.,** Changes in inositol phosphate metabolism are associated with terminal differentiation and neoplasia in mouse keratinocytes, *Carcinogenesis,* 12, 1651, 1991.
25. **Pacini, L., Spinedi, A., Piacentini, M., Limatolo, C., Farrace, M.G., Autuori, F., and Luly, P.,** Calcium-dependence of arachidonic-acid redistribution among phospholipids of cultured mouse keratinocytes, *Biochim. Biophys. Acta,* 1045, 213, 1990.
26. **Tang, W. and Ziboh, V.A.,** Agonist inositol triphosphate-induced release of calcium from murine keratinocytes, a possible link with keratinocyte differentiation, *J. Invest. Dermatol.,* 96, 134, 1991.
27. **Dlugosz, A.A., Mischak, H., Mushinski, J.F., and Yuspa, S.H.,** Transcripts encoding protein kinase C-alpha, delta, epsilon, zeta, and eta are expressed in basal differentiating mouse keratinocytes in vitro and exhibit quantitative changes in neoplastic cells, *Mol. Carcinogenesis,* 5, 286, 1992.
28. **Bollag, W.B., Ducote, J., and Harmon, C.S.,** Effects of the selective protein kinase C inhibitor R031–7549 on the proliferation of cultured mouse epidermal keratinocytes, *J. Invest. Dermatol.,* 100, 240, 1993.
29. **Dlugosz, A.A. and Yuspa, S.H.,** Staurosporine induces protein kinase C agonist effects and maturation of normal and neoplastic mouse keratinocytes in-vitro, *Cancer Res.,* 51, 4677, 1991.
30. **Tong, P.S., Sperry, P.J., Vom Dorp, C., Khalifa, B.A., and Wheeler, L.A.,** Inhibition and down-regulation of protein kinase C activity in epidermal keratinocytes caused increase in DNA synthesis, *Clin. Res.,* 40, 520A, 1992.
31. **Moscat, J., Flemming, T.P., Molloy, C.J., Lopez-Barahona, M., and Aaronson, S.A.,** The calcium signal for Balb/MK keratinocyte terminal differentiation induces sustained alterations in phosphoinositide metabolism without detectable protein kinase C activation, *J. Biol. Chem.,* 264, 1128, 1989.

31a. **Chakravarthy, B.R., Isaacs, R.J., Morley, P., Durkin, J.R., and Whitfield, J.F.,** Stimulation of protein kinase C during Ca^{2+}-induced keratinocyte differentiation, *J. Biol. Chem.,* 270, in press, 1995.

32. **Chakravarthy, B.R., Bussey, A., Whitfield, J.F., Sikorska, M., Williams, R.E., and Durkin, J.P.,** The direct measurement of protein kinase C (PKC) activity in isolated membranes using a selective peptide substrate, *Anal. Biochem.,* 196, 144, 1991.
33. **Nairn, A.C. and Aderem, A.,** Calmodulin and protein kinase C cross-talk: the MARCKS protein is an actin filament and plasma membrane cross-linking protein regulated by protein kinase C phosphorylation and by calmodulin, in *Interactions Among Cell Signalling Systems, Ciba Found. Symp.,* 164, 145, 1992.

34. **Filvaroff, E., Stern, D.F., and Dotto, G.P.,** Tyrosine phosphorylation is an early and specific event involved in primary keratinocyte differentiation, *Mol. Cell. Biol.,* 10, 1164, 1990.
35. **Zhao, Y., Hanafusa, H., Sudol, M., and Krueger, J.,** Altered tyrosine kinase activity of c-Src and c-Yes during calcium-induced keratinocyte differentiation, *J. Cell. Biochem. Suppl.,* 17A, 260, 1993.
36. **Zhao, Y., Sudol, M., Hanafusa, H., and Krueger, J.,** Increased tyrosine kinase activity of c-Src during calcium-induced keratinocyte differentiation, *Proc. Natl. Acad. Sci. U.S.A.,* 89, 8298, 1992.
36a. **Medema, R.H. and Bos, J.L.,** The role of p21 ras in receptor tyrosine kinase signaling, *Crit. Rev. Oncogenesis,* 4, 615, 1993.
37. **Gherzi, R., Sparatore, B., Patrore, M., Sciutto, A., and Briata, P.,** Protein kinase C mRNA levels and activity in reconstituted normal human epidermis relationships to cell differentiation, *Biochem. Biophys. Res. Commun.,* 184, 283, 1992.
38. **Matsui, M.S., Chew, S.L., and Deleo, V.A.,** Protein kinase C in normal human epidermal keratinocytes during proliferation and calcium-induced differentiation, *J. Invest. Dermatol.,* 99, 565, 1992.
39. **Takahashi, H. and Iizuka, H.,** Analysis of the 5′-upstream promoter region of human involucrin gene activation by 12-*O*-tetradecanoylphorbol-13-acetate, *J. Invest. Dermatol.,* 100, 10, 1993.
40. **Hohl, D., Lichti, U., Breitkreutz, D., Steinert, P.M., and Roop, D.R.,** Transcription of the human loricrin gene in-vitro is induced by calcium and cell density and suppressed by retinoic acid, *J. Invest. Dermatol.,* 96, 414, 1991.
41. **Liew, F.-M. and Yamanishi, K.,** Regulation of transglutaminase 1 gene expression by 12-*O*-tetradecanoylphorbol-13-acetate, dexamethasome, and retinoic acid in cultured human keratinocytes, *Exp. Cell Res.,* 202, 310, 1992.
42. **Kim, L.T., Ishihara, S., Lee, C.-C., Akiyama, S.K., Yamada, K.M., and Grinnell, F.,** Altered glycosylation and cell surface expression of beta-1 integrin receptors during keratinocyte activation, *J. Cell Sci.,* 103, 743, 1992.
43. **Jensen, P.J. and Wheelock, M.J.,** Regulation of urokinase plasminogen activator localization in keratinocytes by calcium ion and E-cadherin, *Exp. Cell Res.,* 202, 190, 1992.
44. **Yoneda, K., Imamura, T.F.S., and Ogawa, K.,** Distribution of todrin in the keratinocyte in-vivo and in-vitro, *J. Invest. Dermatol.,* 94, 724, 1990.
45. **Deluca, M., Pellegrin, G., Bondanza, S., Cremona, O., Savoia, P., Cancedda, R., and Marchisio, P.C.,** The control of polarized integrin topography and the organization of adhesion-related cytoskeleton in normal human keratinocytes depend upon number of passages in culture and ionic environment, *Exp. Cell Res.,* 202, 142, 1992.
46. **Fujisawa, H., Muraki, R., and Otsuka, F.,** Effects of interferon-gamma on the formation of cell-cell contact between human epidermal carcinoma cells in low calcium culture, *Acta Dermatol.(Kyoto),* 87, 553, 1992.
47. **Sheu, H.-M., Kitajima, Y., and Yaoita, H.,** Involvement of protein kinase C in translocation of desmoplakins from cytosol to plasma membrane during desmosome formation in human squamous cell carcinoma cells grown in low to normal calcium concentration, *Exp. Cell Res.,* 185, 176, 1989.

48. **Pillai, S., Bikle, D.D., Mancianti, M.-L., Cline, P., and Hincenbergs, M.,** Calcium regulation of growth and differentiation of normal human keratinocytes: modulation of differentiation competence by stages of growth and extracellular calcium, *J. Cell. Physiol.*, 143, 294, 1990.
49. **Grundin, T.G., Roomans, G.M., Forslind, B., Lindberg, M., and Werner, Y.,** X-ray microanalysis of psoriatic skin, *J. Invest. Dermatol.*, 85, 378, 1985.
50. **Menon, G.K., Elias, P.M., and Feingold, K.R.,** Integrity of the permeability barrier is crucial for maintenance of the epidermal calcium gradient, *Br. J. Dermatol.*, 130, 139, 1994
50a. **Menon, G.K. and Elias, P.M.,** Ultrastructural localization of calcium in psoriatic and normal human epidermis, *Arch. Dermatol.*, 127, 57, 1991.
51. **Menon, G.K., Grayson, S., and Elias, P.M.,** Ionic calcium reservoirs in mammalian epidermis: ultrastructural localization by ion-capture cytochemistry, *J. Invest. Dermatol.*, 84, 508, 1985.
52. **Malmqvist, K.G., Carlsson, L.-E., Forslind, B., Roomans, G.M., and Askelsson, K.R.,** Proton and electron microprobe analysis of human skin, *Nucl. Inst. Methods Phys. Res.*, B3, 611, 1984.
53. **Yuspa, S.H., Kilkenny, A., Cheng, C., Roop, D., Hennings, H., Kruszewski, F., Lee, E., Strickland, J., and Greenhalgh, D.A.,** Alterations in epidermal biochemistry as a consequence of stage-specific genetic changes in skin carcinogenesis, *Environ. Health Perspect.*, 93, 3, 1991.
54. **Yuspa, S.H., Kilkenny, A.E., Steinert, P.M., and Roop, D.R.,** Expression of murine epidermal differentiation maskers is tightly regulated by restricted extracellular calcium concentrations in vitro, *J. Cell Biol.*, 109, 1207, 1989.
54a. **Rasmussen, C.D. and Means, A.R.,** Calmodulin is required for cell cycle progression during G1 and mitosis, *EMBO J.*, 8, 73, 1989.
55. **Ben-Ze'ev, A.,** Cytoarchitecture and signal transduction, *Crit. Rev. Eukaryotic Gene Expression*, 2, 265, 1992.
56. **Baxton, R.S. and Magee, A.I.,** Structure and interactions of desmosomal and other cadherins, *Semin. Cell Biol.*, 3, 197, 1992.
57. **McNeill, H., Ryan, T.A., Lewis, R.S., and Nelson, W.J.,** Immediate and early events after the induction of cell-cell contact in polarized epithelial cells, *J. Cell Biol.*, 115, 70a, 1991.
58. **Cook, P.W., Pittelkow, M.R., and Shipley, G.D.,** Growth factor-independent proliferation of normal human neonatal keratinocytes: production of autocrine- and paracrine-acting mitogenic factors, *J. Cell. Physiol.*, 146, 277, 1991.
59. **Guadagno, T.M. and Assoian, R.K.,** G1/S control of anchorage-independent growth in the fibroblast cell cycle, *J. Cell Biol.*, 115, 1419, 1991.
60. **Watt, F.M., Jordan, P.W., and O'Neil, C.H.,** Cell shape controls terminal differentiation of human epidermal keratinocytes, *Proc. Natl. Acad. Sci. U.S.A.*, 85, 5576, 1988.
61. **Adams, J.C. and Watt, F.M.,** Fibronectin inhibits the terminal differentiation of human keratinocytes, *Nature*, 340, 307, 1989.
62. **Adams, J.C. and Watt, F.M.,** Changes in keratinocyte adhesion during terminal differentiation: reduction in fibronectin binding precedes $\alpha 5$ $\beta 1$ integrin loss from the cell surface, *Cell*, 63, 425, 1990.
63. **Yuspa, S.H.,** The pathogenesis of squamous cell cancer: lessons learned from studies of skin carcinogenesis, *Proc. Am. Assoc. Cancer Res.*, 34, 560, 1993.

64. **Schwartz, M.A.,** Transmembrane signalling by integrins, *Trends Cell Biol.*, 2, 304, 1992.
65. **Symington, B.E.,** Fibronectin receptor modulates cyclin-dependent kinase activity, *J. Biol. Chem.*, 267, 25744, 1992.
66. **Guadagno, T.M., Ohtsubo, M., Roberts, J.M., and Assoian, R.K.,** A link between cyclin A expression and adhesion-dependent cell cycle progression, *Science*, 262, 1372, 1993.
67. **Wheelock, M.J. and Jensen, P.J.,** Regulation of keratinocyte intercellular junction organization and epidermal morphogenesis by E-cadherin, *J. Cell Biol.*, 117, 415, 1992.
68. **Lin, C.Q. and Bissel, M.J.,** Multi-faceted regulation of cell differentiation by extracellular matrix, *FASEB J.*, 7, 737, 1993.
69. **Dlugosz, A.A. and Yuspa, S.H.,** Coordinate changes in gene expression which mask the spinous to granular cell transition in epidermis are regulated by protein kinase C, *J. Cell Biol.*, 120, 217, 1993.
70. **Younus, J. and Gilchrest, B.A.,** Modulation of mRNA levels during keratinocyte differentiation, *J. Cell. Physiol.*, 152, 232, 1992.
71. **Tong, P.S., Sperry, J., Vom Dorp, C., Khalifa, B.A., and Wheeler, L.A.,** Inhibition and down-regulation of protein kinase C activity in epidermal keratinocytes caused increase in DNA synthesis, *Clin. Res.*, 40, 520A, 1992.
72. **DiGiovanni, J.,** Multistage carcinogenesis in mouse skin, *Pharm. Ther.*, 54, 63, 1992.
73. **Imamoto, A., Beltran, L., and DiGiovanni, J.,** Differential mechanism for the inhibition of epidermal growth factor binding to its receptor in mouse keratinocytes by anthrones and phorbol esters, *Carcinogenesis*, 11, 1543, 1990.
74. **Nishizuka, Y.,** Studies and prospectives of protein kinase C family for cellular regulation, *Cancer*, 63, 1892, 1989.
75. **Akhurst, R.J., Lee, F., and Balmain, A.,** Localized production of TGFβ mRNA in tumor promoter stimulated mouse epidermis, *Nature*, 331, 363, 1988.
76. **Krieg, P., Schnapke, R., Fürstenberger, G., Vogt, I., and Marks, F.,** TGF-β1 and skin carcinogenesis: antiproliferative effect *in vitro* and TGF-β1 mRNA expression during epidermal hyperproliferation and multistage tumorigenesis, *Mol. Carcinogenesis*, 4, 129, 1991.
77. **Coffey, R.J., Snipes, N.J., Bascom, C.C., Graves-Deal, R., Pennington, C.Y., Weissman, B.E., and Moses, H.L.,** Growth modulation of keratinocytes by transforming growth factors, *Cancer Res.*, 48, 1396, 1988.
78. **Furstenberger, G., Rogers, M., Schnapke, R., Bauer, G., Hofler, P., and Marks, F.,** Stimulatory role of transforming growth factor in multistage carcinogenesis: possible explanation for the tumor-inducing effect of wounding in initiated NMRI mouse skin, *Int. J. Cancer*, 43, 915, 1989.
79. **Mansbridge, J.N. and Hanawalt, P.C.,** Role of transforming growth factor-beta in the maturation of human epidermal keratinocytes, *J. Invest. Dermatol.*, 90, 336, 1988.
80. **Partridge, M., Green, M.R., Landon, J.D., and Feldmann, M.,** Production of TGF-α and TGF-β by cultured keratinocytes, skin, and oral squamous carcinomas: potential autocrine regulation of normal and malignant epithelial cell proliferation, *Br. J. Cancer*, 60, 542, 1989.

81. **Shipley, G.D., Pittelkow, M.R., Wille, J.J., Scott, R.E., and Moses, H.L.,** Reversible inhibition of normal human prokeratinocyte proliferation by type beta transforming growth factor — growth inhibition in serum-free medium, *Cancer Res.,* 46, 2068, 1986.
82. **Pittelkow, M.R., Lindquist, P.B., Abraham, R.T., Braves-Deal, R., Derynck, R., and Coffey, R.J.,** Induction of transforming growth factor-α expression in human keratinocytes by phorbol esters, *J. Biol. Chem.,* 264, 5164, 1989.
82a. **Whitfield, J.F.,** Calcium signals in cell proliferation, differentiation, and death, in *Ion Channels and Ion Pumps,* Foa, P. and Walsh, M.F., Eds., Springer-Verlag, New York, 1994, 19.
83. **Gibbs, S., Lohman, F., Teubel, W., van de Putte, P., and Backendonf, C.,** Characterization of the human *spr2* promoter: induction after UV irradiation or TPA treatment and regulation during differentiation of cultured primary keratinocytes, *Nucleic Acids Res.,* 18, 4401, 1990.
84. **Kartasova, T., van Muijen, G.N.P., van Pelt-Heerschap, H., and van de Putte, P.,** Novel protein in human epidermal keratinocytes: regulation of expression during differentiation, *Mol. Cell. Biol.,* 8, 2204, 1988.
85. **Kartasova, T. and van de Putte, P.,** Isolation, characterization, and UV-stimulated expression of two families of genes encoding polypeptides of related structure in human epidermal keratinocytes, *Mol. Cell. Biol.,* 8, 2195, 1988.
86. **Fornace, A.J.,** Mammalian genes induced by radiation: activation of genes associated with growth control, *Annu. Rev. Genet.,* 26, 507, 1992.
87. **Greif, H., Ben-Chaim, J., Shimon, T., Bechor, E., Eldar, H., and Livneh, E.,** The protein kinase C-related PKC-L(η) gene product is localized in the cell nucleus, *Mol. Cell. Biol.,* 12, 1304, 1992.
88. **Watt, F.M., Boukamp, P., and Hornung, J.,** Effect of growth environment on spatial expression of involucrin by human epidermal keratinocytes, *Arch. Dermatol. Res.,* 279, 335, 1987.
89. **Danks, J.A., Ebeling, P.R., Hayman, J., Chou, S.T., Moseley, J.M., Dunlop, J., Kemp, B.E., and Martin, T.J.,** Parathyroid hormone-related protein: immunohistochemical localization in cancers and normal skin, *J. Bone Miner. Res.,* 4, 273, 1989.
90. **Hayman, J.A., Danks, J.A., Ebeling, P.R., Moseley, J.M., Kemp, B.E., and Martin, T.J.,** Expression of parathyroid hormone related-protein in normal skin and in tumors of skin and skin appendages, *J. Pathol.,* 158, 293, 1989.
91. **Martin, T.J., Moseley, J.M., and Gillespie, M.T.,** Parathyroid hormone-related protein: biochemistry and molecular biology, *Crit. Rev. Biochem. Mol. Biol.,* 1991, 26, 377, 1991.
92. **Atillasoy, E.J., Burtis, W.J., and Milstone, L.M.,** Immunohistochemical colocalization of parathyroid hormone-related protein (PTHrP) in normal human skin, *J. Invest. Dermatol.,* 96, 277, 1991.
93. **Holick, M.F., Nussbaum, S., and Parsons, K.S.,** PTH-like humoral hypercalcemia factor of malignancy may be an epidermal differentiation factor, *J. Bone Miner. Res.,* 3 (Suppl.), S214, 1988.
94. **Milstone, L., Fairley, J., Insogna, K., Stewart, A., and Reuveni, H.,** Parathyroid hormone-like peptides increase intracellular calcium and differentiation of keratinocytes, *J. Cell Biol.,* 107, 73a, 1988.

95. **Orloff, J.J., Ganz, M.B., Ribaudo, A.E., Burtis, W.J., Reiss, M., Milston, L.M., and Stewart, A.F.,** Analysis of PTHRP binding and signal transduction mechanisms in benign and malignant squamous cells, *Am. J. Physiol.*, 262, E599, 1992.
96. **Kaiser, S.M., Laneuville, P., Bernier, S.M., Rhim, J.S., Kremer, R., and Goltzman, D.,** Enhanced growth of a human keratinocyte cell line induced by antisense RNA for parathyroid hormone-related peptide, *J. Biol. Chem.*, 267, 13623, 1992.
97. **Whitfield, J.F., Chakravarthy, B.R., Durkin, J.P., Isaacs, R.J., Jouishomme, H., Sikorska, M., Williams, R.E., and Rixon, R.H.,** Parathyroid hormone stimulates protein kinase C but not adenylate cyclase in mouse epidermal keratinocytes, *J. Cell. Physiol.*, 150, 299, 1992.
98. **Gagnon, L., Jouishomme, H., Whitfield, J.F., Durkin, J.P., MacLean, S., Neugebauer, W., Willick, G., Rixon, R.H., and Chakravarthy, B.,** Protein kinase C-activating domains of parathyroid hormone-related protein, *J. Bone Miner. Res.*, 8, 497, 1993.
99. **Jouishomme, H., Whitfield, J.F., Chakravarthy, B., Durkin, J.P., Gagnon, L., Isaacs, R.J., MacLean, S., Neugebauer, W., Willick, G., and Rixon, R.H.,** The protein kinase-C activation domain of the parathyroid hormone, *Endocrinology,* 130, 53, 1992.
100. **Jouishomme, H., Whitfield, J.F., Gagnon, L., Chakravarthy, B., Durkin, J., and Rixon, R.H.,** Biphasic stimulation of membrane-associated protein kinase C activity in rat osteosarcoma cells by parathyroid hormone-related protein, in *Frontiers in Osteosarcoma Research,* Novak, J. F., Ed., Hogrefe and Huber Publishers, Toronto, 1993, 303.
100a. **Orloff, J.J., Reddy, D., de Papp, A.E., Yang, K.H., Soifer, N.E., and Stewart, A.F.,** Parathyroid hormone-related protein as a prohormone: post translational processing and receptor interactions, *Endocr. Rev.,* 15, 40, 1994.
101. **Regalia, E., Whitfield, J.F., Barisoni, D., and Armato, U.,** Parathyroid hormone-related protein (PTHrP) and the control of the growth and differentiation of normal human keratinocytes, *Proc. 41st Int. Cong. European Tissue Culture Society*, October 9–12, 1994, Verona, Italy, Abstr. Book W3-p8, page 80.
102. **Werkmeister, J.R., Merryman, J.I., McCauley, L.K., Horton, J.E., Capen, C.C., and Rosal, T.J.,** Parathyroid hormone-related protein production by normal human keratinocytes *in vitro, Exp. Cell Res.,* 208, 68, 1993.
103. **Sollberg, S., Rynänen, J., Olsen, D.R., and Uitto, J.,** Transforming growth factor-beta up-regulates the expression of the genes for beta 4 integrin and bullous pemphigoid antigens (BPAG1 and BPAG2) in normal and transformed keratinocytes, *J. Invest. Dermatol.,* 99, 409, 1992.
104. **Pietenpol, J.A., Holt, J.T., Stein, R.W., and Moses, H.L.,** Transforming growth factor β1 suppression of c-*myc* gene transcription: role in inhibition of keratinocyte proliferation, *Proc. Natl. Acad. Sci. U.S.A.,* 87, 3758, 1990.
105. **Pietenpol, J., Stein, R.W., Moran, E., Yaciuk, P., Schlegel, R., Lyons, R.M., Pittelkow, M.R., Munger, K., Howley, P.M., and Moses, H.L.,** TGF-β1 inhibition of c-*myc* transcription and growth in keratinocytes is abrogated by viral transforming proteins with pRB binding domains, *Cell*, 61, 777, 1990.

106. **Merryman, J.I., Rosal, T.J., DeWille, J.W., Farmer, S.J., Werkmeister, J.R., and Capen, C.C.,** Stimulation of parathyroid hormone-related protein production by transforming growth factor-β1 in a squamous carcinoma cell line, *J. Bone Miner. Res.,* 7 (Suppl.), S243, 1992.

106a. **Whitfield, J.F.,** Calcium and cancer, *Crit. Rev. Oncogenesis*, 3, 55, 1992.

107. **Bikle, D.D. and Pillai, S.,** Vitamin D metabolite production and function in keratinocytes, *Ann. N.Y. Acad. Sci.,* 548, 27, 1988.

108. **Janulis, M., Wong, M.-S., and Favus, M.J.,** Structure-function requirements of parathyroid hormone for stimulation of $1,25(OH)_2D_3$ production by rat renal proximal tubules, *Endocrinology*, 133, 713, 1993.

109. **Bittiner, B., Bleehen, S.S., and MacNeil, S.,** 1α,25 Dihydroxyvitamin D_3 increases intracellular calcium in human keratinocytes, *Br. J. Dermatol.*, 124, 230, 1991.

110. **Kim, H.-J., Abdelkadem, N., Katz, M., and McLane, J.A.,** 1,25-Dihydroxy-vitamin-D_3 enhances antiproliferative effect and transcription of TGF-β1 on human keratinocytes in culture, *J. Cell. Physiol.,* 151, 579, 1992.

111. **MacLaughlin, J.A., Cantley, L.C., and Holick, M.F.,** 1,25 Dihydroxy vitamin D-3 increases calcium and phosphatidylinositol metabolism in differentiating cultured human keratinocytes, *J. Nutr. Biochem.,* 1, 81, 1990.

112. **Pillai, S. and Bikle, D.D.,** Role of intracellular free calcium in the cornified envelope formation of keratinocytes, differences in the mode of action of extracellular calcium and 1,25 dihydroxy vitamin D-3, *J. Cell. Physiol.,* 146, 94, 1991.

113. **Dotto, G.P., El-Fouly, M.H., Nelson, C., and Trosko, J.E.,** Similar and synergistic inhibition of gap-junctional communication by *ras* transformation and tumor promoter treatment of mouse primary keratinocytes, *Oncogene*, 4, 637, 1989.

114. **Presland, R.B., Hayclock, P.V., Fleckman, P., Nirunsuksiri, W., and Dale, B.A.,** Characterization of the human epidermal profilaggrin gene, *J. Biol. Chem.*, 267, 23772, 1992.

115. **Eckert, R.L., Yaffe, M.B., Crish, J.F., Murthy, S., Rorke, E.A., and Welter, J.F.,** Involucrin-structure and role in envelope assembly, *J. Invest. Dermatol.,* 100, 613, 1993.

116. **Reichert, U., Michel, S., and Schmidt, R.,** The cornified envelope: a key structure of terminally differentiating keratinocytes, in *Molecular Biology of the Skin: The Keratinocyte*, Darmon, M. and Blumenberg, M., Eds., CRC Press, Boca Raton, FL, 1993, 107.

117. **Ando, Y., Imamura, S., Murachi, T. et al.,** Calpain activates two transglutaminases from porcine skin, *Arch. Dermatol. Res.,* 280, 380, 1988.

118. **Epstein, E.H., Langston, A.W., and Leung, J.,** Sulfation reactions in the epidermis, *Ann. N.Y. Acad. Sci.*, 548, 97, 1988.

119. **Croall, D.E. and DeMartino, G.N.,** Calcium-activated neutral protease (calpain) system: structure, function, and regulation, *Physiol. Rev.*, 71, 813, 1991.

120. **Gavrieli, Y., Sherman, Y., and Ben-Sasson, S.A.,** Identification of programmed cell death in situ by specific labeling of nuclear DNA fragmentation, *J. Cell Biol.*, 119, 403, 1992.

121. **Game, S.M., Huelsen, A., Patel, V., Donnelly, M., Yeudall, W.A., and Stone, A.,** Progressive abrogation of TGF-beta 1 and EGF growth control is associated with tumour progression in ras-transfected human keratinocytes, *Int. J. Cancer*, 52, 461, 1992.

122. **Glick, A.B., Sporn, M.B., and Yuspa, S.H.,** Altered regulation of TGF-beta 1 and TGF-alpha in primary keratinocytes and papillomas expressing v-Has-ras, *Mol. Carcinogenesis,* 4, 210, 1991.
123. **Reiss, M., Stash, E.B., Vellucci, V.F., and Zhou, Z.L.,** Activation of the autocrine transforming growth factor alpha pathway in human squamous carcinoma cells, *Cancer Res.*, 51, 6254, 1991.
124. **Carey, T.E., Laurikainen, L., Nair, T.S., Reinke, T.S., Coling, D.E., Wolf, G.T., Van Waes, C., Liebert, M., and Marcelo, C.,** Regulation of expression and phosphorylation of A9/α6β4 integrin in normal and neoplastic keratinocytes, *Monogr. Natl. Cancer Inst.,* 13, 75, 1992.
125. **Kruszewski, F.H., Hennings, H., Tucker, R.W., and Yuspa, S.H.,** Differences in the regulation of intracellular calcium in normal and neoplastic keratinocytes are not caused by *ras* gene mutations, *Cancer Res.,* 51, 4206, 1991.
126. **Yuspa, S.H. and Morgan, D.L.,** Mouse skin cells resistant to terminal differentiation associated with the initiation of carcinogenesis, *Nature*, 293, 72, 1981.
127. **Yuspa, S.H. and Poirer, M.C.,** Chemical carcinogenesis from animal models to molecular models in one decade, *Adv. Cancer Res.,* 50, 25, 1988.
128. **Rubin, H.L., Parenteau, N.L., and Rice, R.H.,** Coordination of keratinocyte programming in human SCC-13 squamous carcinoma and normal epidermal cells, *J. Cell. Physiol.*, 138, 208, 1989.
129. **Buston, R.S. and Magee, A.I.,** Structure and interaction of desmosomal and other cadherins, *Semin. Cell Biol.,* 3, 157, 1992.
130. **Hennings, H., Lowry, D.T., Robinson, V.A., Morgan, D.L., Fujiki, H., and Yuspa, S.H.,** Activity of diverse tumor promoters in a keratinocyte co-culture model of initiated epidermis, *Carcinogenesis,* 13, 2145, 1992.
131. **Halloran, B.P. and Nissensin, R.A., Eds.,** *Parathyroid Hormone-Related Protein: Normal Physiology and its Role in Cancer,* CRC Press, Boca Raton, FL, 1992.
132. **Campbell, C., Quinn, A.G., Ro, Y.-S., Angus, B., and Rees, J.L.,** p53 mutations are common and early events that precede tumor invasion in squamous cell neoplasia, *J. Invest. Dermatol.,* 100, 746, 1993.
133. **Korsmeyer, S.J., Shutter, J.R., Veis, D.J., Merry, D.E., and Oltvai, Z.N.,** Bcl-2/Bax: a rheostat that regulates an anti-oxidant pathway and cell death, *Semin. Cancer Biol.*, 4, 327, 1993.

5 Calcium as Differentiator and Killer — Colon Cells

I. NORMAL COLON CELLS

The colon epithelium is another example of a cell system in which Ca^{2+} exercises its full range of cycle-promoting, differentiation-triggering, and ultimately killing functions.[1,2]

There are three kinds of differentiated cell in the adult colon epithelium. These are absorptive cells, enteroendocrine cells, and mucus-producing goblet cells.[1–7] These mature, functioning, nonproliferating cells live for only a couple of days before being sloughed off into the fecal stream, and must continuously be replaced by the progeny of a small clone (one to four cells in the mouse) of pluripotent stem cells anchored and thus confined to a specific place or niche at the bottom (cell positions 1 and 2) of each one of millions of monoclonal, flask-shaped or tubular structures known as crypts (Figure 1). The proliferative potential of the pluripotent stem cell is maintained, and the playing of the differentiation cassettes suppressed, by the signals from certain cell surface integrin receptors activated by binding specific ligands in the extracellular matrix of the niche. It is also maintained by signals from receptors for various autocrine, endocrine, juxtacrine, and paracrine factors from blood, neighboring cells, and nerve terminals. Like all cycling cells, these cells express the c-*myc* gene, whose product can be either proliferogenic or apoptogenic depending on the presence of "survival" signalers.[8] Indeed, they protect themselves by producing the Bcl-2 anti-apoptosis protein.[8a,9]

Proliferation of the stem cells is sensitive to the number of cells in their region.[7] Stem cells can detect the death and loss of just a few cells in their region. This loss-triggered signal to the slowly cycling stem cells stimulates them to cycle faster to replace the lost cells.

At the end of its cycle, the stem cell divides into another pluripotent stem cell which stays stuck in the niche and a sister cell which is still proliferatively active, but is no longer indefinitely self-renewable because it must leave the stem cell niche. In the mouse, this cell is the first of about eight transit generations that could result in about 250 differentiated cells. It expresses a set

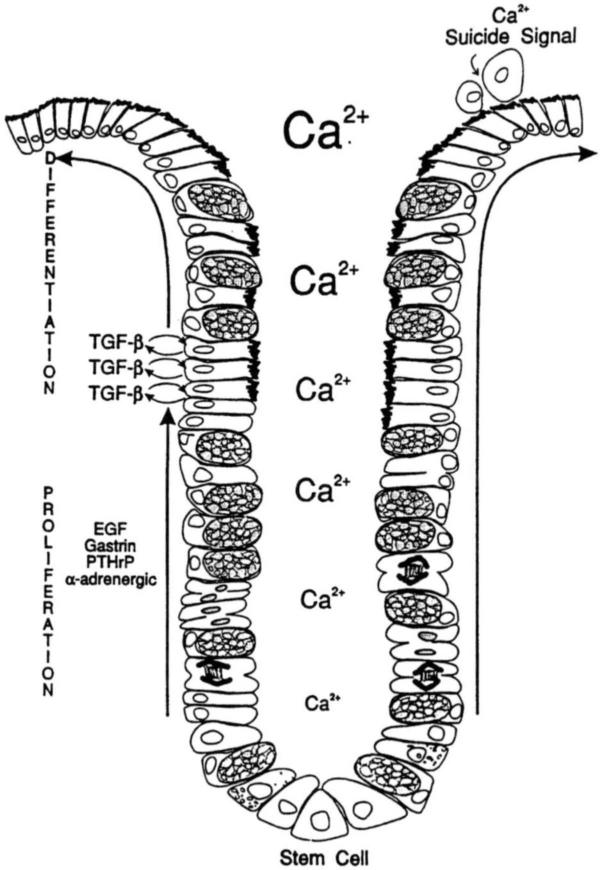

FIGURE 1 A possible model for the control of proliferation, differentiation, and death of colon epithelial cells. The absorptive cells, enteroendocrine cells, and mucus-producing goblet cells are all the monoclonal progeny of the single, self-renewing, pluripotent stem cell anchored to its specific niche at the base of the crypt and expressing cell cycle genes such as c-*myc* and protected from apoptosis by survival factors such as the Bcl-2 protein. Because the responses of colon cell and keratinocyte proliferation and differentiation to Ca^{2+} are very similar, this model includes a Ca^{2+} gradient like the epidermal gradient, with the lowest Ca^{2+} levels being in the proliferative zone in the lower half of the crypt and highest levels being in the differentiation and dying diffpoptosis zones in the upper crypt and the outer flat mucosa facing the Ca^{2+}-rich fecal stream. Such a gradient could be established by the stratification of Ca^{2+}-binding mucins produced by appropriately positioned goblet cells. As the cycling stem cell progeny leave the bottom niche and glide upward along the basal lamina, they reach a point in the Ca^{2+} gradient where their Ca^{2+} sensors fire to signal them to stop cycling and produce the cycle-stopping autocrine/paracrine TGF-βs and their receptors, maybe one or more PTHrP fragments, and a new set of differentiation-specific integrin receptors and other cell-cell adhesion molecules and structures. The cells also stop making the anti-apoptosis Bcl-2 protein and start the apoptogenic mechanisms when they reach the mouth of the crypt and move out onto the flat mucosal surface. From Whitfield.[78]

of integrins specific for the moderate adherence to the extracellular matrix and the generation of signals needed to migrate and maintain the undifferentiated phenotype and a functioning cell cycle engine.[9a] This precursor cell and its progeny, expressing such proliferation-related genes as c-*myc* and c-*myb* and protected by Bcl-2, glide along the basal lamina and up the crypt wall in step with fibroblast-myocyte pericryptal cells of the underlying lamina propria.[1,2] The proliferation of these precursor cells is driven primarily by signals from growth factor autocrine loops and the growth-specific integrins.[1,2,9a,10] This core mechanism is modulated by the ebb and flow of other factors, such as epidermal growth factor (EGF) and gastrin from the upper gastrointestinal tract and noradrenergic agonists from nerve nets enveloping the lower crypts.[1,2] These factors establish a proliferative circadian rhythm and provide input from the diet and other external stimuli.

The productivity of the stem cell-precursor cell populations is immense. Each crypt in the mouse proximal (ascending) colon contains about 100 proliferating cells which generate about 6 cells per hour while in the distal colon there are about 200 proliferating cells per crypt which generate about 21 cells every hour.[3,6]

Limited though it is, the available information indicates that Ca^{2+} signals drive the cycles of colon crypt cells just as they drive the cycles of activated fibroblasts, hepatocytes, basal skin keratinocytes, and other normal cells. Freshly isolated normal human colon cells need about 0.1 m*M* external Ca^{2+} to proliferate optimally while their proliferation, and the proliferation of rat colon cells, is stopped by the high (0.8 to 2.0 m*M*) Ca^{2+} concentrations that can be found in the fecal water.[11–14] Exposure to 2 m*M* Ca^{2+} also inhibits growth factor (EGF, gastrin)- and deoxycholic acid-induced ornithine decarboxylase activity and DNA synthesis in rat colon mucosa cultures.[15] Thus, colon cells probably have Ca^{2+} receptors on their surfaces that function like those displayed on the surfaces of keratinocytes and parathyroid cells,[15a] which can give a reprogramming signal that stops proliferation when the external Ca^{2+} concentration exceeds a critical threshold.

The sensitivity of these cells to external Ca^{2+} is probably modulated by several factors, including $1\alpha,25(OH)_2$ vitamin D_3 which binds to a cell surface receptor and, within a minute, triggers a surge of Ca^{2+} from internal stores, a more prolonged influx of external Ca^{2+}, phosphoinositide breakdown, and membrane-associated protein kinase Cs (PKCs) activity.[16–18] Autocrine/paracrine PTHrP fragments may also have a role as they probably do in keratinocytes, because small intestine cells express PTHrP mRNA, and picomolar concentrations of PTHrP can stimulate Ca^{2+} transport by chicken small intestine cells.[18,20]

The question immediately arises as to whether there is a differentiation-determining Ca^{2+} gradient along the length of the crypt, which is like the epidermis's differentiation-determining gradient described in the previous chapter. To my knowledge no one has looked. However, something must shield the cycling cells in the lower crypts from the apparently millimolar Ca^{2+}

concentrations in the fecal stream which otherwise would sooner or later seep down into the lowest reaches of the crypt and stop cell proliferation. The likely candidates for the anti-Ca^{2+} barrier are the Ca^{2+}-binding chrondroitin sulfates, heparan sulfates, and particularly the sialo- and sulfomucins secreted by the goblet cells concentrated in the lower regions of the crypt.[1,2] A decreasing content of Ca^{2+}-binding mucins and increasing Ca^{2+} concentration from the base to the mouth of the crypt would establish an external Ca^{2+} gradient like that in the epidermis (Figure 1).[1,2]

The proliferation of precursor cells in the colon also requires cAMP signals and a predominance of the proliferation-promoting PKAI isozymes over the differentiation-promoting PKAII isozymes.[21] It appears that the proliferation-promoting EGF receptor signals triggered by EGF or autocrine/paracrine transforming growth factor (TGF)-α increase the expression of PKAIs regulatory RI subunit gene and reduce the expression of the PKAII regulatory subunit gene.[21,22] Autocrine/paracrine PTHrP may be an autocrine/paracrine stimulator of the PKAI isozymes because cells of a colon line express PTH/PTHrP receptors which are coupled to adenylyl cyclase.[20]

At a certain point the upwardly gliding, cycling cell emerges from behind the anti-Ca^{2+} barrier, collides with a superthreshold level of Ca^{2+}, and starts getting different signals from a changing population of integrin receptors for components of the basal lamina, and from cell-cell adhesion receptor molecules.[1,2] This would be analogous to the cycle-stopping lifting of skin keratinocytes off the basal lamina. Signals from the cell surface Ca^{2+} sensor stop the cell from starting a new cycle in the upper crypt by stimulating it to express a TGF-β isoform and its receptor when it reaches the upper half of the crypt.[23,24] This would establish an autocrine loop which changes the expression of the cell's Ca^{2+}-dependent, Ca^{2+}-signaling integrins and other receptors for extracellular components, down regulates TGF-α production, and shifts the balance of the PKAs from the proliferation-promoting PKAIs to the differentiation-promoting PKAIIs.[1,21,22] The differentiating cell shuts down its proliferation-related c-*myc* and c-*myb* genes.[25] A perhaps unexpected part of the differentiation program is a large, sustained increase in the expression of the normal H-Ras and K-Ras (but not N-Ras) proteins that surge only transiently during the growth-division cycles of various cells. Thus, there are large (three- to fivefold) surges in the expression of the H-*ras* gene and smaller increases in the K-*ras* gene expression when Na-butyrate stimulates the differentiation of the 13G and 16E variants of the HT-29 human adenocarcinoma cells into *nontumorigenic*, mucin-secreting goblet cells.[26] A role for Ras proteins in a mature colon cell function is indicated by the fact that colon crypt cells do not contain detectable (with 142–24 EOS antibody that recognizes the highly conserved 96–118 amino acid sequence of H-, K-, and N-Ras proteins) Ras proteins, whereas the mature absorptive and goblet cells on the flat mucosal surface of the colon contain high levels of the proteins.[27] Among the probable functions of Ras proteins in the mature colon cells is to stimulate PKCs and the secretion of things such as enzymes and mucins,[28,29] and, as we learned in Chapter 2, promote protein synthesis. Ras proteins and the proteins that specifically modulate their GDP/GTP exchange (e.g., p21 Ras GDS) and GTPase (p21 Ras GAP) activities work with other small G proteins (e.g., 5 mgp21, 5 mgp25A) and their specific activity-modulating proteins to drive the transport of product-loaded secretory vesicles from the *trans* cisternae of the Golgi apparatus to target sites on the cell membrane, the fusion of the vesicles with the cell membrane, and the ultimate expulsion of their contents from the cell by exocytosis.[29]

To quote Pienta et al.,[30] *"What a cell touches determines what a cell does"*. The rising colon epithelial cell starts binding to the extracellular matrix and its neighbors differently. The Ca^{2+} and TGF-β signals have stopped the cells proliferating and expressing the migration-growth-specific α2 integrins displayed by the proliferating cells lining the base of the crypts, and at the same time they have promoted the expression of differentiation-specific β1 integrins, type E-cadherins, and I-CAMs that couple the cell to the extracellular matrix and its neighbors and are necessary for glandular differentiation.[9a,31–35] These receptors are activated when the cell binds to extracellular matrix components (e.g., collagen I and laminin) and they send signals to the nucleus that stop the expression of the genes encoding cell cycle engine components and trigger the coordinate expression of genes with matrix-specific regulatory elements whose products determine specific cell shape and functions.[34–36] The composition of the extracellular matrix is also determined by products of the fibroblast-myocytes of the pericryptal lamina propria which respond to paracrine factors such as TGF-β from the epithelial cells.[34–36]

An example of how signals from cell surface components may dampen proliferation and promote differentiation is provided by neural cell-adhesion molecules (N-CAMs). Thus, one N-CAM, the placental LAR receptor, appears to have two protein-tyrosine phosphatase domains in its cytoplasmic portion.[37] When expressed on the cell surface and activated by cell-cell adhesion, such receptors could stop cell cycling and promote differentiation by dephosphorylating, and consequently deactivating, autophosphorylated active mitogenic velcroceptor protein tyrosine kinases (YPKs) and their substrates.

Some cells become absorptive enterocytes with apical brush borders facing the colon lumen and bristling with hydrolytic enzymes such as disaccharidases, peptidases, and Na-dependent glucose cotransporters that collect water and nutrients from the passing fecal stream. Their basolateral borders contain α_2- and β-adrenergic receptors, VIP receptors, and outwardly directed Na-independent GluT2 transporters to release the apically absorbed glucose into the blood.[1,2] Other cells become goblet cells which tend to stay in the crypt[4,7] to make, and then use, the Ras-driven machinery to secrete the mucins of a putative Ca^{2+} shield.

Upon emerging from the crypt, the now-mature, functioning cell first becomes part of the hexagonal cuff of cells (the colon's truncated version of the small intestine's villi) that surrounds the mouth of the crypt, and then part of the colon's flat mucosal surface.[5,11] As long as the cell is not neoplastically transformed it will not, indeed it cannot, proliferate because of its adhesion-receptors-induced reprogramming, the functioning TGF-β receptor autocrine loop, and the high Ca^{2+} content of the fecal stream.[1,2] Eventually the cell is sloughed off the basal lamina at an extrusion zone and swept away in the passing fecal stream. Until now, it has not been known whether it is a still healthy, functioning colon epithial cell that is shoved protesting off the basal lamina by younger crypt cells pushing up from below and by cells streaming out of adjacent crypts, or whether terminally differentiating colon cells, like terminally differentiating keratinocytes, kill themselves with a Ca^{2+} surge and apoptosis before being shoved into the fecal stream. It now appears that colon cells are indeed programmed to kill themselves by apoptosis at the appropriate time (Figure 1). In fact, the cells prepare themselves for apoptosis (diffpoptosis) by stopping the production of the anti-apoptosis Bcl-2 protein when they reach the upper half of the crypts.[8a] It seems that suicide does not happen continuously throughout the colon mucosa. Instead, it occurs in waves spreading over the flat colon mucosa.[38] As the cells in a cluster of crypts reach the mouths of the crypts, they trigger the killer Ca^{2+} surge and apoptosis.[38] At first the chromatin fragmentation (as indicated by a procedure consisting of tagging the free 3′-OH ends of the cleaved DNA with biotinylated polydU polymer using terminal deoxyribonucleotide transferase and then staining the labeled ends with avidin-conjugated peroxidase) in the suiciding apoptosing cells is concentrated around the nuclear periphery, but spreads into the nuclear center as the cell moves out onto the flat mucosal surface.[38,39] This chromatin chopping is one of the two mighty pillars of apoptosis. The other pillar, the formation of a tough envelope by a Ca^{2+}-stimulated transglutaminase, is indicated by the observation of Iyengar et al.[40] that human feces contain a substantial number of hardened, *squame-like* colon epithial cells which, as should be expected, still exclude trypan blue.

II. COLON CANCER

Lurking in the fecal stream are carcinogens and tumor promoters such as secondary bile acids, as well as PKCs-stimulating diacylglycerols and free fatty acids from the hydrolysis of dietary fats.[1,2,41] According to Dr. H. Schneider in this laboratory, the secondary bile acids, such as lithocholic acid, also enhance carcinogen-induced mutagenesis in colon cells by potently inhibiting glutathione-*S*-transferase. It would seem that the factor limiting the appearance of colon cancers is the fecal load of the carcinogen enhancers and tumor promoters.

From the start, we must understand that the mutations that initiate the prolonged, multistep carcinogenic process must occur in the self-renewing, niche-bound stem cells in crypt cell positions 1 and 2 but not in the cycling, upwardly migrating transit cells, simply because these amplifying transit cells live for only a few days.[4,7] Cancer is common in the colon, but rare in the small intestine. Why? Potten[7] has suggested an answer to this question which implicates the Ca^{2+}-dependent apoptotic suicide mechanism. Mutagen-damaged colon and small intestine cells may respond to this damage by killing themselves (with their p53-operated chromosome break-monitoring, safety-check mechanism). It happens that the most apoptosis-prone (most altruistic?) cells in the mouse small intestine crypts are around cell position 4, which is right in the middle of the stem cell zone, whereas the most apoptosis-prone colon cells in the mouse colon are in the mid-crypt region (e.g., cell position 11 to 12) far away from the stem cells in crypt cell positions 1 and 2. Therefore, in the small intestine, mutagen-damaged stem cells kill themselves with their apoptogenic $Ca^{2+}/Ca^{2+}\cdot$CaM-activated hydrolytic enzymes and transglutaminases while mutagen-damaged colon stem cells survive to generate dangerous, self-renewing stem cells in which the carcinogenic clonal selection process has been initiated and are primed to respond to tumor promoters.

According to the conventional view, human colorectal cancer cell clones appear in hyperplastic adenomatous polyps.[5,42] However, this "polyps-only" concept may be wrong.[42] Shimoda et al.[43] have reported that 78% of 951 early lesions and advanced colorectal cancers had arisen *de novo* from the flat mucosa rather than from preexisting polyps. Nonpolypoid colorectal carcinomas appear to originate in hyperplastic, distended, flask-shaped (instead of the normal test tube-shaped), branching, bifurcated aberrant crypts.[42,44] Polyps are rare in rodents,[42,45] and their colorectal cancers probably arise *de novo* in similar aberrant crypts.[46,47]

The first precancerous changes are the distortion of crypt morphology, increase in proliferation activity, and extension of the proliferative zone into the upper crypt and even out onto the flat mucosa.[1,2,5,41,42] Another early change that might be associated with altered proliferative activity is a breakdown in the putative mucin Ca^{2+} shield, which is reflected in a change from brown-black sulfomucin staining with high iron diamine-alcian blue to blue sialomucin staining.[42]

The carcinogen-initiated colon cells increase their proliferative activity by becoming more responsive to various combinations of growth factors and by starting to produce, or overproduce, autocrine proliferogens such as platelet-derived growth factor (PDGF) subunits, insulin-like growth factor (IGF-1) and IGF-1 variants, as well as the prime proliferogen, TGF-α.[1] They may also increase the proliferogenic impact of their autocrine TGF-α and exogenous EGF by increasing their display of EGF/TGF-α receptors.[1,48] The importance of a TGF-α autocrine/paracrine loop for the proliferation of at least some colon carcinoma cells is indicated by the ability of a 23-base antisense oligonucleotide that recognizes TGF-α mRNA to reduce TGF-α secretion and proliferation of LIM1215 colon carcinoma cells.[49] Moreover, at some point mutant cells appear which are insensitive to TGF-β and thus cannot downregulate TGF-α production, a key part of the differentiation program.[1] The cells might also start producing or perhaps overproducing PTHrP, which is known to stimulate adenylyl cyclase and proliferation of at least some transformed colon cells, such as human LoVo colon carcinoma cells which therefore must have PTH/PTHrP receptors on their surfaces.[20] Indeed, activation of these receptors might be the reason why primary hyperparathyroidism is associated with colon tumors.[50] The upshot is a persistent expression or overexpression of cell cycle genes such as c-*myc* and the ubiquitous B-*myb*.

It has been suggested that the tumor-promoting proliferative hyperactivity might be a compensatory response to the heightened killing or damaging of mucosal cells by secondary bile acids such as lithocholic acid or free fatty acids.[1,41] The crypt cells would be responding to positive signals, or to a reduced flow of inhibitory signals, coming down to them from the surface along some intercellular communication pathway(s), such as the gap junctions.[1] Reducing the *soluble* secondary bile acid and free fatty acid contents of the fecal stream by increasing the dietary Ca^{2+} load (Ca^{2+} binds avidly to these acids and renders them insoluble) does seem to reduce crypt cell proliferation and the development of colon cancer.[1] Indeed, dietary Ca^{2+} loading, together with certain site-specific cAMP analogs to be discussed below, is believed by some to be a primary tool for the prevention and "differentiation-therapy" of colon cancer.[1,21,22]

The increasingly self-driven, initiated colon cell clones, like initiated keratinocytes, respond only partly or not at all to differentiation signals and thus tend to continue proliferating when they reach the Ca^{2+}-rich upper levels of the crypt or the mucosal surface. This is the crucial change in cellular behavior that opens the way to the formation of polyps, adenomas, and, finally, carcinomas.[1] To take this step, the cells must lose, or become much less responsive to or even unresponsive to, signals from the receptors and mechanism that normally sense an increase in the external Ca^{2+} concentration above a critical threshold and cause normal cells to stop cycling and start differentiating.[7,15a] A clue to how this early change might come about in most (but, please note, not all) polyp-derived colorectal tumors is the finding that 60 to 63% of sporadic tumors, ranging from the smallest analyzable adenomas to the largest malignant carcinomas, come from cells that have suffered loss or mutations of a gene, APC, in the 21 region of the lonq (or "q") arm of chromosome 5 (Figures 2 and 3).[51–59] APC is the gene for a condition known as adenomatous polyposis coli (APC) or familial adenomatous polyposis (FAP), in which the colon becomes progressively carpeted with thousands of polyps, a few of which eventually spawn cancers.[51–59] The deletions or intragenic mutations result in the loss, truncation, or inactivation of the APC protein.[54–56,59] Proof of the key role of the APC gene in the control of intestinal cell proliferation

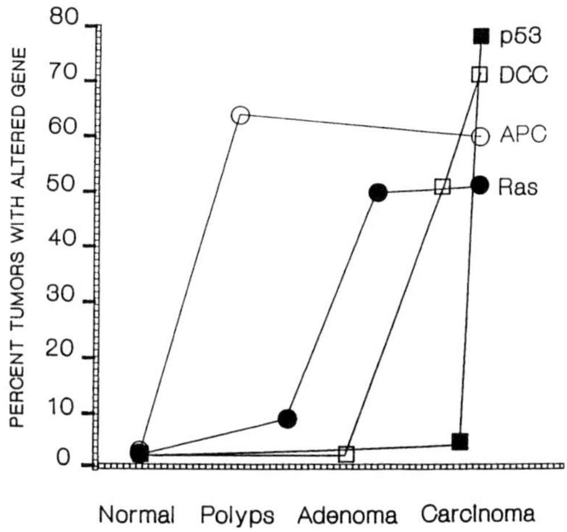

FIGURE 2 Some of the sequential gene mutations and losses that occur at the various stages of colon carcinogenesis. First comes the loss, or loss-of-function mutation, of the APC (adenomatous polyposis coli) gene on chromosome 5q21. The loss of the APC gene product, maybe a surface Ca^{2+} sensor, results in the extension of the cryptal proliferative zone and the consequent piling up of cells into polyps. Then often comes a *ras* gene mutation which results in the appearance of a mutant activated Ras protein which bypasses Ca^{2+} controls, stimulates proliferation, inhibits differentiation, prevents apoptosis, and thus promotes the transition from polyp to adenoma. The late loss of the genes for the surface L-CAM-like DCC cell-adhesion molecule and p53(TP53) appear to be important for the emergence and expansion of carcinoma clones from adenomas. The loss of DCC is symptomatic of the general breakdown of intercellular linkages and communications that are required for mobilization and metastasis. The loss of the important ability of p53 to force a cell to repair chromosome damage or die enables the emergence and expansion of highly malignant mutant clones.

and differentiation is the fact that a germline nonsense mutation of the murine homolog of the human APC gene, mAPC, in a male C57BL/6J mouse has given rise to the *Min* (*M*ultiple *I*ntestinal *N*eoplasia) strain of mouse in which there are multiple tumors extending from duodenum to distal colon.[60–62] These tumors are the products of an enormously increased self-renewal capacity of pluripotent stem cells whose progeny retain, though far less frequently express, the ability to differente along regionally appropriate pathways.[63] Further support for loss or mutation of the APC gene being a frequent early event in colorectal carcinogenesis is the observation of Goyette et al.[63] that inserting a normal human chromosome 5 into human SW480 colorectal adenocarcinoma cells by fusing them with chromosome 5-bearing, colchicine/cytochalasin-generated "microcells" suppressed the tumorigenicity of the cancer cells in athymic nude mice. Tanaka et al.[64] found that microcell-mediated introduction of a normal chromosome 5 into tumorigenic spindle-shaped, anchorage-independent (able to form colonies in soft agar) COKFu human colon carcinoma cells eliminated the cells' tumorigenicity and anchorage-independence, and gave them a normal, epithelial cell shape.

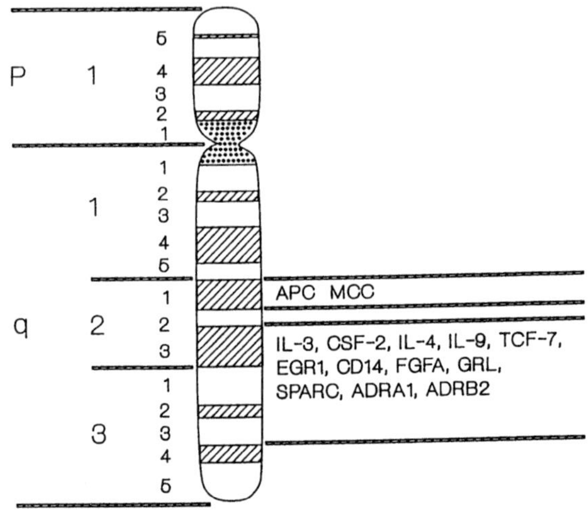

FIGURE 3 The colon polyposis APC and MCC (*M*issing in *C*olon *C*ancer) genes are located in the long arm of human chromosome 5 near to, and maybe in, a cluster of genes coding for various growth factors and receptors.[75] Among these are (in order of appearance from upper left to lower right or centromere to telomere) the genes for interleukin-3 (IL-3), colony-stimulating factor-2 (CSF-2 or granulocyte-macrophage-stimulating factor), interleukin-4 (IL-4), T-cell-specific transcription factor (TCF-7), early growth response factor-1 (EGR1), CD14 antigen (CD14), endothelial growth factor (FGFA), glucocorticoid receptor (GRL), osteonectin (SPARC), α_1- and β-$_2$adrenergic receptors (ADRA1, ADRB2).

So what might the APC protein do and what has Ca^{2+} got to do with this? The currently popular idea is that APC is a tumor suppressor and the familial polyposis results from the loss or inactivation of this protein.[51–59] But how could this be, because familial polyposis is an autosomal dominant disorder and the colorectal cells of a heterozygous familial polyposis patient presumably still have one normal chromosome 5 with a functioning suppressor gene? Maybe one APC gene cannot by itself make enough APC protein to suppress precursor cell proliferation. Or perhaps cycling heterozygous colorectal cells have a high probability of losing their remaining normal APC gene.[57] How might the APC protein suppress tumorigenesis? A clue to the answer is the striking loss of responsiveness to Ca^{2+} that happens during the progressive selection of colon cell clones from near-normality to adenomas and ultimately carcinomas. Thus, by the adenomatous stage, the cells still need about 0.1 m*M* Ca^{2+} to proliferate optimally *in vitro*, but their proliferation-stopping, differentiation-triggering Ca^{2+} switch no longer operates or it operates only sporadically and incompletely.[65–69] For example, the supranormal proliferative activity of primary colon adenoma cells in the presence of 0.05 to 0.1 m*M* Ca^{2+} can be stimulated further by raising the external Ca^{2+} concentration to 2.2 m*M* which would stop the proliferation of normal cells.[67] By the carcinoma stage, the cells no longer need external Ca^{2+} to proliferate optimally.[67–69] But it has been claimed that the proliferation of the cells of some cultured colorectal carcinoma lines (e.g., the CACO-2 line) can still be inhibited by exposure to higher Ca^{2+} concentrations such as 1.8 m*M*.[70] However, this is not really true because the proliferative activity of these cancer cells in the presence of 1.8 m*M* Ca^{2+} was still a large, and very abnormal, 70% of the activity in the presence of 0.25 m*M* Ca^{2+}.[70] Thus, the APC protein might be a sensor involved in the Ca^{2+}-dependent triggering of terminal differentiation. It is a 2844-amino acid polypeptide that is linked to the cytoskeleton by β-catenin which also binds the homophilic cell-adhesion molecule, E-cadherin, to the cytoskeleton.[71,72] E-cadherins are, in fact, differentiation-promoting Ca^{2+}-sensor molecules on epithelial cell surfaces whose external N-terminal domains are linked by Ca^{2+} to the N-terminal domains of E-cadherins on adjacent cells.[72a] Perhaps APC is a similar, Ca^{2+}-sensing, cytoskeleton-coupled signaler. APC has a potential "coiled-coil" structure in its N-terminus, a repeated 20-amino acid sequence in its central region, and a stretch of basic amino acids in its C-terminal region.[71] APC is probably a receptor because a part of it (and the MCC [*M*issing in *C*olon *C*ancer] product of a gene contiguous with the APC gene on chromosome 5q21 [Figure 3]) is homologous to the G protein-binding m3 domain of the Ca^{2+}-mobilizing, PKCs-stimulating cholinergic muscarinic receptor[73] and another part is homologous to the yeast Ral-1 protein that controls Ras protein activity.[51,55,56] APC might be a Ca^{2+} sensor like the enterocyte receptor of Pazianas et al.,[15a] the keratinocyte's putative Ca^{2+} receptor or the G protein-coupled Ca^{2+} sensor on the surface of parathyroid chief cells discussed in the previous chapter. APC might be activated either directly by suprathreshold external Ca^{2+} levels or by Ca^{2+}-dependent binding to specific cell-adhesion molecules such as integrins in the upper crypt. The activated APC would trigger Ca^{2+} surges and a burst of PKCs activity which would stimulate TGF-β expression. This would shift the integrin repertoire and thus stop the cell from cycling and start the terminal differentiation (diffpoptosis) program. The truncation of APC's C-terminus, which is associated with cancer, would disconnect the protein from β-catenin and the cytoskeleton and eliminate or reduce the cell's responsiveness to Ca^{2+} or contact with differentiation zone-specific adhesion molecules. This could be the reason why colon cells from the early to the most advanced stages of carcinogenesis can proliferate at all levels of the crypt and even on the mucosal surface.

Whatever the deletion or loss-of-function mutation of the 5q21 gene(s) on the long arm of chromosome 5 does to upset the Ca^{2+} control of differentiation and proliferation, it should dysregulate the proliferation of other cells and they too should be hypersusceptible to transformation. This seems to be the case.[41] Fibroblasts from FAP or Gardner's syndrome (a related polyposis syndrome) have increased growth potential as indicated by above-normal plating efficiency and saturation density. They are more susceptible to transformation by *N*-methyl-*N′*-nitro-*N*-nitrosoguanidine (MNNG), adeno-SV40 hybrid virus, and SV40 virus. They have above-normal chromosomal instability and, unlike fibroblasts from age-matched normal subjects, they can be transformed by γ-radiation. They are more susceptible than normal fibroblasts to transformation by a mutant activated Ras protein, and, unlike normal fibroblasts, they can be transformed by TPA. The 5q21 loss-of-function mutations or deletion is also associated with tumors of the small intestine, pancreas, thyroid, and liver, as well as osteomas of the jaw, hypertrophy of the retinal epithelium, and proliferation of fibroblasts in the body wall to produce desmoid tumors.[41,57,74]

While the loss of the gene(s) in 5q21 has only limited ramifications, the loss of another region of this fascinating chromosome results in another kind of cancer. Chromosome 5 seems (at least so far) to be different from other chromosomes in having clusters of functionally related genes such as those coding for hematopoietic growth factors and receptors for α_1 and β_2 adrenergic agonists as well as our putative APC Ca^{2+} sensor/receptor (Figure 3).[75] Loss of sequences in the narrow 5q21 region caused by cytotoxic therapies (chemotherapeutic drugs, radiation) triggers iatrogenic myelodysplastic syndrome or acute myeloid leukemia.[75] Indeed, this is a model for the triggering of colon polyp formation by the loss of the 5q21 APC or MCC sequences. Some of the genes in the critical region are interleukin (IL)-3 (5q23–31), IL-4 (5q23–31), IL-5 (5q23–31), IL-9 (5q23–31), colony-stimulating factor (CSF)-2 or granulocyte-macrophage CSF, and CSF-1R (5q33).[75] Thus, the loss, or the reduced production, of a growth-differentiation factor or receptor results in the selective lethal expansion of a pool of cycling mutant precursor cells which cannot differentiate into proliferatively inactive functioning cells or commit suicide (apoptosis) in response to their differentiation failure and overproduction.

However, we must not lose sight of the 30 to 40% of colorectal cancers that may *not* have mutated APC genes (Figure 2). Mutation of this gene leads to the polyps of the polyp-to-cancer sequence, but mutation of some other gene(s) might be responsible for initiating those events leading to non-polyp-derived cancers.[42] One of these other genes may be the HNPCC (*H*ereditary-*N*on-*P*olyposis *C*olon *C*ancer) gene on human chromosome 2.[76] This gene encodes a "mismatch repair" enzyme that removes wrongly paired nucleotides and replaces them with the correct nucleotides.[76] Deletion or loss-of-function mutation of this gene impairs mismatch repair and thus increases the cellular mutation load and, like the loss of p53(TP53) function, amplifies the effectiveness of dietary carcinogens and tumor promoters.[76]

The *average* intracellular Ca^{2+} concentration of colon cancer cells shortly after isolation (130 ± 11 nmol/l in the presence of 0 m*M* external Ca^{2+} or 186 ± 17 nmol/l in the presence of 1 m*M* external Ca^{2+}) is slightly, but significantly, lower than the average concentration in cells freshly isolated from normal mucosa (151 ± 10 nmol/l in the presence of 0 m*M* external Ca^{2+} or 256 nmol/l in the presence of 1 m*M* external Ca^{2+}).[77] By contrast, cells from "normal" colon mucosa adjacent to colon carcinomas have 2.4 to 9.6 times higher concentrations of internal Ca^{2+}, suggesting a higher level of Ca^{2+} signaling.[77] Thus, a larger fraction of these cells may be induced to cycle by paracrine growth factors secreted into the local region by the cancer cells. Alternatively, it might reflect the hyperresponsiveness of the cells in a cancer-prone field to fecally borne PKCs-activating Ca^{2+}-mobilizing tumor promoters. Finally, it might be the result of Ca^{2+} loading due to cell injury by factors coming from the cancer or by fecal secondary bile acids or free fatty acids.[78] At any rate, once they have achieved the cancerous state, cells apparently no longer respond to these Ca^{2+}-elevating factors, whatever they might be.

A commonly observed premalignant change is a 40 to 50% drop in the cellular diacylglycerol content and the total PKCs activity without a change in the fraction of PKC isozymes (about 30%) normally associated with the cell membrane.[1,79,80] However, the membrane-associated fraction rises to 45% in colon cancer cells.[1] Since PKC inhibits proliferation of *normal* colon cells and is part of the differentiation signal,[1,81,82] the depletion of the total enzyme(s) pool could contribute to the resistance of neoplastic colon cells to the signals that stop proliferation and start differentiation.

PKCs stimulators such as long-chain diacylglycerols (especially those with $C_{18.1}$ oleic acid residues), TPA, and secondary bile acids either do not reduce or actually stimulate the proliferation of transforming mucosal cells starting with the early premalignant FAP cell.[81,82] (However, it must be noted that *overexpression* of the β1 isoform of protein kinase C *inhibits* the proliferation of some colon cancer cells.[79]) A clue to the reason for this dramatic conversion of membrane-associated PKC into a mitogen during carcinogenesis is provided by the observation that a burst of membrane-associated PKCs activity, induced by TPA or the abundant fecal diacyglyceride, diolein, to trigger proliferation and sustain the activity of a YPK which catalyzes the phosphorylation of Y residues of a 63-kDa protein in premalignant and malignant colon cells, but not in normal colon cells or in colon carcinoma cells, which are induced to differentiate by PKCs.[83] This protein-tyrosine kinase is probably $p60^{c\text{-}src}$, which is known to be activated early in colon carcinogenesis, and which can indeed be stimulated by TPA-activated PKCs C.[84–86] This YPK and its 63-kDa substrate may be parts of a PKCs-activated mitogenic mechanism which is suppressed in normal cells and in differentiating colon carcinoma cells,[83] possibly via TGF-βs which stop proliferation. Thus, TPA-stimulated PKCs might stop normal cells cycling and promote their terminal differentiation by stimulating the production and secretion of autocrine/paracrine TGF-βs which prevent the initiation of DNA replication by preventing the Rb suppressor protein from being inactivated by hyperphosphorylation by cyclin-dependent PKs near the end of the G_1 build-up. But, colon cells become increasingly less responsive to TGF-βs as they advance toward malignancy.[23] Thus, the increased production and secretion of autocrine/paracrine TGF-βs induced by PKCs-stimulating promoters such as TPA, diolein, and free secondary bile acids selectively promote the expansion of premalignant and malignant clones while suppressing the proliferation and triggering the terminal differentiation of normal cells.[23]

The role of hyperexpressed, hyperactive $p60^{c\text{-}src}$ in colon carcinogenesis is unknown. The enzyme could be separately involved in both promoting proliferation and differentiation. A clue to the mystery is that $p60^{c\text{-}src}$ is hyperexpressed in moderately to well-differentiated colon tumors, but not in more advanced, poorly differentiated colon cancers.[86] Since p60$^{c\text{-}src}$ is activated by hooking onto, among other things, activated EGF/TGF-α velcroceptors with its SH2 domain,[87] it is probably a player in proliferation of the transforming cells, but it may at the same time bolster a faltering differentiation mechanism. In the more advanced stages of carcinogenesis, p60$^{c\text{-}src}$ may no longer be important for proliferation and the reversion of its expression to a normal level may cripple the differentiation mechanism.

PKCs-activating bile acids or fecal diacylglycerols could promote the neoplastic transformation of colon cells in several different ways. Membrane-associated PKCs are key players in secretion.[1] Therefore, tumor promoters may stimulate the secretion of the autocrine mitogens that drive the transforming cells' proliferation. They also may stimulate the secretion of proteases, such as the cysteine proteinases cathepsin B and urokinase, which are all expressed by colon cells in the later stages of neoplastic transformation and which contribute to the selective proliferation of the cells by shutting off gap-junctional cross-communication with neighboring normal cells, and may actually destroy neighboring normal cells and less advanced premalignant cells.[1,41] As we shall see below, a common later step in colon carcinogenesis is the mutation of a H-*ras*, or, more often, a K-*ras*, gene[41] whose wild-type product normally functions in the proliferatively shut-down mature cells on the mucosal surface.[26,27] The mutant or activated Ras proteins can block differentiation of epithelial cells, are mitogenic, anti-apoptotic, and are further activated when phosphorylated by tumor promoter-induced, membrane-associated PKCs activity.[8,88,89] Perhaps one of the most important effects of the membrane-PKCs activity stimulated by tumor promoters is a surge of oxygen radicals generated by PKC-activated NAD(P)H oxidase in the plasma membrane.[89] These radicals generate clastogenic agents that operate through Ca^{2+}-dependent nucleases to promote the chromosome changes and gene deletions that mark the later stages of carcinogenesis.[89]

It appears that cyclic AMP and the PKAs can be important players in colon carcinogenesis. It seems that postzygotic mutations in exon 8 of the gene encoding the adenylyl cyclase-activating $G_s\alpha$ proteins can generate multiple adenomatous gastrointestinal polyps in patients with McCune-Albright syndrome, discussed in Chapter 3.[90] These so-called *gsp* mutations result in the production of $G_s\alpha$ proteins which are constitutively activated and thus maintain supranormal adenylyl cyclase activity because of the elimination of their autoinhibitory GTPase activity. An important factor in the failure of colon cancer cells to respond to differentiation signals and proliferate inappropriately may be the overexpression of the proliferation-promoting PKAIs and the suppression of the differentiation-promoting PKAIIs.[21,22,91] Indeed, Bradbury et al.[91] found that all of 50 colorectal cancer specimens had no detectable RII subunits of PKAIIs while "normal" cells of carcinoma-adjacent and distant mucosa had detectable amounts of both RII and RI subunits. The importance of this abnormal ratio of RI to RII subunit levels for colon cancer cell behavior is indicated by the fact that the proliferation of colon cancer cells is inhibited by 8-Cl-cAMP, a cAMP analog that selectively upregulates the expression of the PKAIIs and downregulates the expression of the PKAIs.[21,22] Furthermore, reducing RI subunit expression with antisense oligonucleotides targeted against the first 21 bases of the human RIα mRNA arrested the proliferation of LS-147T colon carcinoma cells without cytotoxicity, while the subcutaneous injection of the antisense deoxyribonucleotides twice per week reduced the growth of these LS-147T tumor cells by 50% in athymic nude mice.[92] Thus, restoring the normal balance of the expression of PKA isozymes with site-specific cyclic AMP analogs may someday prove to be an effective "differentiation therapy" for colon cancer.

Colon carcinogenesis is particularly amenable to differentiation therapy, because the single layer of colon flat mucosa cells, unlike other parts of the body, are directly bathed in an unfiltered or unprocessed watery medium that can be loaded with the preventative or therapeutic agent. Over the last 25 years a large amount of evidence has accumulated (see Kinsella[41] and Whitfield[78,93] for reasonably recent lists) and is still accumulating that colon carcinogenesis might be prevented by increasing the dietary Ca^{2+} load. One possible reason for this is that a sustained high Ca^{2+} concentration in the fecal water might break through the anti-Ca^{2+} mucin shield in the crypts and reduce the proliferation and promote the differentiation of the earliest initiated, but still responsive, cells before they generate progeny which have lost their responsiveness to external Ca^{2+}. It might also suppress even more strongly any residual proliferation out on the surface mucosa. But the initiated clones of colon cells soon lose their responsiveness to external Ca^{2+}, and when they do, Ca^{2+} probably becomes a potentially dangerous, TPA-like, tumor promoter which, above a critical level, suppresses the proliferation of normal cells without affecting and even stimulating the proliferation of the neoplastic cells. Indeed, oral calcium supplementation (e.g., 1.2 to 1.5 g of calcium carbonate daily) significantly *increases* the DNA-synthetic activity of colon epithelial cells in patients with adenomatous polyps.[94,95] Another reason for the reduction of colon carcinogenesis by dietary Ca^{2+} loading is the formation of insoluble Ca^{2+} complexes with, and thus inactivation of, tumor-promoting secondary bile acids.[78,93]

As mentioned above, variants appear in the transforming colon cell population which have mutant *ras* genes coding for constitutively activated Ras proteins. Thus, the cells in about 10% of early adenomas have a mutant *ras* gene (Figure 2).[24,41,53] This proportion jumps to 50% in the larger, more advanced adenomas, but does not increase further in carcinomas (Figure 2).[53] This indicates that *ras* mutation with the production of activated Ras proteins is not a key event in the ultimate malignant transformation. Indeed, a *ras* mutation is also not essential for the premalignant progression because 50% of colon carcinomas have normal *ras* genes. However, the high fraction of *large* adenomas having mutant *ras* genes indicates that activated Ras proteins promote premalignant growth. How might they do this? Transforming colon cells hyper-express the c-*myc* gene whose product can be either proliferogenic or apoptogenic depending on whether the cell can also make something to prevent apoptosis.[8] Among the anti-apoptosis proteins are activated Ras proteins,[8] which also eliminate the need for external Ca^{2+} to complete the G_1 build-up and stimulate proliferation.[89] Thus, the mutant Ras proteins, unlike their normal counterparts, can potently promote the expansion of transforming cell clones by subverting Ca^{2+} controls, stimulating proliferation, and preventing apoptosis. These activated Ras proteins also stimulate the expression of the c-*fos* proto-oncogene, and the genes coding for autocrine motility factor(s), and collagenase IV which are key parts of the invasiveness/metastasis phenotype.[1,89] A role for *ras* gene mutations and activated mutant Ras proteins in the *optimal* progression of colon carcinogenesis is indicated by the ability of fecal Ca^{2+} loading to prevent mutations of the K-*ras* gene and reduce the number and size of colon cancers induced in male Sprague-Dawley rats by 1,2-dimethylhydrazine.[96]

As the PKCs-stimulating (and other) tumor promoters in the fecal stream relentlessly drive the carcinogenic clonal selection process forward, from one generation of cells to the next, variant cells emerge which no longer display the set of surface molecules which normal colorectal cells use to sense and respond to neighboring cells and their position within the crypt. These variants display antigens on their surfaces that are normally expressed by fetal colorectal cells. Fetal cells display A, B, CEA (carcinoembryonic antigen), H, the Lewis antigens (sialosyl Le^a, Le^b, Le^x, Le^y), and MN-related (T or Gal-β(1–3)-GalNac) antigens on their surfaces.[42] After birth and the establishment of the mature colonic epithelial monolayer and monoclonal crypts, the proximal colon cells continue expressing A, B, H, and Le^b antigens, but they stop expressing the other Lewis antigens and T-antigen while the rest of the colorectal cells stop expressing *all* of these antigens.[42] However, the cells in adult hyperplastic and adenomatous polyps start expressing all of these fetal antigens.[42] Eventually variants emerge which have cut the signal-generating/transmitting links to neighbors and basal lamina. Thus, they don't make the Ca^{2+}-binding type E-cadherin receptors and the L-CAM-like surface glycoprotein DCC, the product of the *dcc* (*D*eleted in *C*olon *C*ancer) gene on the long arm (q21) of human chromosome 18.[53,97,98] These later changes signal impending malignancy and are not involved in the earlier proliferative dysregulation. Thus, the loss of heterozygosity and function of the *dcc* gene is rarely seen in early human colorectal adenomas, but is found in 50% of advanced adenomas and in more than 70% of colorectal carcinomas (Figure 2).[53,97,98]

Another frequent change associated with the conversion to malignancy is "dominant" loss-of-function mutation of the p53 gene.[53,98–100] The loss-of-function mutation is often followed by a second selection for the deletion of the remaining wild-type gene and elimination of suppressor function.[100] The loss of heterogygosity of the p53 gene on the short (p) arm of human chromosome 17 (17p 13 to be more precise) is rarely seen in adenomas at any stage including large adenomas with foci of carcinomas, but is seen in more than 75% of colorectal cancers (Figure 2).[53] As described in Chapter 2, an inactive mutant p53(TP53) can bind to the remaining normal wild-type p53 protein and stop it from entering the nucleus and doing its job of monitoring chromosome breaks and stopping the initiation of DNA replication until the damage is repaired.[100] The loss of p53 function enhances the survival of mutant cells and the emergence of malignant clones,[100] and it is important to understand that it is not involved in the premalignant impairment of terminal differentiation and persistent cycling that combine to produce polyps, aberrant crypts, and benign adenomas.[43,53] Indeed, knocking out the p53 gene does not affect proliferation and morphogenesis in mouse embryos and the animals are completely normal at birth, but they are prone to developing tumors in various tissues.[101]

III. SUMMARY

In many (but by no means all) cases the first, key, proliferation-dysregulating event is a loss-of-function mutation or deletion of a gene on a stem cell's chromosome 5q21, which increases the probability of the stem cell's progeny being able to proliferate in the higher, normally proliferatively inactive higher levels of the crypt. The mutant cells are increasingly less likely to trigger a complete Ca^{2+}-dependent differentiation program (diffpoptosis): the differentiation mechanism has lost its responsiveness to Ca^{2+} and TPA. This leads to hyperplasia and overgrowth, the visible result of which is a benign polyp or small adenoma with a substantial population of noncycling, partially or completely differentiated cells. When cells from small adenomas are planted into culture dishes, they are not yet immortalized and thus cannot proliferate indefinitely, nor can they produce clones from single cells (they can proliferate only when in clumps), and they cannot produce tumors in nude mice.[24] The hyperplasia provides the cycling cells upon which various mutagens and promoters can operate to produce the mutations needed to drive neoplastic development. The appearance of a mutant, activated K-Ras protein in some cells further suppresses the Ca^{2+}-dependent differentiation mechanism and stimulates proliferation. This increases the growth and dysplasia of the adenomas, and the cancers arising from these mutant Ras-driven adenomas would contain more cycling cells and fewer differentiation-competent cells than cancers from adenomas without mutant Ras proteins. The cells from large adenomas are now immortal (have abnormal chromosomes 1) and have largely lost the need for being in clumps to proliferate *in vitro*, but they still cannot produce tumors in nude mice.[24] As the evolution toward malignancy continues cells tend to reduce their chromosomes' G-rich telomeres.[41] However, they may re-express the enzyme, telomerase, which embryonic and stem cells use to maintain their proliferative potential by preventing the progressive telomere reduction that occurs with each cycle and eventually stops further cycling.[102] If telomere reduction is not stopped, it destabilizes the chromosomes, which become prone to loss as well as to fusion-bridge breakage, which results in daughter cells having partly deleted and duplicated chromosomes. Because of this, something happens in some adenoma cells with or without mutant *ras* genes: they lose their functional p53(TP53) genes. This unleashes the generation of deadly variants, because these p53-less cells cannot detect and respond to chromosomal damage by delaying replication until the damage is repaired or apoptotically killing themselves if they cannot repair it. From these mutationally liberated cells comes a shower of mutant clones, among which are the deadly malignant clones that have lost the functions of genes whose products govern cell-cell adhesion and the intracellular communication networks, and can now form tumors in nude mice.[24] However, it must be noted that only 10% of colorectal cells make activated Ras proteins and have lost their APC, *dcc*, and p53 genes. Some may have undergone only two or three of these changes, and some may even have undergone none of them. Clearly, we have glimpsed only the tip of the iceberg. Many other genes may be involved in starting and driving the genesis of colorectal cancers from nonpolyp beginnings.[41,42,103]

Carcinogenesis culminates in the appearance of malignant colon cells that overexpress proliferation-related PKAIs, drive their own proliferation, regardless of the extracellular Ca^{2+} concentration or the availability of exogenous growth factors; cannot effectively differentiate and do not senesce; make and secrete angiogenic and autocrine motility factor(s); and which, like embryonic trophoblast cells and migrating monocytes, have surfaces loaded with the various hydrolases such as cathepsins B and L and specific integrins and other receptors needed to bind to, and chew through, basal laminae, push apart vascular endothelial cells and enter blood vessels, embolize with platelets and other malignant cells, and finally leave the blood vessels and invade and colonize distant organs, especially the liver.[9a,24,104]

REFERENCES

1. **Whitfield, J.F.,**Calcium and cancer, *Crit. Rev. Oncogenesis*, 3, 5, 1992.
2. **Whitfield, J.F.**, Calcium signals in cell proliferation and death, in *Ion Channels and Ion Pumps*, Foa, P. and Walsh, M.F., Eds., Springer-Verlag, New York, 1994, 19.
3. **Gordon, J.I., Schmidt, G.H., and Roth, K.A.**, Studies of intestinal stem cells using normal, chimeric, and transgenic mice, *FASEB J.*, 6, 3039, 1992.
4. **Potten, C.S.**, *Stem Cells,* Churchill Lovingstone, Edinburgh, 1983.
5. **Lev, R.**, *Adenomatous Polyps of the Colon,* Springer-Verlag, New York, 1990.
6. **Sunter, J.P., Appleton, D.R., DeRodriguez, M.S.B., Wright, N.A., and Watson, A.J.**, A comparison of cell proliferation at different sites within the large bowel of the mouse, *J. Anat.,* 129, 833, 1979.
7. **Potten, C.S.**, Significance of spontaneous and induced apoptosis in gastrointestinal tract of mice, *Cancer Metastasis Rev.,* 11, 179, 1992.

8. **Bissonnette, R.P., Shi, Y., Mahboubi, A., Glynn, J.M., and Green, D.R.**, c-myc and apoptosis, in *Apoptosis II*, Tomei, L.D. and Cope, F.O., Eds., Cold Spring Harbor Laboratory Press, Cold Spring Harbor, NY, 1994, 327.

8a. **Hockenbery, D.M., Zutter, M., Hickey, W., Nahm, M., and Korsmeyer, S.J.**, BCL2 protein is topographically restricted in tissues characterized by apoptosis cell death, *Proc. Natl. Acad. Sci. USA*, 88, 6961, 1991.

9. **Korsmeyer, S.J., Shutter, J.R., Veis, D.J., Merry, D.E., and Oltvai, Z.N.**, BCL-2/Bax: a rheostat that regulates an anti-oxidant pathway and cell death, *Semin. Cancer Biol.*, 4, 327, 1993.

9a. **Giancotti, F.G. and Mainiero, F.**, Integrin-mediated adhesion and signaling in tumorigenesis, *Biochim. Biophys. Acta*, 1198, 47, 1994.

10. **Suemori, S., Ciacci, C., and Podolsky, D.K.**, Regulation of transforming growth factor expression in rat intestinal epithelial cell lines, *J. Clin. Invest.*, 87, 2216, 1991.

11. **Friedman, E.A.**, A multistage model of human colon carcinoma development integrating cell culture studies with pathology, *Cancer Invest.*, 3, 453, 1985.

12. **Friedman, E.A.**, Use of tissue culture of human colonic epithelial cells to study mechanisms of calcium protection, in *Calcium, Vitamin D, and the Prevention of Colon Cancer,* Lipkin, M., Newmark, H. L., and Kelloff, G., Eds., CRC Press, Boca Raton, FL, 1991, 159.

13. **Friedman, E.A., Higgins, P.J., Lipkin, M., Shinya, H., and Gelb, A.M.**, Tissue culture of human epithelial cells from benign colonic tumors, *In Vitro*, 17, 632, 1981.

14. **Lipkin, M.**, Neoplasms: aetiology, *Curr. Opinions Gastroenterol.*, 5, 7, 1989.

15. **Arlow, F.L., Walczak, S.M., Moshier, J.A., Pietruk, T., and Majumber, A.P.M.**, Gastrin and epidermal growth factor induction of ornithine decarboxylase in rat colonic explants, *Life Sci.*, 46, 777, 1990.

15a. **Pazianas, M., Maxwell, J.D., Shankar, V.S., Adebanjo, O.A., Lai, F.A., Ventikoraman, A., Huang, V.C.-L., Colston, K.W., Brown, E.M., and Zaidi, M.**, Evidence for a novel calcium-sensing "receptor" on Enterocytes, *J. Bone Miner. Res.*, 9, S220, 1994.

16. **Wali, R.K., Baum, C.L., Sitrin, M.D., and Brasitus, T.A.**, $1,25(OH)_2$ vitamin D_3 stimulates membrane phosphoinositide turnover, activates protein kinase C, and increases cytosolic calcium in rat colonic epithelium, *J. Clin. Invest.*, 85, 1296, 1990.

17. **Civitelli, R., Kim, Y.S., Gunsten, S.I., Fujimori, A., Huskey, M., Avioli, L.V., and Huska, K.A.**, Nongenomic activation of the calcium message system by vitamin D metabolites in osteoblast-like cells, *Endocrinology*, 127, 2253, 1990.

18. **Zhou, L.-X., Nemere, I., and Norman, A.W.**, A parathyroid related peptide induces transcaltachia (the rapid, hormonal stimulation of intestinal Ca^{2+} transport, *Biochem. Biophys. Res. Commun.*, 186, 69, 1992.

19. **Thiede, M.A.**, Expression and regulation of the parathyroid hormone-related protein gene in tumors and normal tissues, in *Parathyroid Hormone-Related Protein: Normal Physiology and its Role in Cancer*, Halloran, B. P. and Nissenson, R. A., Eds., CRC Press, Boca Raton, FL, 1992, 57.

20. **Yu, D., Seitz, P.K., Selvanayagam, P., Rajaraman, S., Townsend, C.M., and Cooper, C.W.**, Effects of parathyroid-related peptide on adenosine 3′,5′-monophosphate and ornithine decarboxylase in a human colonic cell line, *Endocrinology*, 130, 1993, 1992.
21. **Cho-Chung, Y.S. and Clair, T.**, The regulatory subunit of cAMP-dependent protein kinase as a target for chemotherapy of cancer and other cellular dysfunctional-related diseases, *Pharm. Ther.*, 60, 265, 1994.
22. **Cho-Chung, Y.S., Clair, T., Tortora, G., and Yokozaki, H.**, Role of site-selective cAMP analogs in the control and reversal of malignancy, *Pharm. Ther.*, 50, 123, 1991.
23. **Hague, A., Manning, A.M., van der Stappen, J.W.J., and Paraskeva, C.**, Escape from negative regulation of growth by transforming growth factor β and from the induction of apoptosis by the dietary agent sodium butyrate may be important in colorectal carcinogenesis, *Cancer Metastasis Rev.*, 12, 227, 1993.
24. **Williams, A.C., Hague, A., Manning, A.M., van der Stappen, J.W.J., and Paraskeva, C.**, In vitro models of human colorectal cancer, *Cancer Surv.*, 16, 15, 1993.
25. **Torelli, G., Venturelli, D., Colo, A., Zanni, C., Selleri, L., Moretti, L., Calabretta, B., and Torelli, U.**, Expression of c-*myb* protooncogene and other cell cycle-related genes in normal and neoplastic human mucosa, *Cancer Res.*, 47, 5266, 1987.
26. **Augenlicht, L.H., Augeron, C., Yander, G., and Laboisse, C.**, Overexpression of *ras* in mucus-secreting human colon carcinoma cells of low tumorigenicity, *Cancer Res.*, 47, 3763, 1987.
27. **Chesa, P.G., Rettig, W.J., Melamed, M.R., Old, L.J., and Niman, H.L.**, Expression of $p21^{ras}$ in normal and malignant human tissues: lack of association with proliferation and malignancy, *Proc. Natl. Acad. Sci. U.S.A.*, 84, 3234, 1987.
28. **Durkin, J.P., Chakravarthy, B., Mealing, G., Schwartz, J.-L., Tremblay, R., Whitfield, J.F., and Franks, D.J.**, Role of signal transduction events in the proliferative response of cells to a mitogenic viral K-*ras* protein, *Cell Signal.*, 2, 285, 1990.
29. **Takai, Y., Kaibuchi, K., Kikuchi, A., and Kawata, M.**, Small GTP-binding proteins, *Int. Rev. Cytol.*, 133, 187, 1992.
30. **Pienta, K. J., Getzenberg, R.H., and Coffey, S.**, Cell structure and DNA organization, *Crit. Rev. Eukaryotic Gene Expression*, 1, 355, 1991.
31. **Hafez, M.M., Hsu, S., Yan, Z., Winawer, S., and Friedman, E.**, Two roles for transforming growth factor beta 1 in colon enterocytic cell differentiation, *Cell Growth Differ.*, 3, 733, 1992.
32. **Pignatelli, M., Liu, D., Nasim, M.M., Stamp, G.W., Hirano, S., and Takeichi, M.**, Morphoregulatory activities of E-cadherin and beta-1 integrin, *Br. J. Cancer*, 66, 629, 1992.
33. **Stallmach, A., Von Lampe, B., Matthes, E., Bornhöft, G., and Riecken, E.O.**, Diminished expression of integrin adhesion molecules on human colonic epithelial cells during the benign to malignant tumor transformation, *Gut*, 33, 342, 1992.
34. **Pignatelli, M. and Bodmer, W.F.**, Genetics and biochemistry of collagen binding-triggered glandular differentiation in a human colon carcinoma cell line, *Proc. Natl. Acad. Sci. U.S.A.*, 85, 5561, 1988.

35. **Pignatelli, M. and Bodmer, W.F.**, Integrin-receptor-mediated differentiation and growth inhibition are enhanced by transforming growth factor-β in colorectal tumour cells grown in collagen gel, *Int. J. Cancer,* 44, 518, 1989.
36. **Richman, P.J. and Bodmer, W.F.**, Control of differentiation in human colorectal carcinoma cell lines: epithelial-mesenchymal interactions, *J. Pathol.*, 156, 197, 1988.
37. **Shenolikar, S. and Nairn, A.C.**, Protein phosphatases: recent progress, *Adv. Second Messenger Phosphoprotein Res.*, 23, 1, 1991.
38. **Gavrieli, Y., Sherman, Y., and Ben-Sasson, S.A.**, Identification of programmed cell death in situ by specific labeling of nuclear DNA fragmentation, *J. Cell Biol.*, 119, 403, 1992.
39. **Schmidt, G.H., Wilkinson, M.M., and Ponder, B.A.J.**, Cell migration pathway in the intestinal epithelium: an in situ marker system mouse aggregation chimeras, *Cell,* 40, 425, 1985.
40. **Iyengar, V., Albaugh, G.P., Lohani, A., and Nair, P.P.**, Human stools as a source of viable colonic epithelial cells, *FASEB J.*, 5, 2856, 1991.
41. **Kinsella, A.R.**, *Colorectal Cancer, A Scientific Perspective*, Cambridge University Press, Cambridge, 1993.
42. **Shamsuddin, A.M.**, *Diagnostic Assays for Colon Cancer*, CRC Press, Boca Raton, FL, 1992.
43. **Shimoda, T., Ikegami, M., Fujisaki, J., Matsui, T., Aizawa, S., and Ishikawa, E.**, Early colorectal carcinoma with special reference to its development *de novo, Cancer,* 64, 1138, 1989.
44. **Roncucci, L., Stamp, D., Medline, A., and Bruce, R.**, Aberrant crypt foci and microadenomas in the human colon, *Am. J. Gastroenterol.*, 85, 1284, 1990.
45. **Pozharisski, K.M.**, Morphology and morphogenesis of experimental epithelial tumors of the intestine, *J. Natl. Cancer Inst.*, 54, 1115, 1975.
46. **McLellan, E.A. and Bird, R.A.**, Specificity study to evaluate induction of aberrant crypts in murine colon, *Cancer Res.,* 48, 6183, 1988.
47. **McLellan, E.A. and Bird, E.A.**, Aberrant crypts: potential preneoplastic lesions in the murine colon, *Cancer Res.*, 48, 6187, 1988.
48. **Gross, M.E., Zorbas, M.A., Danels, Y.J., Garcia, R., Gallick, G.E., Olive, M., Brattain, M.G., Boman, B.M., and Yeoman, L.C.**, Cellular growth response to epidermal growth factor in colon carcinoma cells with an amplified epidermal growth factor receptor derived from a familial adenomatous polyposis patient, *Cancer Res.*, 51, 1452, 1991.
49. **Sizeland, A.M. and Burgess, A.W.**, Anti-sense transforming growth factor α oligonucleotides inhibit autocrine stimulated proliferation of a colon carcinoma cell line, *Mol. Biol. Cell*, 3, 1235, 1992.
50. **Contreas, C.N., Desai, T.K., and Arlow, F.A.**, Relationship of hormones and growth factors to colon cancer, *Gastroenterol. Clin. North Am.*, 17, 761, 1988.
51. **Bodmer, W.F., Bailey, C.J., Bodmer, J., Bussey, H.J.R., Ellis, A., Gorman, P., Lucibello, F.C., Murday, V.A., Rider, S.M., Scambler, P., Sheer, D., Soloman, E., and Spurr, N.J.K.**, Localization of the gene for familiar adenomatous polyposis in chromosome 5, *Nature*, 328, 614, 1987.
52. **Dunlop, M.G.**, Colorectal cancer genetics, *Semin. Cancer Biol.,* 3, 131, 1992.
53. **Fearson, E.R. and Jones, P.A.**, Progressing toward a molecular description of colorectal cancer development, *FASEB J.*, 6, 2783, 1992.

54. **Kinzler, K.W., Nilbert, M.C., Vogelstein, B., Bryan, T.M., Levy, D.B., Smith, K.J., Preisinger, A.C., Hamilton, S.R., Hedge, P., Markham, A., Carlson, M., Joslyn, G., Groden, J., White, R., Miki, K., Miyoshi, Y., Nishisho, T., and Nakamura, Y.**, Identification of a gene located at chromosome 5q21 that is mutated in colorectal cancers, *Science*, 251, 1366, 1991.
55. **Kinzler, K.W., Nilbert, M.C., Su, L.-K., Vogelstein, B., Bryan, T.M., Levy, D.B., Smith, K.J., Preisinger, A.C., Hedge, P., McKechnie, D., Finnieur, R., Markham, A., Groffen, J., Boguski, M.S., Altschul, S.F., Horii, A., Ando, H., Miyoshi, Y., Miki, Y., Nishisho, I., and Nakamura, Y.**, Identification of FAP locus genes from chromosome 5q21, *Science*, 253, 661, 1991.
56. **Nishisho, I., Nakamura, Y., Miyoshi, Y., Miki, Y., Ando, H., Horii, A., Koyama, K., Utsunomiya, J., Baba, D., Hedge, P., Markham, A., Krush, A.J., Petersen, G., Hamilton, S.R., Nilbert, M.C., Levy, P.B., Boyan, T.M., Preisinger, A.C., Smith, K.J., Su, L.-K., Kinzler, K.W., and Vogelstein, B.**, Mutations of chromosome 5q21 genes in FAP and colorectal cancer patients, *Science*, 253, 665, 1991.
57. **Ponder, R.A.J.**, Inherited predisposition to cancer, *Trends Genet.*, 6, 213, 1990.
58. **Powell, S.M., Zilz, N., Beazer Barclay, Y., Bryan, T.M., Hamilton, S.R., Thibodeau, S.N., Vogelstein, B., and Kinzler, K.W.**, APC mutations occur early during colorectal tumorigenesis, *Nature*, 359, 235, 1992.
59. **Solomon, E., Voss, R., Hall, V., Bodmer, W.F., Jass, J.R., Jeffreys, A.J., Lucibello, F.C., Patel, I., and Rider, S.H.**, Chromosome 5 allele loss in human colorectal carcinoma, *Nature*, 328, 614, 1987.
60. **Moser, A.R., Dove, W.F., Roth, K.A., and Gordon, J.I.**, The *Min* (multiple intestinal neoplasia) mutation: its effect on gut epithelial cell differentiation and interaction with a modifier system, *J. Cell Biol.*, 116, 1517, 1992.
61. **Moser, A.R., Pitot, H.C., and Dove, W.F.**, A dominant mutation that predisposes to multiple neoplasia in the mouse, *Science*, 247, 322, 1990.
62. **Su, L.-K., Kinzler, K.W., Vogelstein, B., Preisinger, A.C., Moser, A.R., Luonzo, C., Gould, K.A., and Dove, W.F.**, Multiple intestinal neoplasia caused by a mutation in the murine homolog of the APC gene, *Science*, 256, 668, 1992.
63. **Goyette, M.C., Cho, K., Fashing, C.L., and Stanbridge, E.J.**, Progression of colorectal cancer is associated with multiple tumor suppression gene defects but inhibition of tumorigenicity is accomplished by correction of any single defect via chromosome transfer, *Mol. Cell. Biol.*, 12, 1387, 1992.
64. **Tanaka, K., Oshimura, M., Kikuchi, R., Seki, M., Hayashi, T., and Miyaki, M.**, Suppression of tumorigenicity in human colon carcinoma 5 or 18, *Nature*, 349, 340, 1991.
65. **Friedman, E.A.**, Familial polyposis colonic epithelial cells respond inappropriately to endogenous modulators of colonic cell growth, in *Familial Adenomatous Polyposis*, Herrera, L., Ed., Alan R. Liss, New York, 1990, 275.
66. **Danes, B.S.**, Effect of increased calcium concentration on *in vitro* growth of human colonic mucosal lines, *Dis. Colon Rectum*, 34, 552, 1991.
67. **Buset, M., Lipkin, M., Winawer, S., Swaroop, S., and Friedman, E.**, Inhibition of human colonic epithelial cell proliferation *in vivo* and *in vitro*, *Cancer Res.*, 46, 5426, 1986.

68. **Guo, Y.-S., Draviam, E., Townsend, C.M., and Singh, P.**, Differential effects of Ca^{2+} on proliferation of stomach, colonic, and pancreatic cancer cell lines, *in vitro, Nutr. Cancer*, 14, 149, 1990.
69. **Swierenga, S.H.H., Auersperg, H., and Wong, K.S.**, Effect of calcium deprivation on the proliferation and ultrastructural of cultured human carcinoma cells, *Cancer Res.,* 43, 6012, 1983.
70. **Cross, H.S., Hubert, C., and Peterlik, M.**, Antiproliferative effect of 1,25-dihydroxyvitamin D_3 and its analogs on human colon adenocarcinoma cells (CACO-2): influence of extracellular calcium, *Biochem. Biophys. Res. Commun.,* 179, 57, 1991.
71. **Rubinfeld, B., Souza, B., Albert, I., Müller, O., Chamberlain, S.H., Masairz, F.R., Munemitsu, S., and Polakis, P.**, Association of the APC gene product with β-catenin, *Science*, 262, 1731, 1993.
72. **Su, L.-K., Vogelstein, B., and Kinzler, K.W.**, Association of the APC tumor suppressor protein with catenins, *Science*, 262, 1734, 1993.

72a. **Birchmeier, W. and Behrens, J.**, Cadherin expression in carcinomas: role in the formation of cell functions and the prevention of invasiveness, *Biochim. Biophys. Acta*, 1198, 11, 1994.

73. **Morley, P., Tsang, B.K., Whitfield, J.F., and Schwartz, J.L.**, The effect of muscarinic cholinergic agonists on intracellular calcium and progesterone production by chicken granulosa cells, *Endocrinology*, 130, 663, 1992.
74. **Nakamura, Y.**, The role of the adenomatous polyposis coli (APC) gene in human cancers, *Adv. Cancer Res.*, 62, 65, 1993.
75. **LeBeau, M.M.**, Deletion of chromosome 5 in malignant myeloid disorders, *Cancer Surv.*, 15, 143, 1992.
76. **Marx, J.**, Gene defect identified in common hereditary colon cancer, *Science*, 262, 1645, 1993.
77. **Edelstein, P.S., Thompson, S.M., and Davies, R.J.**, Altered intracellular calcium regulation in human colorectal cancers and in "normal" adjacent mucosa, *Cancer Res.*, 51, 4492, 1991.
78. **Whitfield, J.F.**, Calcium switches, cell cycles, differentiation, and death, in *Calcium, Vitamin D, and Prevention of Colon Cancer*, Lipkin, M., Newmark, H.L., and Kelloff, G., Eds., CRC Press, Boca Raton, FL, 1991, 31.
79. **Phan, S.C., Morotomi, M., Guillem, J.G., LaGerfo, P., and Weinstein, I.B.**, Decreased levels of 1,2-*sn* diacylglycerol in human colon tumors, *Cancer Res.*, 51, 1571, 1991.
80. **Sauter, G., Nerlich, A., Spengler, U., Kopp, R., and Pfeiffer, A.**, Low diacylgycerol values in colonic adenomas and colorectal cancer, *Gut*, 31, 1041, 1990.
81. **Friedman, E.A., Gillin, S., and Lipkin, M.**, 12-*O*-tetradecanoyl-phorbol-B-acetate stimulation of DNA synthesis in cultured preneoplastic familial polyposis colonic epithelial cells but not in normal colonic epithelial cells, *Cancer Res.*, 44, 4078, 1984.
82. **Friedman, E.A., Urmacher, C., and Winawer, S.**, A model for human colon carcinoma evaluation based on the differential response of cultured preneoplastic, premalignant, and malignant cells to 12-*O*-tetradecanoylphorbol-13-acetate, *Cancer Res.*, 44, 1568, 1984.
83. **Lee, H., Winawer, S., and Friedman, E.**, Signal-tranduction through pp63 tyrosine phosphorylation in human colon carcinoma cell lines, *Proc. Am. Assoc. Cancer Res.,* 51 (Abstr), 450, 1991.

84. **Bolen, J. B., Veillette, A., Schwartz, A. M., DeSeau, V., and Rosen, N.**, Activation of pp60$^{c\text{-}src}$ protein kinase activity in human colon carcinoma, *Proc. Natl. Acad. Sci. U.S.A.*, 84, 2251, 1987.
85. **Cartwright, C.A.A., Meister, A.I., and Eckhast, W.**, Activation of the pp60$^{c\text{-}src}$ protein kinase is an early event in colonic carcinogenesis, *Proc. Natl. Acad. Sci. U.S.A.*, 87, 558, 1990.
86. **Weber, T.K., Steele, G., and Summerhayes, I.C.**, Differential pp60$^{c\text{-}Src}$ activity in well and poorly differentiated human colon carcinomas and cell lines, *J. Clin. Invest.*, 90, 815, 1992.
87. **Brickell, P.M.**, The p60$^{c\text{-}Src}$ family of protein-tyrosine kinases: structure, regulation, and function, *Crit. Rev. Oncogenesis*, 3, 401, 1992.
88. **Medema, R.H. and Bos, J.L.**, The role of p21 ras in receptor tyrosine kinase signaling, *Crit. Rev. Oncogenesis*, 4, 615, 1993.
89. **Whitfield, J.F.**, *Calcium, Cell Cycles, and Cancer*, 1st ed., CRC Press, Boca Raton, FL, 1990.
90. **Weinstein, L.S., Shenker, A., Gejman, P.V., Merino, M.J., Friedman, E., and Spiegel, A.M.**, Activating mutations of the stimulatory G protein in the McCune-Albright syndrome, *N. Engl. J. Med.*, 325, 1688, 1991.
91. **Bradbury, A.W., Miller, W.R., Clair, T., Yokozaki, H., and Cho-Chung, Y.S.**, Overexpressed type I regulatory subunit (RI) of cAMP-dependent protein kinase (PKA) as tumor marker in colorectal cancer, *Proc. Am. Assoc. Cancer Res.*, 31, 172, 1990.
92. **Clair, T., Yokozaki, H., Tortora, G. et al.**, An antisense oligodeoxynucleotide targeted against the type I regulatory subunit (RIα) mRNA of cAMP-dependent protein kinase (PKA) inhibits the growth of LS-147T human colon carcinoma in athymic nude mice, *Proc. Am. Assoc. Cancer Res.*, 51 (Abstr.), 1645, 1991.
93. **Whitfield, J.F.**, Calcium: driver of cell cycles, trigger of differentiation, and killer of cells, in *Cellular and Molecular Targets for Chemoprevention*, Steele, V.L., Stoner, G.D., Boone, C.W., and Kelloff, G.J., Eds., CRC Press, Boca Raton, FL, 257, 1992.
94. **Bostick, R.M., Potter, J.D., Fosdick, L., Grambsch, P., Lampe, P.W., Wood, J.R., Louis, T.A., Ganz, R., and Grandits, G.**, Calcium and colorectal epithelial cell proliferation: a preliminary randomized, double-blinded, placebo-controlled clinical trial, *J. Natl. Cancer Inst.*, 85, 132, 1993.
95. **Kleibeuker, J.H., Welberg, J.W., Mulder, N.H., van der Meer, R., Cats, A., Limburg, A. J., Kreumer, W.M., Hardonk, M. J., and de Vries, E. G.**, Epithelial cell proliferation in the sigmoid colon of patients with adenomatous polyps increases during oral calcium supplementation, *Br. J. Cancer*, 67, 500, 1993.
95a. **Pavelic, Z.P., Pavelic, L., Kuvelkar, R., and Gapany, S.R.**, High c-*myc* protein expression in benign colorectal lesions correlates with the degree of dysplasia, *Anticancer Res.*, 12, 171, 1992.
96. **Lior, X., Jacoby, R.F., Teng, B.B., Jacoby, R., Davidson, N., Sitrin, M., and Bradsitus, T.A.**, K-*ras* mutations in 1,2-dimethylhydrazine-colonic tumors: effects of supplemental dietary calcium and vitamin D deficiency, *Cancer Res.*, 51, 4305, 1991.

97. **Fearson, E.R., Cho, K.R., Nigro, J.M., Kem, S.E., Simons, J.W., Rapport, J.M., Hamilton, S.R., Preisinger, A.C., Thomas, G., Kinzler, K.W., and Vogelstein, B.**, Identification of a chromosome 18q gene that is altered in colorectal cancers, *Science*, 247, 49, 1990.
98. **Nowell, P.C.**, Cytogenic approaches to human cancer genes, *FASEB J.*, 8, 408, 1994.
99. **Baker, S.J., Fearson, E.R., Nigro, J.M., Hamilton, S.R., Preisinger, A.C., Jessup, J.M., Van Tuinen, P., Ledbetter, D.M., Barker, D.F., Nakamura, Y., White, R., and Vogelstein, B.**, Chromosome 17 deletions and p53 gene mutations in colorectal carcinomas, *Science*, 244, 217, 1989.
100. **Baker, S. and Vogelstein, B.**, The p53 gene in human cancer, in *Nuclear Processes and Oncogenes*, Sharp, P.A., Ed., Academic Press, New York, 1992, 105.
101. **Rennie, J.**, Essential but expendable: when do mastergenes regulate cell growth? *Sci. Am.*, 267, 25, 1992.
102. **Rennie, J.**, Immortals enzyme, *Sci. Am.*, 271, 14, 1994.
103. **Peltomaki, P., Sistomen, P., Mecklin, J., Pylkkänen, L., Aaltonen, L., Nordling, S., Kere, J., Jarvinen, H., Hamilton, S.R., Petersen, G., Kinzler, K.W., Vogelstein, B., and de la Chapelle, A.**, Evidence that the MCC-APC gene region in 5q21 is not the site for susceptibility to hereditary nonpolyposis colorectal carcinoma, *Cancer Res.*, 52, 4530, 1992.
104. **Mareel, M.M., De Baetselier, P., and VanRoy, F.M.**, *Mechanisms of Invasion and Metastasis*, CRC Press, Boca Raton, FL, 1990.

6 Lymphocytes

I. T-LYMPHOCYTES

The thymus is colonized by progenitor T-cells mainly at a certain stage of embryonic development (between days 10 and 13 in the mouse), but colonization may continue at a much lower level through-out life.[1,2] The progenitor T-cells display "homing" molecules, such as the murine Joro 75 and Joro 37–5, on their surfaces and thymic vascular cells have addressins such as ICAM-1, VCAM-1, and α_6 integrins on their surfaces.[1,2] When the specifically marked progenitor cells meet the appropriate vascular integrins they bind to them and stop — they have reached "home". The signal from the ligand-bound integrin receptors cause the endothelial cells to part and let the progenitor T-cells enter the perivascular space by diapedesis.[1,2] Driven by migration or chemotactic factors (e.g., β_2-microglobulin), the cells, using their signal-generating integrins, move along matrix proteins such as fibronectin to the basement membrane supporting the subcapsular thymic epithelial cells.[1,2] Like metastasizing cancer cells they recognize and penetrate the basement membrane and establish contact with the subcapsular epithelial or stromal cells.[1,2]

Contact with the stromal cells induces the precursor cells to become rapidly cycling lymphoblasts. Maybe it is a Ca^{2+}-containing signal triggered when the precursor cells' CD2 receptors bind to LFA-3 antigens on the stromal cells that stimulates this conversion.[1] Indeed, these precursors do not express the clonotypic T-cell receptors (TCRs) or the TCR-associated CD3, CD4, or CD8 antigens (Figure 1).[1,2] As the progeny of the proliferating lymphoblasts are pushed deeper into the thymic cortex they start expressing both CD4 and CD8 antigens (they become "double positive") and low levels of the $\alpha\beta$TCR.[1] Those immature cells that do not recognize self-MHCs cannot generate a signal needed to keep them cycling, and they revert to small cells which die by apoptosis,[1,2] a process named by Kerr et al.,[3] but first characterized with rat thymocytes by Whitfield and colleagues.[4–7] Those cells that do recognize self-MHCs enter the killing zone with its self-antigen-presenting cortical, dendritic cells.[2] If the immature cell recognizes self-MHCs and its now-medium-level complement of TCRs recognize a self-antigen nestling in the self-MHC's groove, the TCR-generated signal cascade, which includes a shower of reactive oxygen intermediates (ROIs) from ptdIns(4,5) P_2 breakdown, will trigger apoptosis.[1,2] Actually,

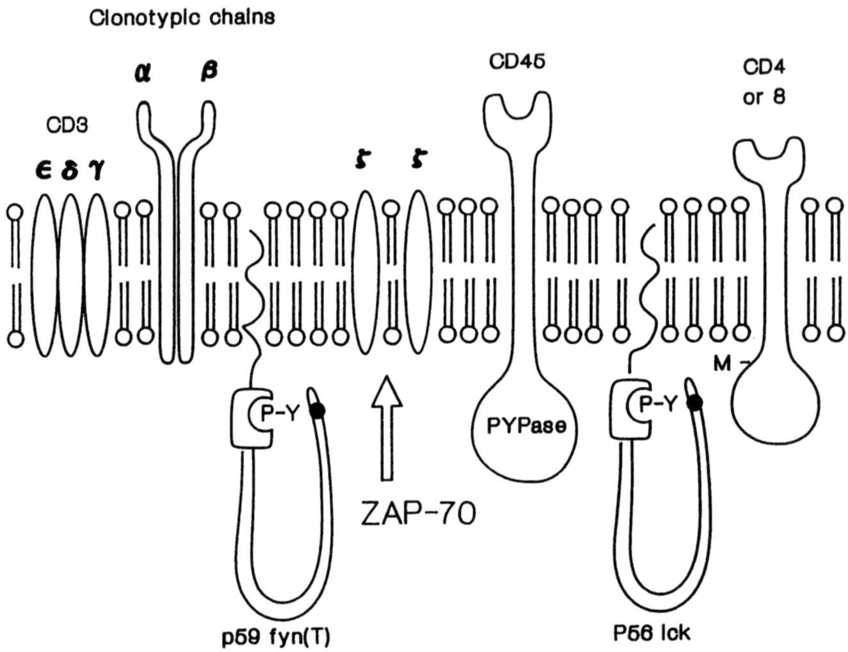

FIGURE 1 The TCR·CD3 complex (α,β clonotypic subunits; γ, δ, and ϵ non-clonotypic CD3 subunits, the accessory $\zeta\zeta$ or $\zeta\eta$ chain subunits, the accessory CD4 or CD8 proteins, and the CD45 PYPase that activates the membrane-associated Src-family YPKs, $p59^{fyn(T)}$ and $p56^{lck}$. In the resting state, the Src-YPKs are curled up in the inactive configuration with their C-terminal pY residues inserted into their own N-terminal SH2 pockets. When a specific, processed antigen (presented by a macrophage or some other APC) binds to the α,β clonotypic subunits, the signal is relayed to the CD3 subunits that somehow activate CD45. CD45 then dephosphorylates and activates $p59^{fyn(T)}$ which phosphorylates, and thus velcroizes, TCR·CD3 components. The ZAP-70 YPK is then activated and phosphorylates the ζ subunits. The accessory, nonclonotypic CD4 or CD8 component also binds to the MHC antigens on the APC. This causes CD45 to dephosphorylate and activate the $p56^{lck}$ YPK which is linked to CD4 or CD8 by a metal (symbolized by M). These several associations and YPK activations result in the formation of a group of modular velcroceptors that assemble the teams of proliferogenic signal enzymes and adaptors.

all immature, double-positive T-cells are sensitive to the apoptogenic actions of glucocorticoids and ionizing radiation.[2] Those immature, double-positive or partly double positive ($CD4^+CD8^-$, $CD4^-CD8^+$) cells, whose TCRs do not recognize any of the self-antigens presented by the dendritic cells, escape the lethal, apoptogenic consequences of premature TCR signaling and pass, or are pushed, safely through the killing zone and crowd into the thymic medulla.[1,2] There they contact the medullary stromal cells and mature into single-positive $CD4^+CD8^-$, or $CD4^-CD8^+$ small T-cells with CD3- and TCR-loaded surfaces.[1,2] Glucocorticoids can no longer trigger apoptosis and the cells are more resistant to apoptogenic, ionizing radiation. Most importantly they are no longer killed by TCR signaling. Instead, the TCR signal now becomes proliferogenic instead of apoptogenic.

What is the reason for this dramatic change in the response to various signals? It appears to be the expression during medullary maturation of a 26-kDa protein, Bcl-2.[2] This protein collects in endoplasmic reticulum membranes, nuclear membranes, but especially in the inner mitochondrial membrane. Bcl-2 may block apoptogenesis by preventing the accumulation of ROIs and the membrane-damaging peroxidation of membrane lipids by these ROIs, which are byproducts of normal receptor signaling (e.g., activation of NAD(P)H oxidase by receptor-activated protein kinases, PKCs) or produced directly by ionizing radiation.[2,8–11] Bcl-2 seems to operate by binding to, and inactivating, the related Bax protein, and it has been suggested that the likelihood of an apoptogenic response is proportional to the levels of Bax/Bax, Bcl-2/Bax, and Bcl-2/Bcl-2 dimers.[11,11a]

The proliferation of thymic lymphoblasts and the production of small lymphocytes is extremely sensitive to Ca^{2+}.[12] Indeed, the story of Ca^{2+} in the cell cycle began with the demonstration by Whitfield and colleagues that external Ca^{2+} concentrations above 1.0 m*M* very rapidly stimulated mitotic activity in freshly isolated rat thymic lymphocytes suspended in a simple, *Tris*-buffered glucose-salts medium, and then reported that raising the Ca^{2+} concentration in the blood did precisely the same thing to thymocytes in the rat.[12] To show this, we first injected animals intraperitoneally (i.p.) with colchicine to stop the thymic lymphoblasts from completing mitosis, and then we measured the subsequent accumulation of blocked mitotic cells, i.e., the flow of cells into mitosis. Then, we injected (i.p.) 1.0 ml of a concentrated Ca^{2+} solution (14 mg of $CaCl_2 \cdot 6H_2O$ per ml) immediately afterwards. Ca^{2+} was injected again at 0.5 h, 1.0 h, and every hour thereafter until 6 h. The resulting train of plasma Ca^{2+} spikes actually quadrupled the flow of thymic lymphoblasts into mitosis. We next reported that the parathyroid hormone (PTH) and PTH-induced Ca^{2+} surges that followed a sharp lowering of the plasma Ca^{2+} concentration with a single injection of a Ca^{2+} chelator such as EDTA or Na_2HPO_4 promptly stimulated the flow of thymus cells into mitosis.[12]

It soon became apparent that these responses were not trivial when we found that Ca^{2+} and PTH controlled the proliferation of thymic lymphoblasts in the rat.[12] PTH maintains the normal blood Ca^{2+} concentration through its high-affinity receptors (receptor I in Figure 5 of Chapter 4), the Ca^{2+}/PKCs signals from which promote the uptake of Ca^{2+} from the nascent urine stream by kidney tubule cells.[13,14] Injecting PTH stimulated the flow of thymus cells into mitosis.[12] During early postnatal development of the rat, overall body growth and thymic mitotic activity were closely correlated to fluctuations in the plasma Ca^{2+} concentration.[12]

If Ca^{2+} and PTH are physiological controllers of the rapid proliferation of the thymus's cortical precursor blast cells, eliminating the hormone by removing the parathyroid glands and thus lowering the plasma Ca^{2+} concentration should cause a rapid thymic atrophy. Indeed, this is exactly what happens.

Perris et al.[15] found that parathyroid removal greatly reduced lymphoblast proliferation, and by 9 d the thymus gland was reduced to a shriveled remnant as the mature small lymphocytes continued to emigrate, but could no longer be replaced by the precursor blast cells. The remnant eventually came to consist of a bare stroma and a few scattered clusters of lymphocytes instead of the great mass of densely packed lymphocytes that normally blankets the stroma.[16]

How could a reduced external Ca^{2+} concentration stop thymic lymphoblast proliferation? The experiments of Whitfield and his colleagues,[12] which showed that raising the Ca^{2+} concentration from 0.6 to 1.25 m*M* stimulated mitotic activity in freshly suspended rat thymic lymphocyte populations, were carried out using a simple, serum-free, *Tris*-buffered glucose-salts solution. Since this solution lacked growth factors, amino acids, and other nutrients and the mitotic response appeared within just 3 or 4 h after isolation, suspension *in vitro,* and addition of Ca^{2+} or PTH, we assumed that we were looking at Ca^{2+} and PTH stimulating the G_2 build-up and initiation of mitosis. But we were astonished to find that raising the Ca^{2+} concentration in either the simple glucose-salts medium or a much richer medium (consisting of the glucose-salts solution plus all of the amino acids, vitamins, and nitrogenous bases of the tissue culture medium 199) stimulated lymphoblasts to initiate DNA replication without directly affecting the G_2 transit into mitosis.[12] Thus, we had stumbled on the fact that the thymus of the 180- to 200-g young adult male rat locks a group (about 50%) of its immature and presumably subcapsular and cortical lymphoblasts into a very unusual quiescent or very slowly transited state in which they have a nearly complete G_1 build-up (i.e., they are treadmilling at, or swimming slowly toward, the G_1/S boundary) and are poised to start replicating DNA upon receipt of an appropriate signal.[12] It turned out that these G_0/late-G_1 thymic lymphoblasts are indeed programmed to maintain pools of the replication enzymes except thymidylate synthase and, I suspect, the cell cycle engines replication-triggering Cdk2·cyclin A PKs which are expressed only in response to a Ca^{2+} signal. The cells probably express the synthase gene, but they promptly destroy either the transcripts or the enzyme product by a mechanism which is inhibited by the Ca^{2+} signal. It seems likely that this is the same late-G_1 signal mechanism that triggers replication in other non-neoplastic cells and was discussed in Chapter 2.

While we were surprised that thymic lymphoblasts could transit S and G_2 phases in just 3 or 4 h, our efforts to find the mechanism that switches on their thymidylate synthase and replication initiators in the late 1960s and early 1970s revealed that thymocytes display an incredible variety of replication-triggering receptors besides the expected cytokine receptors and lectin-binding TCR and other glycoprotein receptors.[12] We were the *first* to glimpse a then-unbelievable new world of growth factors and mitogens! These cells, fresh from the thymus gland, responded to acetylcholine, bradykinin and pancreatic kallikrein, calcitonin, cortisol (10^{-9} to 10^{-8} *M*), cyclic AMP (10^{-8} to 10^{-6} *M*), cyclic GMP, epinephrine and norepinephrine, growth hormone, oxytocin, the adenylyl cyclase-stimulating prostaglandins A and E, PTH, vasopressin, and the very first indicator of the very great things (and hundreds of papers) to come, the potent PKC-activating TPA.[12] Thus, a substantial part of the proliferating or proliferatively stimulable blast cells, which make up 20% of the thymic lymphoid cell population, are the targets of many hormones and other factors and can thus be mobilized rapidly in response to a wide variety of physiological challenges. Evidently their surfaces are bristling with a forest of different hormone receptors, the activation of which can trigger DNA replication and mitosis.

Perhaps the most astonishing response was the stimulation of thymic lymphoblast DNA replication and division by low concentration (10^{-9} or 10^{-8} N) of the glucocorticoid steroid, cortisol, the radiomimetic apoptogenic killer of immature double-positive thymocytes.[1,2,12] Exposing freshly isolated thymic lymphocytes to 10^{-9} M cortisol, in medium containing 0.5 to 1.0 mM Ca^{2+}, caused a large group of lymphoblasts to start replicating DNA within 1 h and then to start entering mitosis about 4 h later.[12] The steroid did nothing when the medium contained no added Ca^{2+}.[12] At the time we suspected that the steroid worked by triggering an influx of external Ca^{2+}. This suspicion has been supported by McConkey et al.,[17,18] who reported that methylprednisolone, a cortisol-mimicking glucocorticoid, caused a cytosolic Ca^{2+} surge. Apparently, the steroid stimulates the expression of a heat-labile, cytosolic protein which stimulates Ca^{2+} influx. When the steroid-induced Ca^{2+} surge exceeds a certain size (e.g., from a basal 85 nM to 320–811 nM), apoptosis is triggered and the chromosomes are chopped into an oligonucleosome "ladder" by a Ca^{2+}-activated endonuclease. Thus, a mitogenic signal can be amplified into an apoptogenic signal.

We focused our efforts on the response to PTH because it is physiologically important. PTH controls the proliferation of thymic lymphoblasts in the animal on two levels: directly by activating receptors on the cell surface and indirectly by determining the level of external Ca^{2+} ions that mediate the signals from the activated receptors.[12–14] Another player in the thymus (and spleen) is PTHrP, which probably operates as an autocrine or paracrine factor.[19] Thus, the PTHrP holoproteins and N-terminal fragments such as PTHrP-(1–40) bind to the same PTH/PTHrP receptors.[20] There are also C-terminal PTHrP fragments with the PTHrP-(107–111) TRSAW motif which are produced by post-translational processing of the holoproteins and bind to a different receptor on lymphocytes and stimulate membrane-associated PKCs activity.[19,20] In fact, it turned out that the response of thymocytes to PTH was the prototype of their responses to most of the other proliferogenic agents.[12]

PTH triggered a prolonged surge of cyclic AMP synthesis which was followed immediately by the onset of DNA replication and entry into mitosis starting 3 or 4 h later.[12] The cyclic AMP surge was expected, because PTH had been shown a couple of years earlier to stimulate adenylyl cyclase in bone and kidney. But there was something odd about this cyclic AMP response to PTH: it was far too slow! The cyclic AMP surge triggered by a direct adenylyl cyclase stimulator such as PGE_1, started almost instantly and took only 5 min to peak while the PTH-induced surge took 20 min to peak.[12] In fact, we found that PTH actually did not stimulate adenylyl cyclase directly. It is now clear that we had stumbled on a different kind of PTH/PTHrP receptor. As discussed in Chapter 4 and illustrated in Figure 5 of Chapter 4, there is now reason to believe that there are two kinds of PTH/PTHrP receptor.[13,14] One of these, RII, independently activates adenylyl cyclase and a phospholipase C that breaks membrane phosphoinositides down into the internal Ca^{2+}-mobilizing Ins(1,4,5)P_3 and PKCs-activating diacylglycerols (DGs) while the other, RI, activates the phospholipase and consequently PKCs, but not adenylyl cyclase. Thymic lymphocytes, like the human and mouse keratinocytes of Chapter 4, express only the RI receptor. Our observations[12] indicated that activating the thymocyte's PTH/PTHrP receptors opens Ca^{2+} channels, but without external Ca^{2+} nothing will happen proliferatively until enough Ca^{2+} is added to the medium. Then, Ca^{2+} surges into the cell through the PTH (PKCs)-opened channels and activates adenylyl cyclase. As discussed in Chapter 2, the next step probably includes the stimulation of PKAs displayed on the thymocyte surface by cyclic AMP released from the cell because unmodified cyclic AMP (which does not enter thymocytes) can bypass Ca^{2+}, bind to high-affinity surface sites, and directly stimulate the initiation of DNA replication.[12]

Clearly, Ca^{2+} and cyclic AMP were the first and second mediators of the signal from the PTH/PTHrP receptors. Thus, raising the external Ca^{2+} concentration above 1.0 m*M*, instead of adding PTH, triggered the same burst of cyclic AMP synthesis, DNA replication, and flow of cells into mitosis as PTH.[12] A powerful, direct adenylyl cyclase stimulator such as PGE_1 did not need any external Ca^{2+} to trigger cyclic AMP synthesis and DNA replication.[12] Adding a small amount (e.g., 10^{-7} *M*) of cyclic AMP or dibutyryl cyclic AMP also triggered DNA replication and the flow of cells into mitosis, but 2′,3′-cyclic AMP and 5′-AMP, which do not activate PKAs, did nothing. Very low (10^{-11} to 10^{-10} *M*) or high (10^{-5} *M*), but not intermediate (10^{-9} to 10^{-7} *M*), concentrations of cyclic GMP (but not 2′,3′-cyclic GMP or 5′-GMP) also triggered the lymphoblast response, the very low ones by a Ca^{2+}-independent mechanism and the high ones by a PTH-like Ca^{2+}-dependent mechanism.[12] But again, the response to cyclic GMP was mediated by a burst of cyclic AMP synthesis.[12]

The cyclic AMP signal that started DNA replication triggered a Ca^{2+}-modulated mechanism which operated during the first 2 min to do something essential (trigger the expression of Cdc2 or cyclin B of the mitosis-promoting Cdc2·cyclin PKs?) for the initiation of mitosis after DNA replication.[12] When the cyclic AMP surge (triggered by PGE_1) was small (a peak value no higher than 140 pmol/10^2 cells) the cells started replicating DNA regardless of whether the external Ca^{2+} concentration was 0 or 1.0 m*M*, but they needed at least 0.5 m*M* of external Ca^{2+} to enter mitosis. On the other hand, when the cyclic AMP surge was much bigger (a peak value of 260 pmol/10^8 cells) the G_2 build-up was blocked when the external Ca^{2+} concentration was higher than 0.2 m*M* during the first 2 or 3 min, but these higher concentrations had no effect when added after 3 min. In either case, the cells stayed in the G_2 phase until the external Ca^{2+} concentration was raised or lowered to the optimal level. When this was done, Cdc2·cyclin A or B PKs presumably began functioning and the cells promptly entered mitosis. The best current guess as to what happened in these ancient experiments (the discovery of the cyclin-dependent PKs was far in the future) is that cyclic AMP (over a wide range of concentrations) triggered DNA replication with or without external Ca^{2+}, but both cyclic AMP and an optimal amount of external Ca^{2+} were needed to trigger the assembly and/or activation of the mitosis-triggered Cdc2·cyclin STPK complexes after chromosome replication was finished. Since cyclic AMP can open, or prolong the opening of, Ca^{2+} channels, the large cyclic AMP surges may have maximally opened (directly or indirectly) the lymphoblast's Ca^{2+} channels. Thus the critical 2 or 3 min after stimulation by a large cyclic AMP surge could have been the time when the Ca^{2+} channels were open and the cell was maximally permeable to Ca^{2+}. The external Ca^{2+} concentration would have been critical for mitosis-induction during this "channel-open" time, because the cell could be loaded with excessively large amounts of Ca^{2+} that would have blocked instead of promoted the postreplication (G_2) accumulation and activation of Cdc2·cyclin A or B STPKs unless, or until, the excess Ca^{2+} was drained from the G_2 cell by incubating it in Ca^{2+}-free medium.

The lesson from all of this is clear: a receptor-induced chain of nonoverlapping, PKCs-induced, Ca^{2+}-induced, and cyclic AMP-induced events led to the stimulation of thymic lymphoblasts. When we caused the first, receptor-induced, event and the third Ca^{2+}-induced event to overlap by adding a proliferogenic amount (1.5 m*M*) of Ca^{2+} along with a receptor activator such as concanavalin A (ConA), PTH, or vasopressin the lymphoblasts did not respond.[12] TPA, a second-event (a burst of membrane-associated PKCs activity) inducer, elicited a powerful, Ca^{2+}-dependent response when the external Ca^{2+} concentration was a nonmitogenic 1.0 m*M*, but there was no response when a mitogenic amount (1.5 m*M*) of Ca^{2+} was added along with TPA.[12] The critical importance of the sequence of the different events is illustrated by our observation that cyclic AMP blocked the response to 1.5 m*M* Ca^{2+} when it was added up to 5 min after Ca^{2+}, but it did not block the response when added at 10 min, because by that time Ca^{2+} had done its job(s) and had kicked the cells into the cyclic AMP-dependent stage.[12] These examples of mediator interference are also indications of the feedback reactions that normally are needed to switch off the signaling mechanisms.

The first of the keys to understanding how the signal triggers DNA replication is our old observation[12] that external cyclic AMP triggered DNA replication from *outside the cell* by binding to high-affinity receptors on the cell surface which are likely to have been the R subunits of PKAs that the cells, like the hepatocytes discussed in Chapter 2, had accumulated on their surfaces during the G_1 build-up and then kept on their surfaces while treadmilling on, or slowly advancing toward, the threshold of the S phase. The second key was the fact that thymidylate synthase was the limiting factor. Thus, the cells could be stimulated to promptly start replicating DNA in Ca^{2+}-free medium when one of a narrow range (5×10^{-5} to 7.5×10^{-5} *M*) of thymidine concentrations was added to the medium. Clearly, those cells lacked only deoxythymidine triphosphate (dTTP), and Ca^{2+} could stimulate the thymidylate synthase activity needed to make the dTMP precursor from dUMP. This was confirmed by three observations:

1. UdR, a precursor of the thymidylate synthase substrate, dUMP, did not stimulate DNA replication in Ca^{2+}-free medium, but it strongly stimulated replication when the medium contained the highest *nonmitogenic* Ca^{2+} concentration, such as 1.0 m*M,* which could stimulate enough thymidylate synthase activity to convert the artificially loaded dUMP pool into enough dTMP which could be phosphorylated to the dTTP needed to trigger replication
2. Proliferogenic Ca^{2+} concentrations reduced the incorporation of exogenous [^{3}H]-thymidine into DNA by stimulating endogenous dTTP synthesis
3. Proliferogenic Ca^{2+} concentrations selectively stimulated the incorporation of exogenous [^{14}C]-formate into the thymine residue of DNA[12]

Thus, the unstimulated cells had been actively expressing all of their replication-related genes, but they selectively destroyed the nascent synthase (perhaps by ubiquitination-induced sensitization to proteases) and the cyclic AMP/PKAs signal might have stopped this destruction. It is also possible that a final intranuclear Ca^{2+}•CaM surge stimulated the assembly of active multienzyme DNA-replicating complexes (replitases) as discussed in Chapter 2.

To recapitulate, it seems that some newborn subcapsular or cortical thymic lymphoblasts have everything they need to replicate DNA except for active thymidylate synthase which becomes the limiting factor that stops them from promptly initiating DNA replication after mitosis. It seems likely that all the lymphoblasts need to start replicating DNA is a Ca^{2+}/cyclic AMP signal from PTH or autocrine/paracrine PTHrP to produce the missing thymidylate synthase and replitases for DNA replication and Cdc2•cyclin PK complexes to initiate mitosis after replication.

These prereplicatively poised, PTH-responsive thymocytes may be mobilized by various trauma such as hemorrhage and partial hepatectomy, which trigger a 1 mg/100 ml drop in the blood Ca^{2+} concentration.[12,16] This drop is caused by a release of calcitonin from the thyroid gland[12] and interleukin-1 (IL-1) from the hepatic Kupffer cells,[21,22] and it is large enough to trigger the maximum release of PTH from the parathyroid chief cells.[23] This postoperative PTH release could be responsible for the stimulation of thymocyte proliferation that follows partial hepatectomy.[24] In other words, unlike the autocrine/paracrine PTHrP, PTH is a wound hormone and thymic lymphoblasts (and spleen lymphocytes) are one of its targets.

The mature single-positive T-cells entering the circulation from the thymic medulla have small nuclei which contain very highly condensed inactive chromatin, a greatly reduced interchromatin matrix and transcript-processing apparatus and an inactive ring-shaped nucleolus.[12] The chromatin condensation is due, in part, to the prevention (apparently by increased lysosomal proteolysis) of certain chromatin-binding non-histone proteins from reaching the nucleus where they could maintain the chromatin in a decondensed active state.[12] The nucleus is wrapped in a thin blanket of cytoplasm containing only a handful of mitochondria, very few polyribosomes, and a poorly developed Golgi complex.[12] Although they make RNA and protein very slowly to maintain pools of essential housekeeping components and surface receptors of various kinds, they do make substantial amounts of 45-S preribosomal RNA which they normally chop into 18-S and 25-S ribosomal RNAs, but they then promptly destroy about 50% of the nascent 18-S RNA. They also transcribe at least one of the cell cycle genes, c-*myb*, but they do not process these transcripts or express c-Myb protein.[12] But, they do make the Bcl-2 protein, which is important for their survival because it prevents the triggering of apoptosis.[2,8–11a,25] Each of these sluggish little cells, with its sparse cytoplasm and handful of mitochondria, is equipped to recognize and bind a specific macrophage-processed and -presented antigen with its clonotypic TCR and to respond to this by growing, proliferating, and becoming a potent secretory cell which produces various cytokines and other things to stimulate or kill other cells.

In Chapter 2 we met the velcroceptors. Each of these is either a monomer (e.g., epidermal growth factor [EGF] and platelet-derived growth factor [PDGF] receptors) with one extracellular ligand-binding domain or a structural heterodimer (e.g., insulin-like growth factor-1 [IGF-1] and insulin receptors) with two extracellular domains, one or two transmembrane domain(s), and one or two cytoplasmic protein tyrosine kinase (YPK) domain(s). Ligand binding causes the monomeric velcroceptors to collect into dimers. The YPK cytoplasmic domain of each component of these ligand-aggregated dimers, or the cytoplasmic YPK domain of each component of ligand-activated structurally heterodimeric receptors, phosphorylates key Y residues on the other. The phosphorylated Y-bearing cytoplasmic motifs are thus auto-velcroized. These velcroized motifs hook a specific team of signaling enzymes and enzyme-adaptors by inserting their phosphorylated pY residues selectively into these enzymes' and adaptors' SH2 pockets. The kinds of signal molecules sticking to the velcroceptor is determined by the sequences flanking the pY residues and the signal molecules' SH2 pockets. Thus, if two receptors have the same Y-containing motif in their cytoplasmic domains, they will hook the same team of SH2-bearing signalers such as PLC-γ_1. This team of signalers sticking to the velcroized receptor motif are phosphorylated by the receptor's YPK domain which sets off a cascade of STPK and dual-function STYPK activities. On the other hand if the receptors have different Y-bearing motifs, they will hook different teams of signalers and thus trigger different signal cascades and cellular responses.

Then there are the *modular velcroceptors*. Each is assembled, *after* ligand binding, from a ligand-binding transmembrane receptor without a cytoplasmic YPK domain, a membrane-attached member of the Src family of YPKs to substitute for a cytoplasmic YPK domain, and a PYPase which activates the Src-YPK to make the active velcroceptor.[6,26a] This ingeniously flexible modularity has been exploited by lymphocytes to assemble different signaler teams and thus generate different signals and stimulate different cellular activities according to the receptors being displayed on the cell surface at particular stages of the cell cycle or differentiation.[26,26a,27,27a]

The T-cell's receptor complex is a modular velcroceptor (Figure 1).[1,26,26a,27,27a] It consists in part of an Ig-like, α,β dimer designed to bind a specific antigen, a transmembrane domain and a cytoplasmic domain *without* YPK activity. Associated with this dimer is the CD3 complex consisting of γ, δ, and ϵ subunits, and the ζζ or ζη chain subunits.[26,26a,27,27a] How can such a kinase-less receptor complex trigger the gene expressions and protein syntheses needed to start a cycle? It can because it associates with, and can somehow activate, one of the Src-family nonreceptor YPKs, p59*fyn(T)* (Figure 1), the level of which is low in immature thymocytes and rises 10-fold as the cells mature and fully express their TCR complexes.[26,27a] Before the αβ dimer binds the macrophage-proffered target antigen, p59*fyn(T)* is in the typical Src YPK inactive "curled" configuration in which the enzyme, attached to a specific membrane receptor protein by its myristoylated N-terminus, blocks its own catalytic domain by curling up to insert the phosphate group of the C-terminal pY^{528} into the more N-terminally sited SH2 pocket (Figure 1).[26,26a,27a] Less than 1% of the membrane-bound p59*fyn(T)* actually binds to TCR·CD3 complexes. By contrast 50 to 90% of the membrane-bound Src-family p59*lck* YPKs bind to the cytoplasmic domains of either CD4 or CD8 TCR coreceptors (Figure 1).[27a] CD4 or CD8 and p56*lck* associate noncovalently, through a metal-mediated linkage to a short motif in the CD4 or CD8 cytoplasmic domain, with conserved pairs of C residues and positively charged amino acids, to the *lck* YPK's 3–32 region containing two C residues.[27a]

T-cells cannot be activated by binding free antigens to their TCRs.[1,26a,27a] Instead, the TCRs must bind to antigens presented to them in the pockets of MHCs I or II on the surfaces of so-called antigen-processing/presenting cells (e.g., B-lymphocytes, dendritic cells).[1,26a,27a] The first step in activation might be the binding of the T-cell's CD2 receptor to LFA-3 antigens on the surface of the antigen-presenting cell.[1] Perhaps this generates a preliminary Ca^{2+} and PKCs signal to mobilize the components of the future velcroceptors and assemble them into functional complexes. The next step is the binding of the TCRα,β dimer to the MHC-borne antigen. The ligand-bound TCR relays the binding signal to the CD3 chains through their associated transmembrane domains. With an integral or complete velcroceptor, such as the EGF receptor, the binding signal directly activates the molecule's cytoplasmic YPK domain. But the ligand-activated TCR·CD3 complex and its associated CD4 or CD8 coreceptors achieve the same end by first associating with, and activating, glycosylated transmembrane PYPases, the CD45s, which make up about 10% of the cell's glycosylated membrane molecules.[1,26,26a,27–27c] The activated CD45s then turn on the YPKs by dephosphorylating their C-terminal, self-SH2-pocketed

Y residues, which causes them to uncurl and expose their catalytic domain to the CD3 and α,β chains.[26,26a] At the same time, another YPK, ZAP-70 associates with the ζζ or ζη chain subunits (Figure 1) which have been phosphorylated by p56*lck* activated by CD4 or CD8 cross-linking.[26b,27a] The upshot of this antigen-triggered shifting of YPks on the inner surface of the T-cell membrane is the assembly of teams of signaling enzymes and SH2/SH3-bearing adaptors and regulators on and around the pY residues of the cytoplasmic domains of the various subunits of the TCR·CD3 complexes and their CD4 or CD8 coreceptors. Among these team members are that star signaler, PLC-γ_1, the 85-kDa regulatory subunit of the functionally obscure, but apparently proliferogenically important, PI3′K, and p120 RasGAP.[27a] When activated (or inhibited as in the case of RasGAP) by phosphorylation by the YPKs, the teams collectively generate the proliferogenic signal as well as other signals that stimulate accessory functions. Of course, we must not forget that a proliferogenic, *polyclonal* signal can be generated by cross-linking CD3 subunits with appropriate antibodies, or by plant lectins such as ConA or phytohemagglutinin, or even in some cases by the combination of a Ca^{2+} ionophore (e.g., A23187 or ionomycin) and a potent PKCs activator such as TPA.[1,12,28] It is noteworthy that these lectins, which have played such a large role in our efforts to understand T-cell activation, bind to the sugar moieties of surface glycoproteins, among which are the external domains of the TCR complex.[26a] Thus, the signal generated by the lectins is in fact a signal from the TCR complex.

As the TCRs nudge onto the antigen in the pockets of the antigen-presenting cells' MHCs, neighboring CD4 or CD8 antigens bind to other specific motifs on the MHCs II or I, respectively.[1] This accessory binding to the MHCs triggers the assembly, and activation by the CD45s, of CD4- or CD8·p56*lck* velcroceptors (Figure 1).[1,26,27b] Either the antigen-responding T-cell needs an incredibly redundant signal to activate itself, or, what seems more likely, TCR·CD3·p59*fyn(T)* velcroceptors and CD4 or CD8·p56*lck* velcroceptors hook different teams of signal molecules and thus evoke different mixes of responses. Thus, for example, the CD4 velcroceptor signal selectively promotes the adhesion of the T-cell to the antigen-presenting cell by activating the constitutively expressed LFA-1 adhesion antigen.[1]

Yet another signal is generated when the T-cell associates with an antigen-presenting B-cell. The T-cell's CD28 receptor glycoprotein is activated by binding to the B-cell's B7/BB1 surface glycoprotein. The signal generated by this binding includes a burst of YPK(s) activity, but not PLC-γ_1 activation and a Ca^{2+} surge or burst of PKC activity, and its job seems to be to supplement the work of the other modular velcroceptor teams by stimulating the production and secretion of certain lymphokines, such as granulocyte-macrophage colony-stimulating factor (GM-CSF), interferon (IFN)-γ, IL-2, and tumor necrosis factor (TNF)-α.[26a]

Members of the signaling team, such as PLC-γ_1, that stick on to the activated TCR·CD3 complexes are activated by phosphorylation on Y residues within 5 s, and DGs and Ins(1,4,5)P_3 from PtdIns(4,5)P_2 breakdown by the activated PLC-γ_1 start appearing between 30 and 40 s.[1,12,26,26a,27,27a] We will learn further on that the same PLC-γ_1-containing signal team sticks onto the activated mIg receptor complexes of B-lymphocytes, because both the TCR and mIg complexes share the same Y-rich motif in their cytoplasmic domains.[1] The Ins(1,4,5)P_3 (IP_3) binds to IP_3 receptors (IP_3Rs) on internal storage vesicles and causes a Ca^{2+} spike which starts and peaks in seconds and lasts for only a minute or so.[29,30] If there is no external Ca^{2+}, the internal Ca^{2+} concentration will still spike, but it then drops prematurely to the starting value.[29,30] However, as discussed in Chapter 2, the emptying of the internal stores causes the release of a coupling factor which opens voltage-insensitive Ca^{2+} channels to admit Ca^{2+} from the medium.[29,30] Thus, if there is enough Ca^{2+} in the medium, activating the TCR complex with antigen or a polyclonal agent (anti-CD3 antibody or phytohemagglutinin) triggers a sharp Ca^{2+} spike which drops to a lower, but still suprabasal plateau, which lasts for as long as 8 h.[29,30] According to Gelfand[29] it is this Ca^{2+} influx and sustained Ca^{2+} plateau, not the initial Ca^{2+} spike, that specifically starts the production of the IL-2 needed to start the G_1 build-up. Thus, it seems that the job of the Ins(1,4,5)P_3-induced emptying of the internal calcium stores is not to produce a Ca^{2+} spike but to release the Ca^{2+} channel-opening coupling factor. But there is another possibility. T-lymphocytes also have IP_3Rs in their plasma membranes.[31] These receptors collect along with the TCR·CD3 and other signaling components into polar "caps" where they are associated with a Ca^{2+} concentration that is higher than elsewhere in the cell and stays high even after the Ca^{2+} level in other parts of the cell starts dropping.[31] Therefore, the receptor-generated Ins(1,4,5)P_3 may do two things: trigger the emptying of internal stores to release channel-coupling factor and activate membrane IP_3R receptor/Ca^{2+} channels to let Ca^{2+} flow into the cell.

How might the in-rushing Ca^{2+} stimulate the all-important expression of IL-2? It seems that this Ca^{2+} activates a phosphatase, calcineurin, which somehow collaborates with the Ras-induced MAPPKs cascade (to be discussed further on) to stimulate the translocation of a T-cell-specific cytoplasmic protein transcription factor to the nucleus where it combines with a resident protein belonging to the AP1 (c-Fos/c-Jun) family to form a complex called NFAT (*N*uclear *F*actor of *A*ctivated *T*-cells) which activates the IL-2 gene by binding to the gene's T-cell-specific enhancer element.[32]

Lymphocytes do not have voltage-sensitive Ca^{2+} channels, and membrane depolarization by incubation in high-K^+ medium actually inhibits Ca^{2+} uptake rather than stimulates it as it does in cells with voltage-sensitive channels.[29,30] This inhibition of Ca^{2+} uptake is probably due to the reduced electrochemical inward driving force.[30] As expected, depolarization inhibits the events leading to IL-2 production and proliferation. The Ca^{2+} surge also activates K^+ channels and the resulting outward flow of K^+ hyperpolarizes (by 5 to 7 mV) the cells. But, this hyperpolarization is not important, because preventing it with the K^+ channel-blocking scorpion toxin, charybdotoxin, does not affect proliferogenesis by TCR·CD3 complex activators or lectins.[29]

The DGs generated by the phosphoinositide breakdown stimulate membrane-associated PKCs which have been translocated to perireceptor, hence DG-rich membrane sites, or which were already attached to the membrane, but in an inactive state waiting for receptor-generated DGs (see below). The role(s) of the PKCs in activation is unknown. But they are needed, because downregulating them (i.e., destroying them by endocellular protease activity) by prolonged pretreatment of the cells with a phorbol ester such as TPA prevents the subsequent initiation of a proliferogenic response.[1] Perhaps one of the reasons for this is that PKCs activate cytoskeletally anchored NFκB·IκB complexes by phosphorylating the IκB inhibitory component which releases p59·p65 NFκB heterodimers that then move into the nucleus where they stimulate the expression of several genes with the so-called κB regulatory element, two of which encode the proliferogenically all-important c-Myc apoptogenic/proliferogenic transcription factor and the IL-2 receptor's β-subunit.[33–35] In the years BVE (before the velcroceptor era) the results of experiments on normal T-cells activated by plant lectins (mainly ConA and phytohemagglutinin), which are polyclonal, TCR-activating proliferogens, and experiments on normal, "immortalized", or transformed T-cell lines treated with PKCs-stimulating phorbol esters such as TPA strongly suggested that the PKCs and Ca^{2+} together mediate T-cell activation, and that the PKCs were responsible for starting the multistep expression of IL-2 receptor subunits, specifically the nonsignaling α(Tac)-subunit.[12,29,30] But things are different in the velcroceptor era, now that the TCR·CD3 complex has been defined and polyclonal proliferogenesis can be triggered by cross-linking antibodies to nonclonotypic TCR components, such as the CD3 subunit.[1,12] While PKCs are key parts of the TCR signal, the events they trigger are actually unknown, because the mimicry of TCR-triggered events by hyperpotent, excessive PKCs stimulation by a phorbol ester such as TPA must no longer be accepted as definitively indicating the participation of the PKCs in these events. For example, since TPA activates Ras proteins in T-cells it was assumed that PKCs are responsible for the Ras response to TCR activation.[36] But while the PKCs' pseudosubstrate inhibitor peptide blocks Ras activation in T-cells by TPA, it does *not* affect TCR-induced Ras activation.[36]

The activation of Ras protein is in fact a key event in T-cell activation.[27,36] Ras proteins can be activated by inhibiting GAP proteins that stimulate the Ras GTP-ase activity and thus reduce the level of active GTP·Ras complexes, or they can be activated by stimulating guanine nucleotide-releasing proteins (GNRPs), such as mSos,[36a] that trigger the replacement of GDP by GTP. The TCR signal does not appear to activate GTP-GDP exchange[36] as might have been expected, because mSos is commonly collected and presented to membrane-associated Ras proteins by Grb-2 SH2/SH3 motifs-bearing adaptor proteins with SH motifs that stick to activated velcroceptors and SH3 motifs that bind Ras proteins.[36a] Actually, T-cells have a permanently turned-on GNRP.[36] Instead, the signal phosphorylates and turns off the Ras-inactivating GAP protein, which causes a build-up of active GTP·Ras complexes.[27,36] These

GTP·Ras proteins form complexes with the Raf-1 STPK and the MAP kinases.[37] As discussed in Chapter 2, this triggers a key cascade. GTP·Ras activates Raf-1, which activates the dual-function STY-MAPPK kinase, which activates the ST-MAPPKs 1 and 2, which phosphorylate and "activate" the ribosomal S6 protein and several "master" protein transcription factors, such as c-Jun and the proliferation-promoting (though apoptogenic if not supported by other proliferogenic signalers[11a,38]) c-Myc.[27] It has already been mentioned that one of the most important of the many things the Ras-MAPPKs cascade does is cooperate with the surging Ca^{2+} and Ca^{2+}-activated calcineurin phosphatase to stimulate the translocation of the T-cell-specific cytoplasmic protein component of the NFAT transcription factor complex to the nucleus where it joins with an AP1 family member to form the complete NFAT complex that activates the IL-2 gene.[32] The GTP·Ras complexes also stimulate a STPK (normally not PKCs, but maybe insulin-activated protamine kinase) that increases the normally low translatability (or low "translational strength") of growth-related messages by phosphorylating the eIF-4E component of the translation-initiation eIF-4 complex.[39–41]

Activation of the TCR triggers a cyclic AMP surge and the translocation of the proliferation-related $PKAI_{\alpha}$,[42] but not $PKAII_{\alpha}$, into TCR·CD3·PKAIα complexes.[43] There are also traces of a small (10 to 20%) increase in the total cellular cyclic AMP content in response to the mitogenic complexing of CD3 subunits by monoclonal antibodies.[12] This increase is small because it is restricted to the cell membrane. Proliferogenic lectins trigger an early cyclic AMP surge in human, murine, and rat T-cells, and the lectin, phytohemagglutinin, triggers within 2 min a patchy immunofluorescence pattern for cyclic AMP along the plasma membrane with no fluorescence in the nuclear membrane.[12,44] The likelihood of this cyclic AMP surge being important for T-cell activation is indicated by the report of Klimpel et al.[45] that the mitogenic lectins, ConA, phytohemagglutinin, or pokeweed mitogen, at optimally mitogenic doses, but not the nonmitogenic wheat germ mitogen and *Agaricus bisporus* lectins, selectively activate PKAI within 30 min. It is also indicated by the results of Shirakawa et al.[46] who showed that C3H/HeJ thymocytes cannot be stimulated to express IL-2 receptors and initiate DNA replication by phytohemagglutinin without an initial 10-min signal from IL-1 receptors. This comitogenic signal is a cyclic AMP surge which peaks at 10 min and subsides completely by 60 to 120 min.[46] It can be replaced completely by an initial 10-min exposure to 1 μM 8-Br-cyclic AMP or dibutyryl cyclic AMP, or by an adenylyl cyclase stimulator such as cholera toxin or forskolin, or by a cyclic nucleotide PDE inhibitor such as aminophylline.[46] Therefore, it is likely that a cyclic AMP/PKAI signal is a player in T-cell activation much as it is in the hepatocyte activation discussed in Chapter 2. One of the roles of the cyclic AMP/PKAI signal is to trigger the mobilization and collection of various TCR components and other things, such as membrane-associated Ca^{2+} influx-stimulating IP_3 receptors, into polar caps.[31,47]

The TCR signal also includes superoxide (O_2^-) and a variety of short-lived oxygen species such as the iron-catalyzed OH$^\bullet$ radicals and longer-lived oxidants which can produce lipid peroxides and Ca^{2+}-dependent, DNA strand-breaking ("clastogenic") agents.[11a,12,48] These *signal-generated* oxidants are responsible for the DNA strand breaks in activated lymphocytes as well as a burst of guanylyl cyclase activity and prostaglandin synthetase activities.[12] This shower of reactive oxygen species probably originates from plasma membrane NAD(P)H oxidase stimulation by DG-activated PKCs.[12,49] These ROI byproducts of signal generation are dangerous, and, like those generated by ionizing radiation, will be apoptogenic unless the cell can inactivate them or repair the damage they cause.[11a,12,50] It seems that this is the purpose of the Bcl-2 protein which the T-lymphocyte maintains or makes after activation if its supply has diminished for some reason.[8–11a,51] The inability to make Bcl-2 to counter the effects of signal-generated oxygen radicals may be at least part of the reason why TCR·CD3 activation kills immature T-cells.[2]

Another contributor to the cell's defense against signal-induced DNA strand breaks and p53-triggered apoptosis might be the Na^+/H^+ antiporter.[12] Receptor-activated PKCs stimulate the Na^+/H^+-antiporter by raising its pH set-point.[12,30] While it seems that the PKCs-activated Na^+/H^+-antiporter does not raise the T-cell's pH_i or affect prereplicative development under normal conditions in the presence of HCO_3^-, the increased influx of Na^+ it causes might serve the important function of promoting the rejoining of signal-induced DNA strand breaks.[12,29,30]

The TCR signal evokes an astonishing response from the sluggish, little resting T-cell. The Ras-triggered PKs cascade activates transcription factors and releases them from their moorings to the cytoskeleton. The signal somehow stops lysosomal proteolysis, and non-histone proteins, such as the activated and liberated transcription factors, now move rapidly into the nucleus.[12] The Ca^{2+} surge triggers the nucleus-specific actomyosin motor in the nuclear matrix which moves the chromatin into the appropriate configuration and position for receiving, and responding to, the incoming proliferogenic transcription factors.[52] Within an hour, histones are phosphorylated, the nucleus, with its repositioning chromatin structure, starts expanding into an active state, gene transcript wastage stops, and there is a massive increase in the expression of as many as 8000 to 10000 genes, among which are the first of the cell cycle genes such as c-*fos*, and c-*jun* c-*myc*.[12] Among the many others are the genes for GM-CSF, growth hormone, IL-3, prolactin, proopiomelanocortin (the precursor of adrenocorticotropin [ACTH] and endorphins), and thyroid-stimulating hormone (TSH), which makes the activated T-cell into a diffuse multipotent "micro-endocrine gland" whose several products modulate the immune response, hemopoiesis, and other functions. This astonishing response is the work of several promiscuous transcription factors, such as AP1 (c-Fos/c-Jun) and NFκB, activated in addition to the NFAT complex by the potent TCR signal.[12] At first glance the repositioning nuclear chromatin seems to decondense, but the total volume of condensed chromatin actually remains the same, and the nuclear decondensation and increase in volume are actually due mainly to expansion of the nuclear matrix and the transcription machinery.[12] Around 6 h there is the first burst of nuclear pore complex production, presumably to organize and facilitate the increased traffic of gene transcripts out of the nucleus and various components including transcription factors into the nucleus.[12] The high ribonuclease activity of the quiescent cell stops, ribosomal RNA wastage stops, and ribosomes accumulate to bring the now dramatically growing cell to the point where it can start the G_1 build-up.[12]

In the cytoplasm, the signal increases energy production, and the phosphorylation of the ribosomal S6 protein and the eIF-4 translation initiation complex by the Ras-triggered PK cascade powerfully enhances protein synthesis and in particular increases the translatability of the growth-related messages which were already in the cytoplasm and are now pouring out of the activated nucleus. The opening of Ca^{2+} channels in the plasma membrane increases glucose uptake, possibly by triggering the formation of active dimeric glucose transporters.[12] The signal also causes the cells to make and display IGF-1 and insulin heterodimeric velcroceptors which are necessary for the G_1 build-up and whose signals (from the team of signalers sticking to the insulin receptor-specific substrate-1, IRS-1, adaptor clinging by its SH2 motifs to each activated receptor) further drive the Ras-induced PK cascade, protein synthesis, and the production of glucose transporters.[1,12,53,54] The increased glucose uptake feeds an increased glycolysis resulting from a signal-activated web of Ca^{2+}/Ca^{2+}•CaM/cyclic AMP-modulated enzyme activities. The Ca^{2+} ions flowing into the cytosol plunge into the cell's few mitochondria where they stimulate ATP production by lowering the Km values (increasing the substrate affinities) of pyruvate dehydrogenase, isocitrate dehydrogenase, and oxoglutarate dehydrogenase.[12] But the mitochondria are at first too few in number to handle the additional load of glycolysis-generated pyruvate which spills over into the lactate pool.[12] However, this will change as the cell starts making the large numbers of mitochondria needed to fuel the massive growth response.

The TCR•CD3 signal must start the cell growing into a state of competence to respond to the signals from a second modular velcroceptor-associated team, to make the second signaler and the parts of its receptor at the appropriate times, and to make something to prevent the cell from being killed by the ROIs from the signaling mechanisms. Unless the cell takes appropriate steps, the ROIs can do several things, among which are the stimulation of the transcription of the c-*fos* and c-*myc* genes and the release of Ca^{2+} from the mitochondria and initiation of Ca^{2+} cycling by the mitochondria.[12,50] The flow of Ca^{2+} into and out of the oxidatively stressed mitochondria dissipates the all-important potential difference across the inner mitochondrial membrane and thus reduces proton flow down the gradient and ATP production by H^+-ATPase. The developing energy shortage eventually causes the Ca^{2+} pumps in the plasma membrane to fail, which results in a cytosolic Ca^{2+} surge that stimulates the transglutaminase, protease, and endonuclease activities that are responsible for apoptosis.[55] Around 6 h the cell starts making the second signaler, IL-2.[1,12] It also increases the expression of the Bcl-2 protein which associates with the inner mitochondrial membrane and to a lesser extent with other membranes and prevents the ROIs from triggering apoptosis.[8–11a]

The high-affinity IL-2 receptor is another modular velcroceptor like the TCR.[47] Among the early genes stimulated by the resting T-cell's TCR·CD3 complex is the gene for the IL-2 receptor's 55-kDa α- (or Tac)-subunit.[12] The α-subunit consists of a 219-residue, IL-2-binding external domain, a 19-residue transmembrane domain, and a very short, 13-residue cytoplasmic domain.[56] This subunit binds IL-2 with a low affinity ($K_D = 10^{-8}$ *M*), and it does not signal.[56] However, it becomes a key part of the high-affinity IL-2 receptor when it associates with the even lower-affinity ($K_D = 10^{-7}$ *M*), 525-residue, 75-kDa β-subunit which surfaces much later, around 24 h.[12,31,56] The β-subunit consists of an external, 215-residue IL-2-binding domain, a 25-residue transmembrane domain, and a 268-residue cytoplasmic domain. Although it is the signaling part of the receptor, the cytoplasmic domain has no PK motifs.[56] But the domain is linked by its "acidic" regions to a Src-family YPK, p56^{lck}, and this "serine-rich" region is critical for the mitogenic signal.[26,56] The third part of the receptor is the 64-kDa γ-subunit which does not bind IL-2 and has a still-obscure signaling role.[56] By 24 h all three components have surfaced and combined to form trimeric, high-affinity ($K_D = 10^{-11}$ *M*) IL-2 receptors or IL-2Rs. When it binds IL-2, the IL-2R, like the TCR·CD3 complex, somehow mobilizes and activates CD45 (Figure 2), the master PYPase that activates the YPK associated with the β-subunit.[1,26,27b,56]

It seems that a much lower level of TCR·CD3 complex signaling is needed to trigger IL-2R expression than is needed to trigger IL-2 production.[12,30,36] Moreover, while a Ca^{2+} surge is needed for IL-2 expression, IL-2R expression (at least the first expressed α-subunit) does not.[29,30] Thus, normal T-lymphocytes exposed to suboptimal concentrations of phytohemagglutinin, which do not cause a Ca^{2+} surge, express and display the α-subunit on their surfaces and ultimately complete the formation of the IL-2Rs, but they do not express the IL-2 gene and therefore cannot start making DNA unless supplied with IL-2 or treated with the Ca^{2+} ionophore A23187.[29,30]

The IL-2R signal drives the G_1 build-up to DNA replication, but despite a lot of effort we still do not know what this signal is.[56] Is it a sort of generic signal, identical to the TCR modular velcroceptor signal, which evokes different responses because the cells have been primed by the TCR·CD3 signal and have a different set of available targets? A major problem has been the use of constitutively activated T-cell lines, some of them neoplastic, instead of normal TCR- or CD3-primed T-cells to determine the nature of the IL-2 signal.[12] According to some, IL-2 does not stimulate adenylyl cyclase, but does stimulate PLC in transformed CT6 cells in serum-free medium.[12] The activated PLC breaks PtdIns(4,5)P_2 down into Ca^{2+}-mobilizing Ins(1,4,5)P_3 and PKCs-stimulating DGs. This signal opens Ca^{2+} channels in response to the emptying of internal stores and reduces adenylyl cyclase activity.[12] But others, working

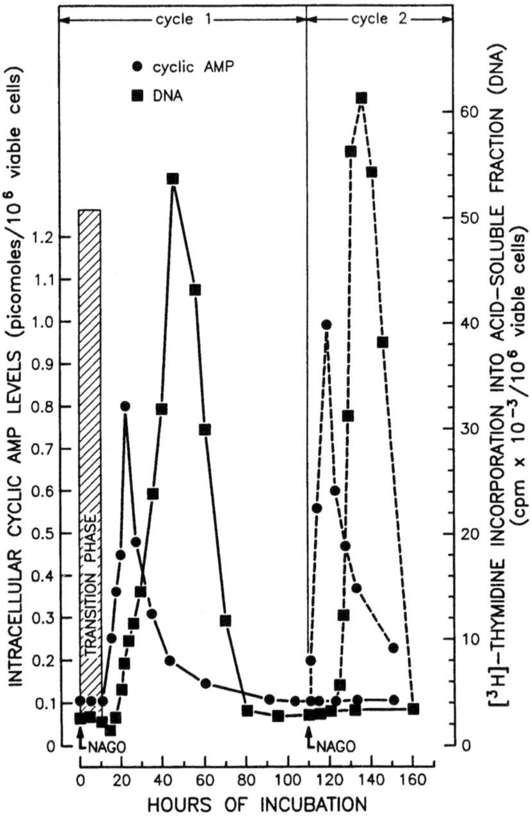

FIGURE 2 The sequential surges of cyclic AMP and DNA synthesis in spleen lymphocytes activated by a combination of neuraminidase and galactose oxidase (NAGO). In the first cycle, there was a 10-h delay before the cyclic AMP surge, which was the time needed for the transition of the quiescent cells to a state of competence to start the G_1 build-up. There was no delay when NAGO was added before the progeny of the first cycle had had a chance to return to the proliferatively inactive state. The data from which these curves were drawn were generously provided by Dr. J.E. Foker (Medical School, University of Minnesota, Minneapolis, MN).[83]

with the same cells in medium containing fetal bovine serum, found that IL-2 does *not* stimulate phosphoinositide breakdown, PKCs, or a Ca^{2+} surge.[12,30] Moreover, downregulating PKCs in cells of one established line by prolonged exposure to TPA does not diminish their responsiveness to IL-2, and mutant, neoplastic T-cells that cannot make their principal PKC isoform can still respond to IL-2 and initiate DNA replication.[12] Koyasu et al.[57] have argued that the signal generated by IL-2 cannot be a simple PLC-triggered Ca^{2+}/PKCs signal, because the combination of the A23187 Ca^{2+} ionophore and the PKCs-activating TPA that can mimic the TCR•CD3 signal in some T-cell lines does not stimulate DNA replication without IL-2. Another indication of a difference is that while IL-2 stimulates the proliferation of TCR·CD3-primed cells, adding IL-2 before activating the TCR·CD3 complex *kills* the cells by triggering apoptosis.[58]

At first sight it seems that the TCR·CD3 and IL-2R signals might in fact be the same because their activation involves the same Src-family PKs and CD45 PYPase. But it is also likely that different teams of signalers hook onto the two receptors. IL-2, like antigen, promptly increases PI3′-K activity, stimulates a Ras-mediated Raf-1/MAP-PKs cascade, and triggers the breakdown of glycosylphosphatidylinositol into glycosylated diacyglycerol and an inositol phosphate glycan.[27,36,56,59] But in agreement with the observations of Mills et al.[60] on neoplastic T-cells, IL-2 does not trigger PtdIns(4,5)P_2 hydrolysis in the IL-2-dependent progeny of phytohemagglutinin-stimulated *normal* human T-cells.[61] Therefore, Il-2R does not hook onto PLC-γ_1 as does TCR·CD3 and thus does *not* generate a Ca^{2+} signal. But the receptor might be linked to a PtdCh-specific PLC or PLD which generates, directly or indirectly, PKC-stimulating DGs without Ca^{2+}-mobilizing breakdown products. Thus, Chakravarthy et al.[62] and Lu et al.[63] have found that IL-2 stimulates a preexisting pool of membrane-associated PKCs in the IL-2-addicted CTLL-2 T-cells and S49T-lymphoma cells without translocating enzymes from the cytosol to DG-rich perireceptor sites in the plasma membrane. There may also be a burst of adenylyl cyclase activity as there is with TCR·CD3 activation, because Wickremasinghe et al.[61] found that IL-2 stimulated a cyclic AMP surge in normal human peripheral blood lymphocytes which peaked between 20 and 30 min at a level 1.6 to 2.3 times higher than the basal value. Moreover, Ebie et al.[64] found that IL-2 strongly stimulated cyclic AMP synthesis in the blood of cancer patients, but not normal subjects (probably because fewer peripheral T-cells from normal subjects are activated and express IL-2Rs).

In Chapter 2 we learned that there are two cyclic AMP surges during the prereplicative development of hepatocytes from quiescence to DNA replication and that the second one of these is apparently universal in normal cells and is essential for the G_1-build-up surge. The same seems to apply to normal T-lymphocytes. Thus, Foker et al.[65] found that the cyclic AMP content of ConA-stimulated human T-cells tripled just before the onset of DNA replication and dropped to the basal level before DNA-synthetic activity peaked. Monahan et al.[66] found that the cyclic AMP content of phytohemagglutinin-stimulated human T-lymphocytes surged between 30 min and 1 h, subsided around 5 h, rose again during the next hour to produce the G_1 surge, and ultimately subsided sometime between 8 and 24 h, which was well before the onset of DNA replication.

According to one scenerio, the G_1 cyclic AMP surge would be stimulated directly or indirectly by IL-2, and it would start the G_1 build-up as proposed in Chapter 2 by stimulating the expression of the G_1 build-up-specific cyclins and cyclin-dependent PKs. But according to Cantrell et al.,[36] for example, T-cells finish making IL-2 within 12 h of activation, but high-affinity IL-2Rs do not surface until 24 to 48 h. Therefore, the G_1 cyclic AMP surge may be induced by some ubiquitous endogeneous factor, as proposed in Chapter 2, and then PKAs directly stimulate the expression of the IL-2R α gene (which has a cAMP-responsive element)[67,68] and indirectly, perhaps via the NFκB transcription factor, stimulates the expression of the IL-2R β gene which has a κB regulatory element.[31] When they appear, the high-affinity receptors would be stimulated by preaccumulated IL-2.[36]

In the activated T-cell, the IL-2R signal stimulates the later, replication-related events of the G_1 build-up. It stimulates the appearance of the p53 protein,[12,56] which is needed to monitor the as yet unreplicated chromosomes for breaks and delay replication until the breaks are repaired or trigger apoptosis if they are irreparable. The signal also stimulates the appearance of the c-Myb protein,[12,56] which as we have seen in Chapter 2, is probably a key player in the coordinate expression of the genes coding for the replication-related enzymes and regulators such as Cdk2-cyclin A PKs, which trigger replication by activating the unwinding of replication origins. The IL-2R signal somehow stimulates the expression of the *cdc2* gene late in the G_1 build-up and the appearance of Cdc2 protein at or near the onset of DNA replication in T-cells activated by anti-CD3-antibody or phytohemagglutinin.[69,70] Activating these cells with phorbol dibutyrate stimulated IL-2R expression, but it did not stimulate IL-2 production, Cdc2 appearance, or DNA replication.[69] But adding IL-2 with the phorbol ester enabled the cells to make Cdc2 and replicate their DNA.[69] While Furukawa et al.[69] and Kim et al.[70] tracked the 34-kDa Cdc2, it seems likely from our discussion in Chapter 2 that the actual replication-triggering cyclin-dependent PK was the 33-kDa Cdk2, which Lucas et al.[71] found to appear along with Cdc2 just before the onset of DNA replication. The key role of the cyclin-dependent PKs at the end of the IL-2R-triggered G_1 build-up has also been demonstrated by Morice et al.[72] using the IL-2-addicted CTLL-2 cytotoxic leukemia cells. Thus, IL-2 stimulated the appearance of cyclic A and E, the assembly of active Cdk2·cyclin A, Cdk2·cyclin E, and Cdc2·cyclin A PK complexes, their entry into the nucleus, and the initiation of DNA replication.[72] Silencing the IL-2R signal with rapamycin completely prevented cyclin A expression, partly prevented cyclin E expression, and blocked the onset of DNA replication.[72] Even more significant is the observation of Furukawa et al.[69] that antisense oligodeoxynucleotides that reduced the increase in *Cdc2* (and presumably Cdk2) gene expression inhibited the initiation of DNA replication in phytohemagglutinin-stimulated human T-cells without stopping the preceding induction of c-*myc* and c-*myb* expression.

The appearance of the mitosis-specific Cdc2 alongside the G_1-specific Cdk2 and the accumulating cyclin A (for the formation of replication-triggering Cdk2-cyclin complexes) and p53 protein poses a serious threat to the T-cell should it be stopped from starting replication and reaching later stages of the cycle where these potent proteins are appropriately used, processed, and eliminated. Thus, preventing late-G_1 T-cells from initiating replication, with excess thymidine, for example, causes cyclin A and p53 (and c-Myc) specifically to continue accumulating to very high levels by 20 h or so when they trigger a lethal, apoptogenic mechanism.[73] The mechanism might be the accumulation of enough active Cdc2·cyclin A PKs to trigger an apoptosis-like, premature mitotic catastrophe (see Chapter 2) consisting of nuclear membrane breakdown, chromatin condensation, and eventual cleavage of the chromatin into a "ladder" of oligonucleosomes.[73] This may be one of the ways by which p53 kills cells with irreparably damaged chromosomes.

The continued expression of high-affinity IL-2Rs needs the presence of antigen-presenting cells and TCR·CD3 signals.[1] Without these, the high-affinity receptors bind the rapidly diminishing IL-2, are internalized with their bound IL-2, degraded, and replaced by only the nonsignaling IL-2R α subunits. However, if the still-activated/competent cells are exposed to IL-2 before losing too many of their high-affinity receptors they can restore their IL-2R complement (IL-2 stimulates IL-2R expression) and immediately start the G_1 build-up to DNA replication. But eventually they become quiescent and will not start cycling unless activated by TCR·CD3 signals (or lectin-triggered signals). The signal from the high-affinity IL-2Rs is a general cell cycle gene activator which can stimulate early cell cycle genes such as c-*myc* in T-cells of established lines constitutively expressing high-affinity IL-2Rs as well as genes specific for the later G_1 build-up in normal TCR·CD3-activated T-cells.

In Chapter 2 we learned that rat hepatocytes and other cells accumulate CaM during their prereplicative development and that this CaM plays a key role in the initiation of DNA replication. It seemed that part of this CaM collects in the prereplicative nucleus. Then, at the end of the G_1 build-up there is a surge of Ca^{2+} from nuclear envelope stores. The resulting nuclear Ca^{2+}·CaM complexes promote the assembly and activation of DNA replication complexes (replitases). Similar things may happen in T-cells,[74] but the evidence is sparse. Between 6 and 10 h after exposure to phytohemagglutinin, human T-cells equally increase the expression of their CaM I and II genes, but they increase the expression of the CaM III gene much more. By 24 h the CaM I and II transcripts peak at twice the basal value, but the CaM III transcripts continue accumulating until around 70 h when their level has reached nine times the basal value. CaM starts accumulating about the same time as its genes' transcripts, and its level peaks around 48 h at a level which is about twice the starting value. As in regenerating rat liver, the nuclear CaM content also rises along with the appearance of 170- and 180-kDa CaM-binding proteins. As discussed in Chapter 2, this accumulated CaM is probably a key part of replication complexes.

In normal fibroblasts and hepatocytes, the final stage of the G_1 build-up is especially sensitive to the external Ca^{2+} concentration. As we have learned in the previous chapters, this dependence could be due to a need for signals from surface receptors such as the Ca^{2+}-binding integrins to stimulate the expression of the cyclin A needed for the Cdk2·cyclin A PKs that activate DNA replication complexes. But does this also apply to T-lymphocytes? The short answer is —

no one knows. But the responsiveness of freshly isolated G_0/late-G_1thymic lymphoblasts to Ca^{2+} suggests that it does. Indeed, these thymic cells might display Ca^{2+} receptors like those on the surfaces of parathyroid chief cells,[74a] because they are set to respond maximally (with DNA replication) to a very small increase of the extracellular Ca^{2+} concentration (from 1.0 to 1.2 m*M*).[12] More support for a late G_1 Ca^{2+}-dependence is the observation of Birx et al.[75] that Ca^{2+} channel blockers prevented phytohemagglutinin-stimulated human T-cells from initiating DNA replication when added *after* the IL-2R signal and during the last 24 h of the prereplicative period. And Blitstein-Willinger and Diamanstein[76] also found that the Ca^{2+} channel blocker verapamil blocked a late-G_1 step in ConA- and phytohemagglutinin-stimulated T-cells.

As discussed in Chapter 2, rat hepatocytes accumulate, and display, PKAs on their surfaces during the G_1 build-up and it seems that accumulated cyclic AMP is released from the cell to stimulate some critical final G_1 events by activating the surface PKAs. The same may apply to T-cells. Thus, thymocytes stalled in late G_1, like late-G_1 hepatocytes, display PKAs on their surfaces and they can be induced to start replicating DNA by external, unmodified cyclic AMP, which cannot enter the cells.[12]

Finally, although we have focused on the modular velcroceptors that drive T-lymphocyte growth, proliferation, and the expression and secretion of cytokines that affect the functions of accessory cells, we must not lose sight of the many other receptors and signal teams that modulate the T-cell response to antigen and contact with other cells.[1,27a] Thus, for example, there are the glycophosphatidylinositol-linked surface molecules such as human CD59, CD55, CD48, CD24, and CD14 as well as murine thy-1 and Ly-6 that can associate with the principal $p56^{lck}$ signal YPK.

II. B-LYMPHOCYTES

The resting mature B-lymphocyte has on its surface a receptor complex similar to the T-cell's TCR·CD3 complex.[1,26,59,77–79] This signaling receptor consists of mIgM (membrane-associated IgM) which is associated with three other transmembrane proteins, the 32- to 34-kDa Igα, the 37- to 39-kDa Igβ, and the 35-kDa Igγ. Igα is homologous to the T-cell's CD3γ, δ, and ϵ chains (Figure 1). It also has a 60-residue cytoplasmic tail with Y residues arranged in a motif which is shared by the Ig molecules of the TCR. This should immediately alert us to the likelihood of a modular velcroceptor which can assemble the same basic signal team as the T-cell's TCR·CDE receptor complex. Indeed, the B-cell's mIgM associates with the Src-family Y-PKs, $p53^{lyn}$, $p56^{lyn}$, $p55^{blk(B)}$ which are not expressed in T-cells.[1,26,59]

Specific antigen binding to mIgM or the polyclonal cross-linking of mIgM by anti-Ig antibody sends a signal through to the complex which somehow activates the ubiquitous CD45 transmembrane phosphatase isoform(s) that activates the membrane-bound receptor-associated YPKs.[27a,59] The YPKs phosphorylate the key motifs on the cytoplasmic domains of the receptor molecules which then hook on to the members of their signal teams, which are activated by the associated YPKs. One of these team members is the familiar PLC-γ_1 which, when phosphorylated by the YPKs, breaks PtdIns(4,5)P_2 down into the Ca^{2+}-mobilizing Ins(1,4,5)P_3 and the PKC-stimulating DGs.[59,77,78] As in T-cells, the Ins(1,4,5)P_3 promptly triggers the release of Ca^{2+} and coupling factor from internal stores. The stores then refill and the initial surge starts subsiding, but the coupling factor opens membrane Ca^{2+} channels to start a prolonged flow of external Ca^{2+} into both cytoplasm and nucleus which holds the intracellular Ca^{2+} concentration at a lower, but still suprabasal, level.[30,79,80] The result is the same biphasic Ca^{2+} response as occurs in activated T-cells.[59] The Ca^{2+} surge is essential, and, together with the DG-stimulated PKCs, contributes to the events that start the B-cell growing.[77,78] The GTP·Ras-induced cascade of PKs stimulates key cell cycle genes and protein synthesis. As in other cells, the early signal-activated genes are c-*fos*, c-*jun*, and the apoptogenically dangerous,[50a] but yet proliferogenically important, c-*myc*. Another gene is *egr*-1 (early growth response-1), whose product stimulates the production of the anti-apoptogenic Bcl-2 protein.[59] In immature (mIgM$^+$, mIgD$^-$) B-cells, the *egr*-1 gene cannot respond to the mIgM signal because its promoter is hobbled by heavy methylation, and the stimulated cell may kill itself by (c-Myc-promoted?) apoptosis.[50] In mature B-cells, the *egr*-1 promoter is hypomethylated, which is why the *egr*-1 gene can respond to the first signal and the cell can grow and proliferate rather than self-destruct.[59]

One of the B-cell-stimulating cytokines produced by activated T-cells is IL-4, whose receptor is a modular velcroceptor that operates by activating a cytosolic YPK and associating with a 170-kDa phosphoprotein adaptor which is phosphorylated by the YPK and collects the receptor's signaling team like (and indeed can be replaced by) the phosphorylated IRS-1 adaptor of the heterodimeric insulin receptor.[54] IL-4 can activate human tonsillar B-cells which have been primed by a submitogenic amount of anti-mIgM antibody.[81] This signal consists of an initial (within 1 min) Ca^{2+} surge (and probably a burst of PKCs activity) followed 15 min later by a cyclic AMP surge which lasts for at least 30 min.[81] Chelating the surging Ca^{2+} with BAPTA prevents the cyclic AMP surge and the activation of the cell, as indicated by the expression of the CD23 antigen, the extracellular domain of which, when cleaved, acts as an autocrine factor that drives the cell further towards activation and differentiation.[1,81] Treating the resting B-cells with a Ca^{2+} ionophore and a PKCs-stimulating phorbol ester and then dibutyryl cyclic AMP mimicked the IL-4 signal and stimulated the expression of CD23 antigen.[81] It is important to note that the signaler sequence is critical. Thus, treating the cells with dibutyryl cyclic AMP *before* IL-4 *prevented* them from responding to the cytokine, despite the fact that a cyclic AMP surge is a necessary part of the normal signal. This is a dramatic and salutary reminder of the importance of signaler sequence, and warns against the far-too-common mistake of concluding that cyclic AMP is a physiological signal blocker from its inhibitory action when added or increased out of context and sequence.

Among the key events stimulated by the first signal is the increased expression of MHC II (Ia) genes stimulated by the NF-BRE transcription factor.[1,59] This results in the loading of the cell surface with MHC II (Ia) antigens.[1,59,77,78] Binding of the accumulated MHC II molecules (bearing processed antigen) to a helper T-cell or even just to a T-cell membrane, or cross-linking the accumulated MHC II molecules on IL-4/anti-Ig-stimulated cells with anti-MHC II antibodies drives the growing cells into the S phase.[59,77,78] According to Cambier et al.[77] and Coggeshall et al.[78] the signal from the MHC II (Ia) receptor is a cyclic AMP surge. The binding of the MHC II-antigen complex to the helper T-cell's TCR•CD3•CD4 complex also activates the T-cell as discussed above. The T-cell's CD28 receptor, activated by binding to the B-cell's B7/BB1 antigen, collaborates with the TCR signal to stimulate the production and secretion of cytokines such as IL-2, which regulates B-cell cycling and antibody secretion, as well as IL-4 and IFN-γ, which stimulate heavy chain switching.[1,26a,59]

Activated B-cells, like activated hepatocytes and T-cells, generate either one or two cyclic AMP surges. Thus, stimulating mouse spleen B lymphocytes with lipopolysaccharides triggers a cyclic AMP surge within 30 min, and this surge either lasts until 12 h, or it subsides and is followed by a second surge, the G_1 surge, which peaks by 12 h.[82] This surge is important because inactivating it, by microinjecting monoclonal anti-cyclic AMP antibody (by fusing the cells with antibody-loaded erythrocytes) when the surge is at its peak, nearly halves the DNA-synthetic response.[82] A dramatic example of the later G_1 cyclic AMP surge is presented in Figure 2. In mouse spleen B-cells activated by a combination of galactose oxidase and neuraminidase, there was a large cyclic AMP surge which began around 10 h, peaked at eight times the quiescent value at 20 h, and had returned to the starting value by the time the DNA-synthetic activity peaked at 40 h.[83] When the newborn spleen cell progeny were immediately restimulated before becoming quiescent, they generated this large G_1 cyclic AMP surge without delay and started replicating DNA soon afterwards.[83] Human RPMI 8866 lymphoma cells also generate a splendid cyclic AMP surge during during their brief G1 build-up (Figure 3).[84] In this example, between the second and third hour of the 4-h G_1 phase following the block, something massively (69-fold) increased the total fluoride-stimulable adenylyl cyclase activity, reduced the low-Km PDE activity by 44%, and caused a fivefold rise in the total cellular cyclic AMP content. The adenylyl cyclase activity and cyclic AMP content then crashed down to the starting level just before the onset of DNA replication (Figure 3).

Eventually, the growing cell, presumably driven by the cell cycle engine's parade of cyclins and cyclin-dependent PSTAIRE motif-bearing STPKs, approaches the end of the G_1 build-up. In other cells this means entry into the

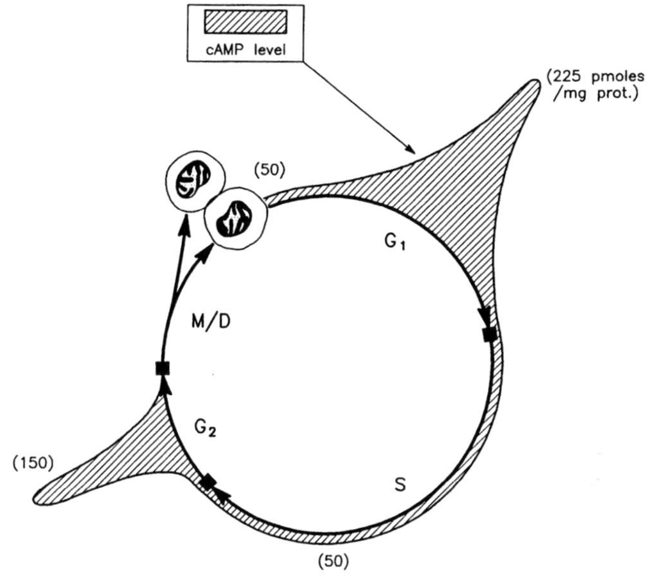

FIGURE 3 Human RPMI 8866 lymphoma cells generate dramatic cyclic AMP transients during the G_1 and G_2 phases of their cycle.[84] The numbers in brackets are cyclic AMP concentrations.

second, Ca^{2+}-dependent, adhesion (integrin)-dependent, Cdk2-cyclin A-dependent stage of prereplicative development. This may also apply to B-cells, because these cells are sensitive to the Ca^{2+} channel blocker, D-600, in later G_1.[78]

After replicating their DNA, B-cells generate a second cyclic AMP surge.[84] We may assume that the initiation of B-cell mitosis involves this cyclic AMP surge as well as the assembly and activation of the Cdc2-cyclin A and Cdc2-cyclin B STPK complexes, as discussed in Chapter 2.

III. LEUKEMIAS AND LYMPHOMAS

When the TCR·CD3 or mIgM complex stops signaling, a normal T- or B-cell or its daughter shuts off its cell cycle genes (e.g, c-*myc*), destroys the cyclin-dependent PKs of its cell cycle engines, shrinks down into quiescence, and, depending on accessory cells and signals, may self-destruct by the Ca^{2+}-dependent apoptosis. If a cell should free itself from the need for antigen-activated receptors to be activated, it and its progeny will have generated the first of several mutations and the clonal selections leading to malignancy. For example, LSTRA murine lymphoma cells express a high level of mutant, hyperactive $p56^{lck}$ which eliminates the need for receptor-stimulated CD45 protein phosphatase to switch this YPK on or to start the prereplicative program.[85,86] If a T-cell is infected with HTLV-1 virus, the constitutively expressed viral Tax protein will activate the cell's NFκB transcription factor which, in turn, will stimulate c-*myc*, the signaling-competent IL-2Rβ, and other growth-related genes.[26a,34] Some mutant T-cells start the transformation process by constitutively expressing IL-2Rs, but they still need the first Ca^{2+} signal to express their IL-2 gene.[12] Thus, they are IL-2 addicts who will die unless they can get their "drug" to stave off death and to proliferate.[12,87] Without IL-2 they stop making the anti-apoptogenic protein Bcl-2 and self-destruct with apoptosis.[88] However, one of the constitutive IL-2R-expressing cells may mutate to constitutive Il-2 production. The resulting double-mutant T-cell clone does not need the TCR·CD3 signal: the autocrine IL-2 and the Il-2R signals will keep the cell alive and cycling.

B-cells have an active recombinase to rearrange their Ig genes, but sometimes the c-*myc* gene gets caught up in the process and is translocated to the Ig heavy chain locus where its expression is constitutively stimulated by the Ig heavy chain gene's enhancer.[89] However, a deregulated c-*myc* gene by itself can be apoptogenic (probably through a p53(TP53)-mediated mechanism) without help from other cell cycle genes (such as *egr*-1) and the second signalers and their receptors.[59,80a,90] Thus, the single-mutant "rearranged" cells are more prone to die by apoptosis.[91] But, if the *bcl*-2 gene should also be translocated to the Ig heavy chain locus, the resulting constitutive production of the anti-apoptosis Bcl-2 protein will prevent the cell from killing itself and thus enhance the expansion of the doubly mutant clone in which the subsequent loss of p53 (or TP53) function (see Chapter 2) will allow the emergence and expansion of further mutant clones which are increasingly liberated and hence increasingly dangerous.[10,89,91–93]

When c-*myc* and other "master" cell cycle genes are constitutively active, or one of the Src-family YPKs (e.g., $p56^{lck}$) is a continuously signaling hyperactive mutant, lymphocytes no longer need antigen-triggered TCR·CD3 or mIgM signals and Ca^{2+} influx to be activated. In one artificial, but heuristic, example, the potent activity of a v-Src YPK stimulates a T-lymphocyte to constitutively make and secrete IL-2.[94] Once cells can do without the antigen-triggered (or its equivalent) signal for activation, they may lose, over subsequent generations, the ability to respond to antigen-activated mIgM or mIgD receptor complexes with a burst of YPK activity and/or a Ca^{2+} surge. Thus, they can safely get rid of a now-useless signaling mechanism.[95] But, sooner or later they also lose the need for the second Ca^{2+} signal to finish the G_1 build-up. As discussed in Chapter 2, this second Ca^{2+} signal is probably generated by adhesion-activated integrin receptors, and it is needed to produce the active Cdk2·cyclin A PKs that are one of the initiators of DNA replication. Shirakawa et al.[96,97] reported that constitutively activated human HPB-ATLL cells could not proliferate in IL-2-free, Ca^{2+}-deficient medium, probably because they still had the normal T-cell's need for a high external Ca^{2+} concentration to finish the

G_1 build-up. Il-2-dependent HT-2 cells and cloned cytotoxic murine T-cells constitutively expressing functional IL-2 receptors still need external Ca^{2+} in the post-IL-2 stage of the G_1 build-up: reducing the external Ca^{2+} concentration stops these cells from initiating DNA replication after exposure to IL-2.[98,99] The IL-2-dependent cells of some human and murine lines need the first Ca^{2+} surge to make IL-2 when stimulated by lectins or monoclonal antibodies to specific surface antigens, but they do not need external Ca^{2+} to initiate DNA replication when the autocrine loop is bypassed by adding IL-2.[12] Murine L1210 leukemia cells do not need the early Ca^{2+} surge to activate themselves and they do not need IL-2 to proliferate, but they still cannot complete the G_1 build-up in Ca^{2+}-deficient medium.[100] The T-cells of Whatman et al.[101] still respond to phytohemagglutinin by expressing IL-2Rs, but they do not need external Ca^{2+} to do so, nor do they need external Ca^{2+} to start the G_1 build-up and initiate DNA replication in response to addition of IL-2. When transforming cells have accumulated enough mutations to spontaneously produce key cell cycle components such as the cell cycle engine's cyclin-dependent PKs, they no longer need receptor-generated Ca^{2+} signals, and they can lower their external Ca^{2+} requirement to a bare minimum. Thus, the cells of various T-cell lines can proliferate normally without IL-2 in severely Ca^{2+}- and Mg^{2+}-deficient media, and their ability to proliferate under such extreme conditions is correlated to their leukemogenicity (oncogenicity). For example, "wild-type" L5178 Y murine lymphoma cells have a much-reduced external Ca^{2+} requirement and no longer need exogenous proliferogens, but they still cannot proliferate in medium containing 0.05 m*M* Ca^{2+} or less.[102] However, there are further mutants of these cells which do proliferate optimally in such severely Ca^{2+}-deficient media.[102] These are far more oncogenic than the "wild-type" lymphoma cells.[102] Other leukemia cells, such as the human HPB-ALL leukemia cells of Shirakawa et al.,[96,97] can proliferate as rapidly in IL-2-free medium containing only 0.01 m*M* Ca^{2+} as they can in medium containing 5 m*M* Ca^{2+}. Another example of ultimately liberated T-cells which have dispensed with all their Ca^{2+}-dependent cell cycle switches are the IL-2-independent human Molt-4-cells whose proliferation, unlike that of normal T-cells, is unaffected by external Ca^{2+} deprivation and Ca^{2+} channel blockers.[75]

REFERENCES

1. **Austyn, J. M. and Wood, J. K.**, *Principles of Cellular and Molecular Immunology*, Oxford University Press, Oxford, 1993.
2. **McCarthy, N. J., Smith, C. A., and Williams, G. T.**, Apoptosis in the development of the immune system: growth factors, clonal selection and *bcl*-2, *Cancer Metastasis Rev.*, 11, 157, 1992.
3. **Kerr, J. F. K., Willie, A. H., and Currie, A. H.**, Apoptosis, a basic biological phenomenon with wider implications in tissue kinetics, *Br. J. Cancer*, 26, 239, 1972.

4. **Perris, A. D. and Whitfield, J. F.**, Dissolution of the condensed chromatin structures of isolated thymocyte nuclei and the disruption of deoxyribonucleoprotein by inorganic phosphate and a phosphoprotein, *Exp. Cell Res.*, 49, 359, 1968.
5. **Whitfield, J. F., Brohée, H., and Youdale, T.**, The effects of X-irradiation on the nuclear structure of rat thymocytes, *Prog. Biochem. Pharmacol.*, 1, 209, 1965.
6. **Whitfield, J. F., Kellerer, S., Brohée, H., and Youdale, T.**, The Feasibility of a New Procedure for Biological Dosimetry, Euratom Report EUR 2505.e, Presses Academiques Europeennes, Brussels, 1965.
7. **Whitfield, J. F., Perris, A, D., and Youdale, T.**, Destruction of the nuclear morphology of thymic lymphocytes by the corticosteroid cortisol, *Exp. Cell Res.*, 52, 349, 1968.
8. **Butke, T. M. and Sandstrom, P. A.**, Oxidation stress as a mediator of apoptosis, *Immunol. Today*, 15, 7, 1994.
9. **Hockenbery, D. M., Oltvai, Z. N., Yin, X. M., Milliman, C. L., and Korsmeyer, S. J.**, Bcl-2 functions in an antioxidant pathway to prevent apoptosis, *Cell*, 75, 241, 1993.
10. **Korsmeyer, S. J.**, Programmed cell death: *Bcl*-2, in *Important Advances in Oncology*, Devita, V. T., Hellman, S., and Rosenberg, S.A., Eds., J.B. Lippincott, Philadelphia, 1993, 19.
11. **Korsmeyer, S. J., Shutter, J. R., Veis, D. J., Merry, D. E., and Oltvai, Z. N.**, Bcl-2/Bax: a rheostat that regulates an anti-oxidant pathway and cell death, *Semin. Cancer Biol.*, 4, 327, 1993.
11a. **Bredesen, D. E.**, Neuronal apoptosis: genetic and biochemical modulation, in *Apoptosis II*, Tomei, L. D. and Cope, F. O., Eds., Cold Spring Harbor Laboratory Press, Cold Spring Harbor, NY, 1994, 397.
12. **Whitfield, J. F.**, *Calcium, Cell Cycles, and Cancer*, 1st ed., CRC Press, Boca Raton, FL, 1990.
13. **Jouishomme, H., Whitfield, J. F., Chakravarthy, B., Durkin, J. P., Gagnon, L., Isaacs, R. J., MacLean, S., Neugebauer, W., Willick, G., and Rixon, R. H.**, The protein kinase-C activation domain of the parathyroid hormone, *Endocrinology*, 130, 53, 1992.
14. **Jouishomme, H., Whitfield, J. F., Gagnon, L., MacLean, S., Isaacs, R., Chakravarthy, B., Durkin, J., Neugebauer, W., and Rixon, R.**, Further definition of the protein kinase C activation domain of the parathyroid hormone, *J. Bone Miner. Res.*, 943, 1994.
15. **Perris, A. D., Weiss, L. A., and Whitfield, J. F.**, Parathyroidectomy and the induction of thymic atrophy in normal, adrenalectomized and orchidectomized rats, *J. Cell. Physiol.*, 76, 141, 1970.
16. **Whitfield, J. F., Rixon, R. H., MacManus, J. P., and Bulk, S. D.**, Calcium, cyclic adenosine 3′,5′-monophosphate, and the control of cell proliferation: a review, *In Vitro*, 8, 257, 1973.
17. **McConkey, D. J., Hartzell, P., Nicotera, P., and Orrenius, S.**, Calcium-activated DNA fragmentation kills immature thymocytes, *FASEB J.*, 3, 1843, 1989.
18. **McConkey, D. J., Hartzell, P., Nicotera, P., Hartzell, P., Belloma, G., Willie, A. H., and Orrenius, S.**, Glucocorticoids activate a suicide process in thymocytes through an elevation of cytosolic Ca^{2+} concentration, *Arch. Biochem. Biophys.*, 269, 365, 1989.

19. **Whitfield, J. F., Isaacs, R. J., Chakravarthy, B., Durkin, J. P., Morley, P., Neugebauer, W., Williams, R., Willick, G., and Rixon, R. H.**, C-terminal fragments of parathyroid hormone-related protein, PTHrP-(107–111) and (107-139), and the N-terminal PTHrP-(1–40) fragments stimulate membrane-associated protein kinase C activity in rat spleen lymphocytes, *J. Cell. Physiol.*, 158, 518, 1994.
20. **Orloff, J. J., Reddy, D., dePapp, A. E., Yang, K. H., Soifer, N. E., and Stewart, A. F.**, Parathyroid hormone-related protein as a prohormone: post translational processing and receptor interactions, *Endocr. Rev.*, 15, 40, 1994.
21. **Boyce, B. F., Yates, A. J. P., and Mundy, G. R.**, Bolus injections of recombinant human interleukin-1 cause transient hypocalcemia in normal mice, *Endocrinology*, 125, 2780, 1989.
22. **Whitfield, J. F.**, Calcium signals in cell proliferation, differentiation, and death, in *Ion Channels and Ion Pumps*, Foa, P. P. and Walsh, M. F., Eds., Springer-Verlag, New York, 1994, 19.
23. **Habener, J. F., Rosenblatt, M., and Potts, J. T.**, Parathyroid hormone: biochemical aspects of biosynthesis, secretion, action, and metabolism, *Physiol. Rev.*, 64, 985, 1984.
24. **Desser-Wiest, L. and Desser, H.**, The proliferation activity of the rat thymocytes after partial hepatectomy, *Exp. Pathol.*, 18, 167, 1990.
25. **Reed, J. C., Miyashita, T., Cuddy, M., and Cho, D.**, Regulation of p26-bcl-2 protein levels in human peripheral blood lymphocytes, *Lab. Invest.*, 67, 443, 1992.
26. **Brickell, P. M.**, The pp60$^{c\text{-}Src}$ family of protein-tyrosine kinases: structure, regulation, and function, *Crit. Rev. Oncogenesis*, 3, 401, 1992.

26a. **Danielian, S. and Fagard, R.**, *Protéines Tyrosine Kinases et Signalisation Cellulaire*, Les Editions INSERM, Paris, 1993.

26b. **Chan, A. C., Desai, D. M., and Weiss, A.**, The role of protein tyrosine kinases and protein tyrosine phosphatases in T cell antigen receptor signal transduction, *Annu. Rev. Immunol.*, 12, 555, 1994.

27. **Medema, R. H. and Bos, J. L.**, The role of p21 ras in receptor tyrosine kinase signalling, *Crit. Rev. Oncogenesis*, 4, 615, 1993.

27a. **Penninger, J. M., Wallace, V. A., Kishihara, K., and Mak, T. W.**, The role of p56^{lck} and p59^{fyn} tyrosine kinases and CD45 protein tyrosine phosphatase in T-cell development and clonal selection, *Immunol. Rev.*, 135, 183, 1993.

27b. **Barclay, A. N., Ed.,** Structure and function of the leucocyte common antigen CD45, *Biochem. Soc. Trans.*, 20, 161, 1992.

27c. **Trowbridge, I. S. and Thomas, M. L.**, CD45: an emerging role as a protein tyrosine phosphatase required for lymphocyte activation and development, *Annu. Rev. Immunol.*, 12, 85, 1994.

28. **Hume, D. A. and Weidemann, M. J.**, *Mitogenic Lymphocyte Transformation*, Elsevier/North Holland, Amsterdam, 1980.
29. **Gelfand, E. W.**, The ionic and biochemical basis for T-cell activation and proliferation, *Clin. Immunol. Immunopathol.*, 61, S1, 1991.
30. **Grinstein, S. and Dixon, S. J.**, Ion transport, membrane potential and cytoplasmic pH in lymphocytes: changes during activation, *Physiol. Rev.*, 69, 417, 1989.
31. **Khan, A. A., Steiner, J. P., Klein, M. G., Schneider, M. F., and Snyder, S. H.**, IP_3 receptor: localization to plasma membrane of T-cells and cocapping with the T-cell receptor, *Science*, 257, 815, 1992.

32. **Woodrow, M., Ceipstone, N. A., and Cantrell, D.**, p21[ras] and calcineurin synergize to regulate the nuclear factor of activated T-cells, *J. Exp. Med.*, 178, 1517, 1993.
33. **Bours, V., Franzoso, G., Brown, K., Park, S., Azareno, V., Tomita-Yamaguchi, M., Kelly, K., and Siebenlist, U.**, Lymphocyte activation and the family of NF-KB transcription factor complexes, *Curr. Topics Microbiol. Immunol.*, 182, 411, 1992.
34. **Duyao, M. P., Kessler, D. J., Spicer, D. B., and Sonenshein, G. E.**, Transactivation of the c-*myc* gene by HTLV-1*tax* is mediated by NFKB, *Curr. Topics Microbiol. Immunol.*, 182, 421, 1992.
35. **Shirakawa, F. and Mizel, S. B.**, In vitro activation and nuclear translocation of NF-KB catalyzed by cyclic AMP-dependent protein kinase and protein kinase C, *Mol. Cell. Biol.*, 9, 2424, 1989.
36. **Cantrell, D. A., Graves, J. D., Izquierdo, M., Lucas, S., and Downward, J.**, T lymphocyte activation signals, *Ciba Found. Symp.*, 164, 208, 1992.
36a. **Bar-Sagi, D.**, The Sos (son of sevenless) protein, *Trends Endocrinol. Metab.*, 5, 165, 1994.
37. **Moodie, S. A., Willumsen, B. M., Weber, M. J., and Wolfman, A.**, Complexes of Ras·GTP with Raf-1 and mitogen-activated protein kinase-kinase, *Science*, 260, 1658, 1993.
38. **Wyllie, A. H.**, Apoptosis and the regulation of cell numbers in normal and neoplastic tissues: an overview, *Cancer Metastasis Rev.*, 11, 95, 1993.
39. **Frederickson, R. M. and Sonenberg, N.**, Signal transduction and the regulation of translation initiation, *Semin. Cell Biol.*, 3, 107, 1992.
40. **Rhoads, R. E.**, Regulation of eukaryotic protein synthesis by initiation factors, *J. Biol. Chem.*, 268, 3017, 1993.
41. **Smith, M. R., Jaramille, M., Liu, Y.-L., Dever, T. E., Merrick, W. C., Kung, H.-F., and Sonenberg, N.**, Translation initiation factors induce DNA synthesis and transform NIH 3T3 cells, *New Biol.*, 2, 648, 1990.
42. **Cho-Chung, Y. S. and Clair, T.**, The regulatory subunit of cAMP-dependent protein kinase as a target for chemotherapy of cancer and other cellular dysfunctional-related diseases, *Pharmacol. Ther.*, 60, 265, 1993.
43. **Skålhegg, B. S., Taskén, K., Hanson, V., Huitfeld, H. S., Jahnsen, T., and Lea, T.**, Location of cAMP-dependent protein kinase type I with the TCR-CD3 complex, *Science*, 263, 84, 1994.
44. **Harper, J. F., Haddox, M. K., Johanson, R. A., Hanley, R. M., and Steiner, A. L.**, Comparison of second messenger action: immunocytochemical and biochemical evidence, *Vitam. Horm.*, 42, 197, 1985.
45. **Klimpel, G. R., Byus, C. V., Russell, D. H., and Lucas, D. O.**, Cyclic AMP-dependent protein kinase activation and induction of ornithin decarboxylase during lymphocyte mitogenesis, *J. Immunol.*, 123, 817, 1979.
46. **Shirakawa, F., Yamashita, U., Cedid, M., and Mizel, S. B.**, Cyclic AMP — an intracellular second messenger for interleukin 1, *Proc. Natl. Acad. Sci. U.S.A.*, 85, 8201, 1988.
47. **Kammer, G. M., Boehm, C. A., Rudolph, S. A., and Schultz, L. A.**, Mobility of the T lymphocyte surface molecules CD3, CD4, and CD8: regulation by a cAMP-dependent pathway, *Proc. Natl. Acad. Sci. U.S.A.*, 85, 792, 1988.

48. **Schreck, R., Richter, P., and Baeuerle, P. A.**, Reactive oxygen intermadiates as apparently widely used messengers in the activation of the NF-κB transcription factor and HIV-1, *EMBO J.*, 10, 2247, 1991.
49. **Bastian, N. R. and Hibbs, J. B., Jr.**, Assembly and regulation of NADPH oxidase and nitric oxide synthase, *Curr. Opinion Immunol.*, 6, 131, 1994.
50. **Richter, C.**, Pro-oxidants and mitochondrial Ca^{2+}: their relationship to apoptosis and oncogenesis, *FEBS Lett.*, 325, 104, 1993.
51. **Reed, J. C., Tsujimoto, Y., Alpers, J. D., Croce, C. M., and Nowell, P. C.**, Regulation of *bcl-2* proto-oncogene expression during normal human lymphocyte proliferation, *Science*, 236, 1295, 1987.
52. **De Boni, U.**, The interphase nucleus as a dynamic structure, *Int. Rev. Cytol.*, 150, 149, 1994.
53. **Reiss, K., Porcu, P., Sell, C., Pietrzkowski, Z., and Baserga, R.**, The insulin-like growth factor 1 receptor is required for the proliferation of hematopoietic cells, *Oncogene*, 7, 2243, 1992.
54. **Varticoski, L., Harrison-Findik, D., Keeler, M. L., and Susa, M.**, Role of PI3-kinase in mitogenesis, *Biochim. Biophys. Acta*, 1226, 1, 1994.
55. **Frost, P., Hart, I., and Kerbel, R., Eds.**, Apoptosis, *Cancer Metastasis Rev.*, 11, 94, 1993.
56. **Minami, Y., Kuno, T., Miyazaki, T., and Taniguchi, T.**, The IL-2 receptor complex: its structure, function, and target genes, *Annu. Rev. Immunol.*, 11, 245, 1993.
57. **Koyasu, S., Suzuki, G., Asano, Y., Osawa, H., Diamantstein, T., and Yahama, I.**, Signals for activation and proliferation of murine T-lymphocyte clones, *J. Biol. Chem.*, 262, 4689, 1987.
58. **Lenardo, M. J.**, Interleukin-2 programs mouse αβ T-lymphocytes for apoptosis, *Nature*, 353, 858, 1991.
59. **Cushley, W. and Harnett, M. M.**, Cellular signalling mechanisms in B lymphocytes, *Biochem. J.*, 292, 313, 1993.
60. **Mills, G. B., Stewart, D. J., Mellors, A., and Gelfand, E. W.**, Interleukin 2 does not induce phosphatidylinositol hydrolysis in activated T-cells, *J. Immunol.*, 136, 3019, 1986.
61. **Wickremasinghe, R. G., Mire, A. R., and Hoffbrand, A. V.**, Interleukin-2 binding to activated human T-lymphocytes triggers generation of cyclic AMP but not inositol phosphates, *FEBS Lett.*, 220, 52, 1987.
62. **Chakravarthy, B., Whitfield, J. F., and Durkin, J. P.**, Inactive membrane protein kinase C, a possible target for receptor signaling, *Biochem. J.*, 304, 809, 1994.
63. **Lu, Y., Tremblay, R., Jouishomme, H., Chakravarthy, B., and Durkin, J. P.**, Evidence that the activation of an inactive pool of membrane-associated protein kinase C is linked to the interleukin 2-dependent survival of T-lymphocytes, *J. Immunol.*, 155, 1495, 1994.
64. **Ebie, N., Harris, J. E., Anderson, K. M., Wolter, J., and Braun, D. P.**, Cyclic nucleotide responses in peripheral blood mononuclear cells of cancer patients following stimulation with interleukin 2 (IL-2) or prostaglandin E (PGE), *Proc. Annu. Meet. Am. Assoc. Cancer Res.*, 27, 345, 1986.
65. **Foker, J. E., Malkinson, A. M., Sheppard, J. R., and Wang, T.**, Studies of cyclic AMP metabolism in proliferating lymphocytes, in *The Molecular Basis of Immune Cell Function*, Kaplan, J. G., Ed., Elsevier/North Holland, Amsterdam, 1979, 57.

66. **Monahan, T. M., Marchand, N. W., Fritz, R. R., and Abell, C. W.**, Cyclic adenosine 3′:5′monophosphate levels and activities of related enzymes in normal and leukemic lymphocytes, *Cancer Res.*, 35, 2540, 1975.
67. **Narumiya, S., Hirata, M., Nanba, T., Nikaido, T., Taniguchi, Y., Tagaya, Y., Okada, M., Mitsuya, H., and Yodoi, J.**, Activation of interleukin-2 receptor gene by forskolin and cyclic AMP analogues, *Biochem. Biophys. Res. Commun.*, 143, 753, 1987.
68. **Vitté-Mony, I., Goy, A., Guizani, L., and Bertoglio, J. H.**, Cyclic AMP upregulates mRNA and surface expression of IL-2Rα p55(Tac) in normal human NK cell clones, *Mol. Immunol.*, 29, 593, 1992.
69. **Furukawa, Y., Piwnica-Worms, H., Ernst, T. J., Kanakura, Y., and Griffin, J. D.**, cdc 2 expression at the G_1 to S transition in human T lymphocytes, *Science*, 250, 805, 1990.
70. **Kim, Y. H., Proust, J. J., Buchholz, M. J., Chrest, F. J., and Nordin, A. A.**, Expression of the murine homologue of the cell cycle control protein $p34^{cdc2}$ in T lymphocytes, *J. Immunol.*, 149, 17, 1992.
71. **Lucas, J. J., Terada, N., Szepesi, A., and Gelfand, E. W.**, Regulation of synthesis of $p34^{cdc2}$ and its homologues and their relationship to p110 Rb phosphorylation during cell cycle progression of normal human T cells, *J. Immunol.*, 148, 1804, 1992.
72. **Morice, W. G., Brunn, G. J., Wiederrecht, G., Siekierka, J. J., and Abraham, R. T.**, Rapamycin-induced inhibition of $p34^{Cdc2}$ kinase activation is associated with G_1/S-phase growth arrest in T lymphocytes, *J. Biol. Chem.*, 268, 3734, 1993.
73. **Gazitt, Y. and Erdos, G. W.**, Fluctuations and ultrastructural localization of oncoproteins and cell cycle regulatory proteins during growth and apoptosis of synchronized AGF cells, *Cancer Res.*, 54, 950, 1994.
74. **Colomer, J., Agell, N., Engel, P., Alberola-Ila, J., and Bachs, O.**, Calmodulin expression during proliferative activation of human T lymphocytes, *Cell Calcium*, 14, 609, 1993.

74a. **Brown, E. M.**, Extracellular sensing, regulation of parathyroid cell function, and role of Ca^{2+} and other ions as extracellular (first) messengers, *Physiol. Rev.*, 71, 371, 1991.

75. **Birx, D. L., Berger, M., and Fleisher, T. A.**, The interference of T cell activation by calcium channel blocking agents, *J. Immunol.*, 133, 2904, 1984.
76. **Blitstein-Willinger, E. and Diamantstein, T.**, Inhibition by isoptin (a calcium antagonist) of the mitogenic stimulation of lymphocytes prior to the S-phase, *Immunology*, 34, 303, 1978.
77. **Cambier, J. C., Justement, L. B., Newell, M. K., Chen, Z. Z., Harris, L. K., Sandoval, V. M., Klemsz, M. J., and Ransom, J. T.**, Transmembrane signals and intracellular "second messengers" in the regulation of quiescent B-lymphocyte activation, *Immunol. Rev.*, 95, 37, 1987.
78. **Coggeshall, K. M., Monroe, J. G., Ransom, J. T., and Cambier, J. C.**, Mechanisms of transmembrane signal transduction during B cell activation, in *B-Lymphocyte Differentiation*, Cambier, J. C., Ed., CRC Press, Boca Raton, FL, 1986, 1.
79. **Parker, D. C.**, T cell-dependent B cell activation, *Annu. Rev. Immunol.*, 11, 331, 1993.

80. **Yamada, H., Mizuguchi, J., and Nakanishi, M.**, Antigen receptor-mediated calcium signals in B cells as revealed by confocal fluorescence microscopy, *FEBS Lett.*, 284, 249, 1991.

80a. **Bissonnette, R. P., Shi, Y., Mahboubi, A., Glynn, J. M., and Green, D. R.**, c-myc and apoptosis, in *Apoptosis II*, Tomei, L. D. and Cope, F. O., Eds., Cold Spring Harbor Laboratory Press, Cold Spring Harbor, NY, 1994, 327.

81. **Michell, R. H., Conroy, L. A., Finney, M., French, P, J., Bunce, C. M., Anderson, K., Baxter, M. A., Brown, G., Gordon, J., Jenkinson, E. J., Lord, J. M., Kirk, C. J., and Owen, J. J.**, Inositol lipids and phosphates in the proliferation and differentiation of lymphocytes and myeloid cells, *Ciba Found. Symp.*, 164, 2, 1992.

82. **Ohara, J., Sugi, M., Fujimoto, M., and Watanabe, T.**, Microinjection of macromolecules into normal murine lymphocytes by means of cell fusion. II. Enhancement and suppression of mitogenic responses by microinjection of monoclonal anti-cyclic AMP into B-lymphocytes, *J. Immunol.*, 129, 1227, 1982.

83. **Whitfield, J. F., Boynton, A. L., MacManus, J. P., Rixon, R. H., Sikorska, M., Tsang, B., and Walker, P. R.**, The roles of calcium and cyclic AMP in cell proliferation, *Ann. N.Y. Acad. Sci.*, 339, 216, 1980.

84. **Millis, A. J. T., Forrest, G. A., and Pious, D. A.**, Cyclic AMP-dependent regulation of mitosis in human lymphoid cells, *Exp. Cell Res.*, 83, 335, 1974.

85. **Casnellie, J. E., Harrison, M. L., Hellstrom, K. E., and Krebs, E. G.**, A lymphoma cell line expressing elevated level of tyrosine protein kinase activity, *J. Biol. Chem.*, 258, 10738, 1983.

86. **Gacon, G., Gisselbrecht, S., Piau, J. P., Boissel, J. P., Tolle, J., and Fisher, S.**, High level of tyrosine protein kinase activity in a murine lymphoma cell line induced by Moloney leukemia virus, *EMBO J.*, 1, 1579, 1982.

87. **Duke, R. C. and Cohen, J. J.**, IL-2 addiction: withdrawal of growth factor activates a suicide program in dependent T cells, *Lymphokine Res.*, 5, 289, 1986.

88. **Deng, G. and Podack, E. R.**, Suppression of apoptosis in a cytotoxic T-cell line by interleukin 2-mediated gene transcription and deregulated expression of the protooncogene bcl-2, *Proc. Natl. Acad. Sci. U.S.A.*, 90, 2189, 1993.

89. **Nowell, P. C.**, Cytogenetic approaches to human cancer genes, *FASEB J.*, 8, 408, 1994.

90. **Debbas, M. and White, E.**, Wild-type p53 mediates apoptosis by E1A, which is inhibited by E1B, *Genes Dev.*, 7, 546, 1993.

91. **Sachs, L. and Lotem, J.**, Control of programmed cell death in normal and leukemic cells: new implication for therapy, *Blood*, 82, 15, 1993.

92. **Campana, D., Coustan-Smith, E., Manabe, A., Buschle, A., Buschle, M., Raimondi, S. C., Behm, F. G., Ashmun, R., Arico, M., Biondi, A., and Pui, C. H.**, Prolonged survival of B-lineage acute lymphoblastic leukemia cells is accompanied by overexpression of bcl-2 protein, *Blood*, 81, 1025, 1993.

93. **Cotter, F. E.**, Molecular pathology of lymphomas, *Cancer Surv.*, 16, 157, 1993.

94. **O'Shea, J. J., Ashwell, J. D., Bailey, T. L., Cross, S. L., Samelson, L. E., and Klausner, R. D.**, Expression of v-src in a murine T-cell hybridoma results in constitutive T-cell receptor phosphorylation and interleukin 2 production, *Proc. Natl. Acad. Sci. U.S.A.*, 88, 1741, 1991.

95. **Michel, F., Merle-Béral, H., Legac, E., Michel, A., Debré, P., and Bismuth, G.**, Defective calcium response in B-chronic lymphocyte leukemia cells, *J. Immunol.*, 150, 3624, 1993.
96. **Shirakawa, F., Yamashita, U., Oda, S., Chiba, S., Eto, S., and Suzuki, H.**, Calcium dependency in the growth of adult T-cell leukemia cells *in vitro*, *Cancer Res.*, 46, 658, 1986.
97. **Shirakawa, F., Yamashita, U., Tanoka, Y., Oda, S., Chiba, S., Eto, S., and Suzuki, H.**, Regulation of adult T-cell leukemia (ATL) cell growth by calcium *in vitro*, in *Growth and Differentiation of Cells in Defined Environment*, Murakami, H., Yamane, I., Barnes, D. W., Mather, J. P., Hayashi, I., and Sato, G. H., Eds., Springer-Verlag, New York, 1985, 459.
98. **Harris, D. T., Kozumbo, W. J., Testa, J. E., Cerrutti, P. A., and Cerottini, J.-C.**, Molecular mechanisms involved in T cell activation. III. The role of extracellular calcium in antigen-induced lymphokine production and interleukin 2-induced of cloned cytotoxic T-lymphocytes, *J. Immunol.*, 140, 921, 1988.
99. **Le Grue, S. J.**, Interleukin-2 stimulus-response coupling is calcium independent, *Lymphokine Res.*, 6, 1, 1987.
100. **Cory, J. G., Cater, G. L., and Karl, R. C.**, Calcium ion-dependent proliferation of L1210 cells in culture, *Biochem. Biophys. Res. Commun.*, 145, 556, 1987.
101. **Whatman, N. P., Crespo, L., Davis, B., Poiesz, B. J., and Zamkoff, K. W.**, Differential effect on fresh and T cells of PHA-induced changes in free cytoplasmic calcium: relation to IL-2 receptor expression, IL-2 production, and proliferation, *Cell. Immunol.*, 111, 158, 1988.
102. **Brennan, J. K., Lichtman, M. A., Chamberlain, J. K., and Leblond, P.**, Isolation of variant lymphoma cells with reduced growth requirements for extracellular calcium and magnesium and enhanced oncogenicity, *Blood*, 47, 447, 1976.

7 Overview and Afterword: a Final Tribute to Ca^{2+}

Although a lot has been said about PKAs, PKCs, and the cyclin-dependent PKs of the cell cycle engine, the real hero of our five stories has been Ca^{2+}, who, operating through a few known and very many still unknown Ca^{2+}-binding signal-transducing proteins, is a principal player in the triggering of the major events in the cell cycle, as well as the terminal differentiation programs of normal epithelial cells. The Ca^{2+} story is immensely complicated because the influence of the ion reaches into all corners and directly or indirectly affects all functions of the cell. Indeed, the real Ca^{2+} story, with its certain to be breath-taking relevations and insights into the cell cycle engine and switches and what goes wrong with them to produce cancers, awaits us in the next millennium.

A Ca^{2+} spike and oscillations, together with bursts of PKA and PKC activities, triggered by velcroceptors, modular velcroceptors, or, in some cells, 7-transmembrane domain serpentine receptors, start normal quiescent cells expressing sets of early cell cycle genes, growing, accumulating CaM, and repositioning their nuclear structures to be able to start the actual G_1 build-up to chromosome replication when the signal is given to do so. Next comes the G_1 cyclic AMP surge with its emphasis on activating the growing cell's nucleus-seeking, gene activating PKAII which contributes to the G_1 build-up by starting a cascade of gene expressions which culminates in the appearance of cyclin-dependent PKs such as Cdk2·cyclin D or E and Cdk4·cyclin D that inactivate Rb suppressor protein and stimulate the accumulation of replication enzymes and initiators. Near the end of the G_1 build-up there is a second train of Ca^{2+} oscillations that starts at the nuclear envelope and spreads throughout the cell. Some of the surging Ca^{2+} rushes into the nucleus along with some of the accumulated CaM, Ca^{2+}·CaM-binding proteins, and Cdk2·cyclin A PKs to help assemble, and start the operation of, the DNA replicating enzyme complexes on the chromosomes' replication origins. Around the same time, the cell releases or actively secretes accumulated cyclic AMP and CaM, which, by stimulating surface PKs, send essential signals that supplement the internal Ca^{2+} and cyclic AMP signals that start replication. After Ca^{2+} and the Cdk2·cyclin A PKs start replication, the Ca^{2+} oscillations stop and the cyclin AMP level drops to the basal starting level. When the cell's RCC-1-associated sensors tell it that the chromosomes have been completely replicated, there is a final cyclic AMP transient, signals start coming from emerging HA receptors, and inactive (because of phosphorylation-blocked catalytic domains), mitosis-initiating Cdc2·cyclin A or B complexes start building up to a critical threshold level. Then the cyclic AMP surge crashes, there is a Ca^{2+} spike, and the Cdc2·cyclin PKs are driven into the nucleus where their catalytic domains are unblocked by the Cdc25 dual-function phosphatase and they activate a group of mechanisms that break up the nuclear envelope and lamina, condense the chromosomes, and build the mitotic spindle. When each chromosome's kinetochore has attached to a bundle of spindle microtubules extending from the polar centrosome and all of the chromosomes have lined up along the spindle equator, another cell sensor recognizes and acknowledges the completion of congression, and the cell triggers a fifth Ca^{2+} spike which activates a protease that inactivates the Cdc2·cyclin PKs by destroying their cyclin components. Otherwise, a persistence of these PKs would lock the cell in mitosis. This fifth Ca^{2+} signal also activates the topoisomerase that uncouples the oppositely pulling sister chromatids by cutting their residual replication-produced tangles and starts the spindle motors. This initiates elongation of the dividing cell and the anaphase movement of its chromosomes to the spindle poles. Finally, Ca^{2+} drives the tightening of the subsurface belt of actomyosin filaments that pinches the elongating cell into two new cells.

But Ca^{2+} does a lot more than trigger key parts of the cell cycle. Cells such as colon epithelial cells and skin keratinocytes are armed with surface sensors that measure the external Ca^{2+} concentration and respond when it exceeds a critical threshold. In keratinocytes these sensors might function like the Ca^{2+} receptors on osteoclasts or parathyroid gland chief cells. When activated, these sensors trigger the hydrolysis of membrane phosphoinositides, the emptying of internal Ca^{2+} stores, the influx of external Ca^{2+}, bursts of PKCs, and YPK activities. This signal somehow stops cell cycle initiation and starts the differentiation (diffpoptosis) program which sooner or later ends in death by an apoptosis-like process triggered by a massive Ca^{2+} influx.

During the multistep carcinogenic process, the clonal variation and selection for proliferative independence, apoptosis resistance, and malignancy facilitated by the appearance of mutant Ras proteins and the loss of the mutation-opposing monitor, p53(TP53), there is a progressive shedding of the need for exogenously triggered Ca^{2+} signals to start and then to complete the G_1 build-up to chromosome replication as normally Ca^{2+}-dependent events become constitutively active or are bypassed. However, transforming cells and the ultimate malignant cells never lose the need for endogenous Ca^{2+} transients, CaM, and Ca^{2+}·CaM-binding proteins to replicate DNA and to start prophase and anaphase and to drive cytokinesis. Transforming colon cells and keratinocytes also lose their Ca^{2+} sensors or become increasingly unresponsive to their signals to stop cycling, to start differentiating, and ultimately to apoptose. One of the reasons why cancer cells, such as colon and mammary cancer cells, cannot respond effectively to differentiation signals is their greatly reduced expression of the differentiation-related RII subunits of the nucleus-seeking, gene-activating PKAIIs and their continued, or enhanced, expression of the proliferation-related RI subunits of the PKAIs.

Abbreviations

AFB	Aflatoxin B
Amino acids	(only the single-letter code is used in this book)
	A, Ala, alanine
	C, Cys, cysteine
	D, Asp, aspartic acid
	E, Glu, glutamioacid
	F, Phe, phenylalanine
	G, Gly, glycine
	H, His, histidine
	I, Ile, isoleucine
	K, Lys, lysine
	L, Leu, leucine
	M, Met, methionine
	N, Asn, asparagine
	P, Pro, proline
	Q, Gln, glutamine
	R, Arg, arginine
	S, Ser, serine
	T, Thr, threonine
	V, Val, valine
	W, Trp, tryptophan
	Y, Tyr, tyrosine
APC	Adenomatous polyposis coli (Chapter 5)
APC	Antigen-presenting cell (Chapter 6)
ASV	Avian sarcoma virus or RSV
BAPTA	1,2-*bis* (2-aminophenoxy) ethane *N,N,N′,N′*-tetraacetic acid
β-ARK	β-Adrenergic receptor kinase
bFGF	Basic fibroblast growth factor
BZIP	Basic leucine zipper domain
$[Ca^{2+}]_i$	Internal ionic calcium concentration
CACF	Ca^{2+} channel-activating factor
CaM	Calmodulin
cAMP	Cyclic AMP, cyclic adenosine 3′,5′-monophosphate
CaMPK	CaM-dependent protein kinase
CaMRE	Calmodulin-responsive element
CANPs	Calcium-dependent neutral protease(s)

ConA	Concanavalin A
CAP	Catabolite gene-activating protein
CEA	Carcinoembryonic antigen
CEN	Centromere
CENP	Centromere protein
CKI,CKII	Casein kinases I and II
Cmd	Calmedin
CRE	Cyclic AMP-responsive element
CSF	Cytostatic factor
C subunits	Common catalytic subunit of cAMP-dependent protein kinases (PKAs)
d	Day(s)
DCC	Surface glycoprotein deleted in colon cancer
DG	DAG or diacylglycerol
DNA lig	DNA ligase
$1\alpha,25(OH)_2D_3$	1α,25-dihydroxy vitamin D_3
ECM	Extracellular matrix
eEF-2	Eukaryotic elongation factor-2
EGF	Epidermal growth factor
egr-1	Early growth response gene-1
EGTA	Ethylene-*bis*-(oxyethylene nitrino) tetraacetate acid, a highly specific Ca^{2+} chelator
eIF-4	Eukaryotic initiation factor-4
EPU	Epidermal proliferation unit
ER	Endoplasmic reticulum
ERK	Early-response protein kinase (see MAPK)
FAK	Focal adhesion kinase, a protein tyrosine kinase (YPK)
FAP	Familial adenomatous polyposis
GADD-45	Growth arrest DNA-damage protein-45
GAP	GTPase-activating protein
GEF	Guanine nucleotide exchange factor
GH	Growth hormone
GIP	GTPase-inhibiting protein
GLP	Glucagon-like protein
GNRP	Guanine nucleotide-releasing protein
G protein	GTP-binding/activated protein
GRP	Glucagon-related intestinal protein
h	Hour(s)
HA	Head activator
HAR	Head activator receptor
hCG	Human chorionic gonadotropin
HGF	Hepatocyte growth factor
HGFR	HGF receptor
HNPCC	Hereditary nonpolyposis colon cancer

HPX	70% partial hepatectomy
HTLV	Human T-cell leukemia virus-1
IBMX	Isomethyl butyl xanthine
Ig	Immunoglobulin
IGF-1 and -2	Insulin-like growth factors 1 and 2
IL	Interleukin
IL-2R	Interleukin-2 receptor
INCENPS	Internal centromere proteins
$Ins(1,4,5)P_3$	Inositol-1,4,5-*tris*-phosphate
i.p.	Intraperitoneal
IP_3	$Ins(1,4,5)P_3$ (see above)
IPPK	Insulin-stimulated protamine protein kinase
IP_3R	IP_3 receptor (see above)
IRS	Insulin receptor-specific substrate-1
K1,5,10,14	Keratin isotypes
KGF	Keratinocyte growth factor
LAP	Laparotomy
L-CAM	Type L-cell adhesion molecule
LH	Luteinizing hormone
LPS	Lipopolysaccharide substance
MAPK	Mitogen-activated protein kinase
MAPKK	Mitogen-activated protein kinase kinase
MCB	MluI cell cycle box
MCC	Missing in colon cancer
MCH	Master-of-ceremonies helicase
MHC	Major histocompatibility complex antigen or protein
mIg	Membrane immunoglobulin
MLCK	Myosin light-chain kinase
MPF	Mitosis — or maturation — promoting factor
MNNG	*N*-methyl-*N′*-nitro-*N*-nitrosoguanidine
MTOC	Microtubule-organizing center
NLS	Nuclear localization signal sequence
NPC	Nuclear pore complex
NFAT	Nuclear factor of activated T-cells
Nucleic acid bases	A, adenine
	C, cytosine
	G, guanine
	T, thymine
	U, uracil
Ori	DNA replication origin
PA	Phosphatidic acid
PAP	Phosphatidic acid phosphohydrolase
PARP	Poly(ADP-ribose) polymerase

PCNA	Proliferating cell nuclear antigen
PDE	Cyclic nucleotide phosphodiesterase
PGEs	Prostaglandins E
PI3'K	Phosphatidylinositol-3′-kinase
PK	Protein kinase
PKA	Cyclic AMP (cAMP)-dependent protein kinase
PKAI	Type I PKA
PKAII	Type II PKA
PKCs	Protein kinase C isoforms
PLA_2	Phospholipase A_2
PLC	Phospholipase C
PLD	Phospholipase D
Polα or Pola	DNA polymerase-α
Polδ	DNA polymerase-δ
PPase	Protein phosphatase
PRL	Prolactin
PtdCh	Phosphatidylcholine
PtdCh-PLC	Phosphatidylcholine-specific phospholipase C
PtdIns(4,5)P_2	Phosphatidylinositol-4,5-*bis*-phosphate
pS	Phospho-serine
PSTYPase	Protein STY phosphatase
PTH	Parathyroid hormone
PTHrP	Parathyroid hormone-related protein
pT	Phospho-threonine
PTX	Parathyroidectomy or pertussis toxin
PVA	Pemphigus vulgaris antigen
pY	Phosphotyrosine
PYPase	Protein-tyrosine phosphatase
Ran	Ras-related nuclear protein
RAS	Yeast Ras proteins
Ras	Animal cell Ras proteins
Rb or RB	Retinoblastoma susceptibility protein
RCC1	Regulator of chromose condensation
RF-A	Replication factor-A, single-strand DNA-binding protein
RF-C	Replication factor-C
ROI	Reactive oxygen intermediate
RR	Ribonucleoside diphosphate reductase or ribonucleotide reductase
RSV	Rous (or avian) sarcoma virus
R subunit	PKA regulatory subunit
Sa. cerevisiae	*Saccharomyces cerevisiae*
Sc. pombe	*Schizosaccharomyces pombe*
SCaBP	Skin Ca^{2+}-binding protein

SH2, SH3	Sarc-homologous region 2 or 3
Spr	Small proline-rich protein
SRE	Serum-responsive element
STOP	Stable microtubules only protein
STYPK	Serine/threonine/tyrosine-phosphorylating protein kinase
STPPase	pS/pT-protein phosphatase
STYPPase	pS/pT/pY-protein phosphatase
T_3	Triodothyronine
T_4	Tetraiodothyronine or thyroxine
Tag	SV40 T-antigen
TCR	T-cell receptor
TG	Thyroglobulin
TGase	Transglutaminase
TGF-α	Tumor growth factor-α
TGF-β	Tumor growth factor-β
TK	Thymidine kinase
7-TM receptors	Receptors with seven transmembrane domains
TPA	12-0-Tetradecanoylphorbol-13-acetate
TPTX	Thyroparathyroidectomy
TP53	Tumor-suppressor protein p53
TRH	Thyrotropin-releasing hormone
TS	Thymidylate synthase
TSH	Thyroid-stimulating hormone
5'-UTR	5′-untranslated region (of mRNA)
VIP	Vasoactive intestinal peptide
YPK	Protein tyrosine kinase

Index

A

B

C

F

G

T